LINEAR ALGEBRA
A Concrete
Introduction

LINEAR ALGEBRA
A Concrete
Introduction

LINEAR ALGEBRA
A Concrete
Introduction

Dennis M. Schneider

Manfred Steeg

Frank H. Young

Knox College

MACMILLAN PUBLISHING CO., INC.
New York

COLLIER MACMILLAN PUBLISHERS
London

Copyright © 1982, Macmillan Publishing Co., Inc.

Printed in the United States of America.

Macmillan Publishing Co., Inc.
866 Third Avenue, New York, New York 10022

Collier Macmillan Canada, Inc.

Library of Congress Cataloging in Publication Data

Schneider, Dennis M.
  Linear algebra: a concrete introduction.

  Includes index.
  1. Algebras, Linear. I. Steeg, Manfred.
II. Young, Frank H. (Frank Hood), 1939-
III. Title.
QA184.S37   512'.5   81-8160
ISBN 0-02-476810-3   AACR2

Printing: 1 2 3 4 5 6 7 8   Year: 2 3 4 5 6 7 8 9

Preface

Linear algebra is deeply rooted in analytic geometry and the theory of systems of linear equations. Until the last twenty or thirty years, students were first introduced to some of the concepts of linear algebra in courses on analytic geometry and the theory of equations. When they were subsequently exposed to an abstract presentation of linear algebra they were equipped to deal with it because they were already familiar with some·of the concrete problems that gave birth to the subject.

Recently, all of this has changed. Analytic geometry has been integrated into courses on the calculus. This has resulted in a substantial increase in the amount of material that needs to be covered in calculus courses. And if material needs to be omitted because of time constraints, it is usually the material on analytic geometry. Also courses on the theory of equations have long since vanished from the curriculum. Thus most students enter a course on linear algebra equipped only with some elementary facts about vectors in R^2 and R^3. Although vectors certainly provide some motivation for studying abstract vector spaces, it is not enough. Students do not see the need for such abstract concepts as linear independence, spanning, bases, and dimension arising simply out of the study of vectors in R^2 and R^3. However, they do see the need for these concepts arising out of concrete problems in analytic geometry and systems of linear equations. For these reasons we feel that most students will learn more abstract linear algebra from a concrete approach based on the theory of linear equations and analytic geometry than they would from the (now traditional) abstract approach.

In view of these remarks, we chose as our point of departure the theory of systems of linear equations. In Chapter 1 we completely develop the Gaussian elimination process and thereby teach the student how to solve any system of linear equations. This chapter is also the natural place to introduce vectors and matrices in order to represent systems of equations. (*The material in this chapter should be covered as rapidly as possible.*) In Chapter 2 we use concrete problems concerning linear systems to motivate the abstract concepts of linear algebra (in *n*-dimensional Euclidean space only). We also stress the interplay between these concepts and the geometry of two- and three-dimensional space. In Chapter 3 we again return to the theory of linear systems, but this time use inconsistent systems to motivate the need to extend the concepts of length and angle to higher-dimensional spaces. Chapter 4 (which requires calculus and may be omitted) extends the concepts of Chapters 2 and 3 to the function space setting. Chapter 5 introduces the notion of a linear transformation and discusses the relationship between these transformations and matrices.

It is not until Chapter 6 (which may be omitted) that we finally give an abstract definition of a vector space and an inner product space. The student should at this point be prepared to appreciate how these definitions provide a single conceptual framework for dealing with problems in linear algebra.

In Chapter 7 we briefly discuss determinants and their relationship to the geometry of space. Eigenvalues and eigenvectors are discussed in Chapter 8. Our discussion leads naturally to the problem of diagonalizing a matrix and the spectral theorem for symmetric matrices. In Sections 8.7 through 8.10 (which require calculus and may be omitted) we apply the theory of linear algebra to systems of differential equations. Finally, in Chapter 9 we discuss some numerical techniques that are useful for solving problems in linear algebra with a computer.

The instructor should note that the text deals exclusively with real vector spaces. Except for a brief remark in Chapter 8, all matters concerning complex numbers are left to the appendix. In the appendix, we define complex numbers and develop their arithmetic. We then point out (via examples and exercises) that all of the material in Chapters 1 and 2 extends to C^n with no change. After motivating a definition of an inner product on C^n we show (again via examples and exercises) that the material in Chapter 3 extends to C^n with no change. The extension of the material in Chapter 8 to C^n (where it belongs) now follows immediately.

The applications that are presented are an important part of the text. They provide the student with a sense of the vast scope and rich nature of the subject. We believe that the theory and applications of linear algebra illuminate each other. Not only does a knowledge of the theory help one to understand the applications, but a knowledge of the applications helps one to understand the theory. The applications that we have chosen are real, not artificial. They are taken primarily from biology, economics, sociology, circuit theory, and data analysis. Each example is carefully motivated, explained, and developed.

Every linear algebra text must contain exercises. They are vital because it is through working the exercises that the student actually confronts the subject and completes the learning process. Exercises will be found in the textual material to encourage the student to read the book in an active rather than passive manner. The problems at the end of each section begin with some fairly routine exercises. More substantial problems follow these. To assist the student, answers for computational exercises and for those problems marked with an asterisk are provided at the end of the book. All problems involving calculus are so indicated.

Since we have found that at this stage most students find set theoretical notation and the sigma notation more of a hinderence than a help, we have avoided their use in the text. We have also avoided using mathematical induction.

The book is organized so that it may be used for both a calculus-based and a non-calculus-based course. Only a very basic understanding of calculus is required for a calculus-based course. If calculus is not required, Chapter 4 and Sections 8.7 through 8.10 must be omitted. In the remainder of the book, the few problems and examples that require calculus are clearly marked.

Although we believe that there are strong pedagogical reasons for covering the topics in the book in the order in which they are presented, the instructor does have many options. For example, after covering Chapters 1, 2, and 3 there are essentially five options available: Chapter 4; Chapter 5; Chapter 6; Chapters 7 and 8; Chapter 9. We have noted the exact dependencies in the following chart.

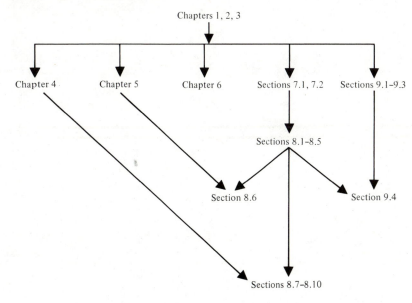

To our students at Knox College and to the students at Monmouth College who have suffered through using Xerox copies of this book in the various stages of its development, we wish to express our sincere thanks. Special thanks are due to those students, colleagues (especially George Converse and Lyle Welch at Monmouth College), and the reviewers who have made many valuable suggestions and criticisms. We also wish to express our gratitude to Betsy Kelly, Mavis Meadows, and Jonathan Young for their excellent typing of the manuscript. Finally, we wish to thank the entire staff at Macmillan for their effort.

D. M. S.
M. S.
F. H. Y.

Contents

Inconsistent

Function S

Systems of Linear Equations and (

Linear Tra

Fundamental Concepts of Linear Al

Abstract V

LINEAR ALGEBRA
A Concrete
Introduction

Systems of Linear Equations and Gaussian Elimination

Linear algebra is a subject of crucial importance to mathematicians and users of mathematics. Applications of linear algebra are found in subjects as diverse as economics, physics, sociology, and engineering. Workers in these fields, as well as mathematicians, statisticians, computer scientists, and management consultants, use linear algebra to express ideas, solve problems, and model real activities.

Linear algebra has its beginnings in the study and solution of systems of linear equations. Before studying linear algebra as an abstract mathematical subject, it is necessary for the student to have some understanding and appreciation of the concrete origins of the subject. We therefore devote this introductory chapter to the theory of systems of linear equations, introducing necessary terminology and notation as well as describing applications and techniques for computation. *We recommend that the material in this chapter be covered as rapidly as possible. Some or all of the applications in Section 1.8 may be postponed until Chapter 8.*

1.1 SYSTEMS OF LINEAR EQUATIONS

The equation $y = mx + b$ is familiar to mathematics students as an equation that represents a nonvertical straight line. It is an example of what we call a linear equation. A linear equation in the two variables x_1 and x_2 is an

1

equation that can be written in the form

$$a_1 x_1 + a_2 x_2 = b,$$

where a_1, a_2, and b are numbers. In general, a **linear equation in the n variables** $x_1, x_2, \ldots, x_n$ is an equation that can be written in the form

$$a_1 x_1 + a_2 x_2 + \cdots + a_n x_n = b,$$

where the **coefficients** $a_1, a_2, \ldots, a_n$ and the **constant term** b are numbers. We adopt the usual notation of using subscripts because this makes it easier to understand and manipulate equations involving several variables (and we do not have to worry about running out of letters). The following are examples of linear equations:

$$x_1 + 7x_2 = 3, \qquad x_1 - 3x_2 + x_4 = \frac{5}{2},$$
$$0.5x_1 = 3x_2 - 7, \qquad x_1 + x_2 + \cdots + x_n = 4.$$

Some examples of equations that are not linear are:

$$x_1^2 + x_1 x_3 = 5, \qquad \frac{1}{x_1} + x_2 + x_3 = 7,$$
$$e^{(x_1)} + x_2 = \frac{1}{2}, \qquad \frac{x_1 + x_2}{x_3 + x_4} = x_5 + 7.$$

Any equation that contains a power of a variable (x_i^r, where $r \neq 1$) or a product of two or more variables (e.g., $x_i x_j$) is not a linear equation.

A **solution** of a linear equation is a collection of values for the variables such that when these values are substituted for the variables, the equation is true. For example, a solution of $x_1 + x_2 = 0$ is $x_1 = 0$, $x_2 = 0$. Another solution of this equation is $x_1 = 1$, $x_2 = -1$. Solving a linear equation involves finding values (numbers) for the variables that make the equation true. Since these values are initially unknown, we often refer to the variables as unknown quantities or **unknowns**. "Variable" and "unknown" are interchangeable terms.

EXERCISE 1   Verify that the indicated values are solutions of the given linear equations.

(a) $3x_1 + 2x_2 = 7$, $x_1 = 1, x_2 = 2$
(b) $3x_1 + 2x_2 = 7$, $x_1 = -3, x_2 = 8$
(c) $x_1 - x_2 = 5$, $x_1 = 2, x_2 = -3$

Frequently, we have more than one equation involving the same variables. For example,

$$\begin{aligned} x_1 + x_2 + x_3 &= 4 \\ 3x_1 - 2x_2 + 2x_3 &= 6 \end{aligned} \tag{1}$$

is a **system** of linear equations in the three variables x_1, x_2, and x_3. This system has two equations and three unknowns. In general, a **system of linear equations** (also called a **linear system**) in the variables $x_1, x_2, \ldots, x_n$ consists of a finite number of linear equations in these variables. The general form of a system of m equations in n unknowns is

$$
\begin{aligned}
a_{11}x_1 + a_{12}x_2 + \cdots + a_{1n}x_n &= b_1 \\
a_{21}x_1 + a_{22}x_2 + \cdots + a_{2n}x_n &= b_2 \\
&\vdots \\
a_{m1}x_1 + a_{m2}x_2 + \cdots + a_{mn}x_n &= b_m.
\end{aligned}
$$

We will call such a system an $m \times n$ (m by n) linear system. The double subscripting of the coefficients has been arranged so that the first subscript refers to the equation and the second subscript refers to the variable. In other words, a_{ij} is the coefficient of x_j in the ith equation. In (1), $a_{12} = 1$ and $a_{21} = 3$. Similarly, b_i is the constant term in the ith equation. In (1), $b_1 = 4$ and $b_2 = 6$. Note also in (1) that $m = 2$ and $n = 3$. We can determine m and n (once we are given the system of equations) by counting equations and variables, respectively.

EXERCISE 2   Given the linear system

$$
\begin{aligned}
4x_1 + 2x_2 - 3x_3 + 4x_4 &= 2 \\
5x_1 - 3x_2 + 4x_3 - 2x_4 &= -7 \\
x_1 + x_2 - x_3 + 3x_4 &= 2.
\end{aligned}
$$

What is the value of a_{31}? of a_{21}? of a_{12}? of a_{24}? of b_3? This is an $m \times n$ system. What do m and n equal?

It should come as no surprise that a solution of a system of linear equations is a collection of values for the variables which makes *all* the equations true. When these values are substituted for the variables, every single one of the equations is a true statement. System (1) has $x_1 = 2$, $x_2 = 1$, $x_3 = 1$ as a solution because $2 + 1 + 1 = 4$ *and* $3 \cdot 2 - 2 \cdot 1 + 2 \cdot 1 = 6$. (The student who is actively reading this book should have just asked if there are any other solutions. Are there?) However, $x_1 = 4$, $x_2 = 0$, $x_3 = 0$ is not a solution of system (1) despite the fact that it is a solution of the first equation of the system. Since $3 \cdot 4 - 2 \cdot 0 + 2 \cdot 0 = 12 \neq 6$, these values do not satisfy the second equation and thus cannot be a solution of the system.

EXERCISE 3   Which of the following are solutions of system (1)?

(a) $x_1 = 1$, $x_2 = 1$, $x_3 = 1$
(b) $x_1 = -2$, $x_2 = 0$, $x_3 = 6$
(c) $x_1 = 1$, $x_2 = 1$, $x_3 = 2$

A system of linear equations may not have any solutions. For example, consider the system

$$x_1 + 3x_2 = 17$$
$$x_1 + 3x_2 = 24.$$

If this system were to have a solution, 17 would have to equal 24, since both are equal to $x_1 + 3x_2$. This equality is clearly impossible. A linear system that does not have any solutions is called **inconsistent**. A linear system that does have solutions is called **consistent**.

EXERCISE 4 Find a value for b such that the following system is inconsistent.

$$2x_1 + 3x_2 = 4$$
$$4x_1 + 6x_2 = b$$

In the next section we discuss applications of linear systems in several disciplines.

PROBLEMS 1.1

*1. Which of the following equations are linear?
(a) $3x_1 - 2x_2 + x_3 = 0$ (b) $2x_1 + 3x_2 + x_4 = x_3 - 8$
(c) $x_1 x_2 + x_3 = 2$ (d) $2x_1 + x_2^2 - x_3 = x_4$
(e) $\dfrac{x_1 - x_2}{x_3} = x_4$ (f) $x_1 + x_2 - x_5 = 4$

2. Identify the linear equations.
(a) $3x = 5y - z$ (b) $5x_1 = 3x_2 - 1$
(c) $x_1 + x_7 = 0$ (d) $\log(x_1) = 4$
(e) $\dfrac{1}{x_1} + x_2 = -52$ (f) $\dfrac{x_2}{x_4} = \dfrac{x_3}{x_1}$

*3. Which of the following are solutions of the equation $2x_1 - 5x_2 + x_3 = 3$?
(a) $x_1 = 1, x_2 = -1, x_3 = -3$ (b) $x_1 = 10, x_2 = 2.5, x_3 = -4.5$
(c) $x_1 = 1, x_2 = -1, x_3 = -4$ (d) $x_1 = \dfrac{2}{9}, x_2 = -\dfrac{5}{9}, x_3 = -\dfrac{2}{9}$

4. Which of the following are solutions of the equation $3x_1 - 2x_2 + 4x_3 + x_4 = 0$?
(a) $x_1 = 1, x_2 = 1, x_3 = 1, x_4 = 4$
(b) $x_1 = 1, x_2 = 2, x_3 = 1, x_4 = 3$
(c) $x_1 = 2, x_2 = 1, x_3 = -1, x_4 = 0$
(d) $x_1 = 2, x_2 = 1, x_3 = -1, x_4 = 1$

*5. Determine which of the following are solutions of the system

$$x_1 + 2x_2 - 2x_3 = 3$$
$$-x_1 + x_2 - 5x_3 = 0$$
$$3x_1 + 3x_2 + x_3 = 6.$$

*Asterisks indicate problems which have answers in the answers section.

(a) $x_1 = 1, x_2 = 1, x_3 = 0$ (b) $x_1 = -7, x_2 = 8, x_3 = 3$
(c) $x_1 = 9, x_2 = 0, x_3 = 3$ (d) $x_1 = 3, x_2 = 0, x_3 = 0$
(e) $x_1 = 9, x_2 = -6, x_3 = -3$ (f) $x_1 = 1, x_2 = 2, x_3 = 2$

6. Determine which of the following are solutions of the system

$$3x_1 - 2x_2 - 3x_3 - 2x_4 = -1$$
$$x_1 + x_2 - x_3 + x_4 = 8$$
$$2x_1 + 3x_2 + x_3 = 21.$$

(a) $x_1 = 1, x_2 = 1, x_3 = 1, x_4 = -1$
(b) $x_1 = 5, x_2 = 3, x_3 = 2, x_4 = 2$
(c) $x_1 = 2, x_2 = 6, x_3 = -1, x_4 = -1$
(d) $x_1 = 0, x_2 = -2, x_3 = 1, x_4 = 1$

7. Explain why the system

$$x_1 + x_2 + 2x_3 = 1$$
$$3x_1 + 3x_2 + 6x_3 = 2$$

cannot have any solutions.

*8. Find three different values for b that will make the following system inconsistent.

$$x_1 + x_2 + 2x_3 = 1$$
$$3x_1 + 3x_2 + 6x_3 = b$$

9. For which values of b will the following system have solutions?

$$x_1 + 2x_2 - x_3 = 4$$
$$2x_1 + 4x_2 - 2x_3 = b$$

*10. Given the system

$$2x_1 + x_2 - x_3 = 4$$
$$x_1 - x_2 + 3x_3 = 2,$$

find an equation with the property that when it is included in the system the resulting system of three equations is inconsistent.

11. Verify that an infinite number of solutions of the system

$$3x_1 - 2x_2 - 3x_3 - 2x_4 = -1$$
$$x_1 + x_2 - x_3 + x_4 = 8$$
$$2x_1 + 3x_2 + x_3 = 21$$

is given by $x_1 = 3 + t, x_2 = 5 - t, x_3 = x_4 = t$, where t is any number.

12. Verify that an infinite number of solutions of the system

$$x_1 + x_3 + x_4 = 7$$
$$x_1 + x_2 - x_4 = 4$$
$$x_2 - x_3 - 2x_4 = -3$$

is given by $x_1 = 2 + t - s, x_2 = 3 - t + 2s, x_3 = 4 - t, x_4 = 1 + s$, where s and t are arbitrary numbers.

13. Verify that a solution of

$$a_{11}x_1 + a_{12}x_2 = b_1$$

$$a_{21}x_1 + a_{22}x_2 = b_2$$

is given by

$$x_1 = \frac{b_1 a_{22} - b_2 a_{12}}{a_{11}a_{22} - a_{12}a_{21}}$$

$$x_2 = \frac{a_{11}b_2 - a_{21}b_1}{a_{11}a_{22} - a_{12}a_{21}}$$

provided that $a_{11}a_{22} - a_{12}a_{21} \neq 0$.

1.2   SOME EXAMPLES OF SYSTEMS OF LINEAR EQUATIONS

Most students of high school algebra have been subjected to problems such as this:

Find three numbers whose sum is 20 and such that (1) the first plus twice the second plus three times the third equals 44 and (2) twice the sum of the first and second minus four times the third equals -14. [This problem is found (together with 19 others involving digits, water tanks, boats, work, and freight trains) in H. B. Fine, *A College Algebra*, Ginn & Co. 1904, pp. 150–152.]

This problem is equivalent to finding the solution of the following system of linear equations (where x_1, x_2, and x_3 are the three numbers we are trying to find).

$$\begin{aligned} x_1 + x_2 + x_3 &= 20 \\ x_1 + 2x_2 + 3x_3 &= 44 \\ 2x_1 + 2x_2 - 4x_3 &= -14 \end{aligned}$$

Problems such as this, although of interest to professional (or habitual) problem solvers, are not important applications of linear equations. They give practice in translating English into the language of mathematics as well as practice in computation but do not give the student adequate motivation for studying the mathematics that is being used. In this section we give several practical and important examples where systems of linear equations arise naturally.

Electric Circuits

Most people think of electricity as something that "flows" through wires. Indeed, it is usually convenient to think of electricity as electrons flowing through wires. When we think of something flowing we naturally think of

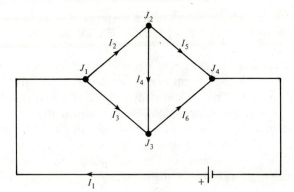

Figure 1.1

the "pressure" behind the flow and the "quantity" of substance flowing. For electrical circuits, the "pressure" behind the electrons is measured in volts and the "quantity" of electrons flowing, called the current, is measured in amperes or amps. For the sake of simplicity we will consider only direct-current (dc) circuits, circuits in which the electricity travels in one direction in each wire.

Let us consider the electric circuit given by Figure 1.1. This circuit has four junctions, places where many wires come together (J_1, J_2, J_3, and J_4). There are also six branches with currents $I_1, I_2, \ldots, I_6$ and one source of electricity in the branch from J_4 to J_1. Each branch of this circuit has been (arbitrarily) assigned an arrow indicating a direction of flow. The actual direction of flow will be given by the sign of the current in that branch. A positive current will mean a flow in the direction of the arrow; a negative current will flow in the opposite direction.

We can use an ammeter to measure both the direction and the amount of current flowing in each wire of this circuit. If the currents in all the wires that come together at a junction are added, it is found that the sum is zero. This is not too surprising—it expresses the fact that the substance (the electrons) which is flowing is not being created or destroyed at the junction. In brief, what goes in must equal what comes out. This is one of two basic laws regarding electric circuits which were first formulated by G. R. Kirchhoff in 1845.

At J_1 we have I_1 amps flowing in and $I_2 + I_3$ amps flowing out. By Kirchhoff's law, $I_1 - I_2 - I_3 = 0$. This is a linear equation with currents as the variables. There will be one equation for each junction. Looking at all four junctions we get the following four equations.

$$
\begin{aligned}
I_1 - I_2 - I_3 \qquad\qquad\quad &= 0 \qquad (\text{junction } J_1)\\
I_2 \quad\; -I_4 - I_5 \qquad &= 0 \qquad (\text{junction } J_2)\\
I_3 + I_4 \quad\; -I_6 &= 0 \qquad (\text{junction } J_3)\\
-I_1 \qquad\qquad\quad +I_5 + I_6 &= 0 \qquad (\text{junction } J_4)
\end{aligned}
$$

We conclude that the values of the currents flowing in the six branches of the circuit of Figure 1.1 must satisfy (be a solution of) this system of linear equations.

It is clear that this system of equations does not have a unique solution. We could double the current in each branch and still satisfy Kirchhoff's law at each junction. Thus a knowledge of the physical problem (the electric circuit has given us information about a mathematical problem (the system of equations). The reverse is also possible. In particular, the theory developed in Chapter 2 will enable us to determine the minimum number of currents that must be measured before the equations above give us complete knowledge of all currents in the circuit.

Pricing Goods in a Closed Economy

Suppose that there are three people in a closed economy, for example, three astronauts orbiting the earth in a space station. No goods enter or leave the station. Suppose that each of these three astronauts is producing one product and that each product is divided up among the three astronauts. Let us be specific about how the goods produced in our simple economy are distributed. Suppose that astronaut 1 produces a certain quantity of item A each day, which is divided up so that $\frac{2}{5}$ goes to astronaut 1, $\frac{1}{5}$ to astronaut 2, and $\frac{2}{5}$ to astronaut 3. Let astronaut 2 produce item B and let $\frac{1}{4}$ of the amount produced be given to astronaut 1 and an equal amount to astronaut 3. Finally, let astronaut 3 produce item C, with $\frac{1}{2}$ of the production going to astronaut 1 and $\frac{1}{4}$ to astronaut 2. We now have complete information about the way the goods in this economy are distributed.

This small community does not need money to accomplish the exchange of goods. But we usually have a closed economy with a much larger number of producers. For example, the earth is a closed economy which consists of a very large number of producing units. Such communities require some common medium of exchange (i.e., money). We then have a way to measure the relative value of each of the goods produced and we can assign prices to the products.

Suppose that the value of the goods consumed by one of the astronauts is greater than the value of the goods produced by that astronaut. This astronaut's supply of money will decrease every day. This cannot continue indefinitely. If our economy is to be stable, if it is to be in a state of equilibrium, we must reject this possibility. Similarly, to have a stable economy, we must reject the possibility of an astronaut consuming goods less valuable than those produced. Thus, in a stable economy, the value of the goods produced by each astronaut must equal the value of the goods consumed by that astronaut.

Assuming that our mini-economy is stable, let us attempt to assign prices to the goods being produced. Let p_i be the value (price) of the goods produced by astronaut i in one day. At the end of each day the first astronaut has $\frac{2}{5}$ of his own production, $\frac{1}{4}$ of the second astronaut's production, and $\frac{1}{2}$ of the third astronaut's production. The value of these goods is $\frac{2}{5}p_1 + \frac{1}{4}p_2 + \frac{1}{2}p_3$. But the value of the goods produced by the first astronaut is p_1. Thus the prices must satisfy the equation

$$\frac{2}{5}p_1 + \frac{1}{4}p_2 + \frac{1}{2}p_3 = p_1.$$

A similar equation can be found for each astronaut and we have the following system of equations in the variables p_1, p_2, p_3.

$$\frac{2}{5}p_1 + \frac{1}{4}p_2 + \frac{1}{2}p_3 = p_1 \quad \text{(astronaut 1)}$$

$$\frac{1}{5}p_1 + \frac{1}{2}p_2 + \frac{1}{4}p_3 = p_2 \quad \text{(astronaut 2)} \tag{1}$$

$$\frac{2}{5}p_1 + \frac{1}{4}p_2 + \frac{1}{4}p_3 = p_3 \quad \text{(astronaut 3)}$$

One solution to this system is $p_1 = 5$, $p_2 = 4$, $p_3 = 4$. Another solution is $p_1 = 10$, $p_2 = 8$, $p_3 = 8$. But from the point of view of economics these two solutions are essentially the same. Merely doubling all prices does not change the relative values of the goods. Another solution of the system is $p_1 = p_2 = p_3 = 0$. Such a solution has no economic meaning. It does not measure the relative values of the goods in a meaningful way.

Two important questions must now be asked. First, given a closed economy, does there always exist a meaningful set of prices for the goods which will yield a stable economy, that is, an economy in which the value of the goods consumed by each individual equals the value of the goods produced by that individual? Second, if such a set of prices exists, is any other set of prices satisfying the same conditions in some sense the same? These questions are of critical importance to the economist. In Chapter 2 we use linear algebra to show that the answer to each of these questions is indeed affirmative.

Curve Fitting in Analytic Geometry

In elementary analytic geometry it is shown that the equation of a nonvertical straight line in the xy-plane is of the form $y = mx + b$. The constants m and b are the slope and y-intercept of the line. There are occasions when the equation of the line is unknown (we do not know m and

b) but some points on the line are known. If the points (x_i, y_i), $i = 1, 2, \ldots, n$, are all on the line, then for each $i = 1, 2, \ldots, n$, we have

$$y_i = mx_i + b.$$

This is a system of n equation in the two unknowns m and b. (The reader will note our sudden change of notation. The standard notation of analytic geometry is so familiar that it is pedagogically unsound to alter it.) Note that n can be any positive integer, including 1.

To make our example more concrete, let us take three points in the plane: $(0,0)$, $(1,1)$, and $(2,1)$. Then our system of equations is

$$0 = m \cdot 0 + b$$
$$1 = m \cdot 1 + b$$
$$1 = m \cdot 2 + b.$$

Since the first equation says that $b = 0$, we may substitute that value for b in the last two equations. They then become $m = 1$ and $2m = 1$. This is clearly impossible. We conclude that no straight line can pass through these three points.

If, on the other hand, we take the two points $(2,3)$ and $(4,7)$, we get the equations

$$3 = m \cdot 2 + b$$
$$7 = m \cdot 4 + b.$$

The unique solution is $m = 2$, $b = -1$. These results agree with our geometric intuition. Three points do not necessarily lie on a straight line, whereas two points *always* determine a unique line.

Suppose that we ask another question about the three points $(0,0)$, $(1,1)$, and $(2,1)$. Can a quadratic curve, one of the form $y = ax^2 + bx + c$, be fitted to these points? If so, then the following equations must be satisfied:

$$0 = a \cdot 0 + b \cdot 0 + c$$
$$1 = a \cdot 1 + b \cdot 1 + c$$
$$1 = a \cdot 4 + b \cdot 2 + c.$$

The first equation says that $c = 0$, so we have

$$c = 0$$
$$a + b = 1$$
$$4a + 2b = 1.$$

There is a unique solution; $a = -\frac{1}{2}$, $b = \frac{3}{2}$, $c = 0$. So we can fit a unique quadratic polynomial to these three points, namely $y = -\frac{1}{2}x^2 + \frac{3}{2}x$.

In this example we have seen some interesting connections between geometric ideas (lines, curves) and algebraic ideas (systems of equations). In subsequent chapters we convert the geometry of Euclidean space into the algebra of systems of equations. We also convert the algebra of systems of equations into the geometry of Euclidean space. The net result will be a more thorough understanding of both.

One further note. We saw above that a system of equations with no solution could arise naturally when studying a certain problem. In Chapter 3 we discuss how such inconsistent systems arise in other situations. We also consider the question of what can be done with such systems.

PROBLEMS 1.2

1. (These problems are for students who enjoy word problems. They are taken from Fine's book.) In each of the following problems certain unknowns satisfy a system of linear equations. Identify the unknowns and exhibit the correct system. Do not solve the system.

   *(a) Find a number of two digits from the following data: (1) twice the first digit plus three times the second equals 37; (2) if the order of the digits be reversed, the number is diminished by 9.

   *(b) The sum of three numbers is 51. If the first number be divided by the second, the quotient is 2 and the remainder is 5; but if the second number be divided by the third, the quotient is 3 and the remainder 2. What are the numbers?

   (c) In a certain number of three digits, the second digit is equal to the sum of the first and the third, the sum of the second and the third digits is 8, and if the first and third digits are interchanged, the number is increased by 99. Find the number.

   *(d) A gave B as much money as B had; then B gave A as much money as A had left. Finally, A gave B as much money as B then had left. A then had $16 and B $24. How much had each originally?

   (e) Find the fortunes of three men, A, B, and C, from the following data: A and B together have p dollars; B and C, q dollars; C and A, r dollars.

   *(f) Given three alloys of the following composition: A, 5 parts (by weight) gold, 2 silver, 1 lead; B, 2 parts gold, 5 silver, 1 lead; C, 3 parts gold, 1 silver, 4 lead. To obtain 9 ounces of an alloy containing equal quantities (by weight) of gold, silver, and lead, how many ounces of A, B, and C must be taken together?

   (g) Two vessels, A and B, contain mixtures of alcohol and water. A mixture of 3 parts from A and 2 parts from B will contain 40% of alcohol; and a mixture of 1 part from A and 2 parts from B will contain 32% of alcohol. What are the percentages of alcohol in A and B, respectively?

2. What system of equations must the currents in each circuit satisfy?

   *(a)   (b)

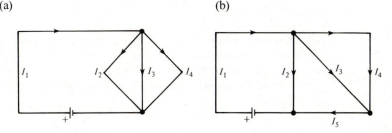

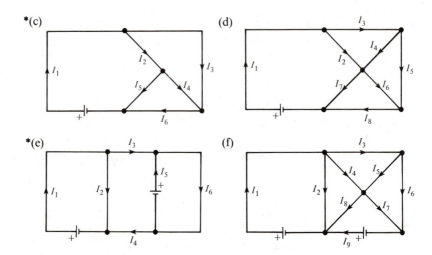

3. Kirchhoff's second law states that the sum of the voltage drops around any closed circuit must be zero. In our more intuitive language this says that the sum of the "pressures" around a circular path must be zero. In essence this law prohibits perpetual flow of electricity unless there is a source for the electricity (e.g., a battery).

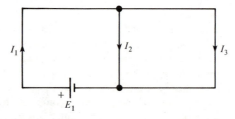

   The voltage E and the current I in any branch of a circuit satisfy the relation $E = IR$ (Ohm's law), where R is the resistance (in ohms) in the branch. For example, consider the circuit pictured. This circuit has three closed loops; the loop consisting of the branches with currents I_1 and I_2, the loop consisting of the branches with currents I_2 and I_3, and the loop consisting of the branches with currents I_1 and I_3. If we let R_i be the resistance in the branch with current I_i, we have the equations

$$R_1 I_1 + R_2 I_2 - E_1 = 0$$
$$R_2 I_2 - R_3 I_3 \quad\quad = 0$$
$$R_1 I_1 + R_3 I_3 - E_1 = 0,$$

each of which corresponds to one of the three closed loops in the circuit. Notice that the voltage from the battery (E_1) has been subtracted because the voltage increase caused by the battery is a negative voltage drop.

   For each of the circuits in Problem 2, find the additional equations satisfied by the currents if:

*(a) All batteries produce 1 volt and all branches have a resistance of one ohm.

*(b) All batteries produce 1 volt and the branch with current I_j has a resistance of j ohms.

4. The prices in a closed economy satisfy the following system of equations:

$$\frac{1}{2}p_1 + \frac{1}{3}p_2 + \frac{1}{4}p_3 = p_1$$

$$\frac{1}{3}p_1 + \frac{1}{3}p_2 + \frac{1}{2}p_3 = p_2$$

$$\frac{1}{6}p_1 + \frac{1}{3}p_2 + \frac{1}{4}p_3 = p_3.$$

*(a) What fraction of production does producer 1 retain? sell to producer 2? sell to producer 3?
 (b) What fraction of production does producer 2 retain? sell to producer 1? sell to producer 3?
 (c) What fraction of production does producer 3 retain? sell to producer 2? sell to producer 1?

5. Consider a closed economy consisting of five producers. Suppose that producer i produces one unit of item i each day and the daily production is distributed according to the following table.

Goods Consumed by:	Item 1	Item 2	Item 3	Item 4	Item 5
Producer 1	$\frac{1}{2}$	0	$\frac{1}{5}$	0	$\frac{1}{12}$
Producer 2	$\frac{1}{8}$	$\frac{1}{3}$	$\frac{1}{5}$	$\frac{2}{3}$	$\frac{1}{3}$
Producer 3	$\frac{1}{8}$	$\frac{4}{9}$	0	0	$\frac{1}{4}$
Producer 4	$\frac{1}{8}$	$\frac{1}{9}$	$\frac{2}{5}$	0	$\frac{1}{4}$
Producer 5	$\frac{1}{8}$	$\frac{1}{9}$	$\frac{1}{5}$	$\frac{1}{3}$	$\frac{1}{12}$

Let p_i, $i = 1, 2, 3, 4, 5$, be the price for one unit of item i. What system of linear equations must these prices satisfy in order to have a stable economy?

*6. Four people stranded on a desert island form a closed economy. The number in the ith row and the jth column of the following table represents the fraction of the ith person's daily production that is consumed by the jth person. What system of equations do the prices satisfy?

	1	2	3	4
1	$\frac{1}{3}$	$\frac{1}{6}$	$\frac{1}{3}$	$\frac{1}{6}$
2	$\frac{1}{5}$	$\frac{2}{5}$	$\frac{1}{5}$	$\frac{1}{5}$
3	$\frac{1}{6}$	$\frac{1}{6}$	$\frac{1}{2}$	$\frac{1}{6}$
4	$\frac{1}{8}$	$\frac{1}{4}$	$\frac{1}{8}$	$\frac{1}{2}$

7. Explain why in a stable economy we must reject the possibility of an individual consuming goods less valuable than those produced by that individual.

8. Verify that $p_1 = -10$, $p_2 = -8$, $p_3 = -8$ is a solution of linear system (1) in the text. Discuss the economic meaning of these prices.

9. The following system of equations cannot be the equations satisfied by the prices in a stable mini-economy. Why?

$$\frac{1}{2}p_1 + \frac{1}{2}p_2 + \frac{1}{2}p_3 = p_1$$

$$\frac{1}{2}p_1 + \frac{1}{3}p_2 + \frac{1}{4}p_3 = p_2$$

$$\frac{1}{2}p_1 + \frac{1}{4}p_2 + \frac{1}{4}p_3 = p_3$$

*10. Suppose that we have an n-person stable economy. Then the prices for the goods satisfy the following linear system.

$$\begin{aligned}
a_{11}p_1 + a_{12}p_2 + \cdots + a_{1n}p_n &= p_1 \\
a_{21}p_1 + a_{22}p_2 + \cdots + a_{2n}p_n &= p_2 \\
&\ \ \vdots \\
a_{n1}p_1 + a_{n2}p_2 + \cdots + a_{nn}p_n &= p_n
\end{aligned}$$

Show that the value of

$$a_{1j} + a_{2j} + a_{3j} + \cdots + a_{nj}$$

does not depend on j. What is its value?

11. Consider a cubic curve $y = ax^3 + bx^2 + cx + d$. What system of linear equations must a, b, c, and d satisfy so that the curve passes through the points $(0,1)$, $(1,2)$, $(3,-1)$, $(-1,-1)$?

*12. What equations must a, b, and c satisfy if the quadratic polynomial $y = ax^2 + bx + c$ is to pass through the points $(0,-2)$, $(1,6)$, $(-2,0)$, $(4,66)$?

*13. What equations must m and b satisfy if the points $(1,-1)$, $(5,7)$, $(-2,-7)$, and $(0,-3)$ are to lie on the straight line $y = mx + b$?

14. What equations must m and b satisfy if the points $(1,5)$ and $(-1,1)$ are to lie on the straight line $y = mx + b$?

1.3   SOLVING SYSTEMS—BACK SUBSTITUTION, EQUIVALENT SYSTEMS

It is usually not sufficient to know that the answer to a problem satisfies a linear system. We need to find the answer by solving the linear system. We begin this section by exhibiting some systems which are in a form that makes them easy to solve. We then investigate some basic operations that will transform a linear system into this special form without changing its solutions.

EXAMPLE 1   Consider the linear system

$$x_1 + x_2 = 4$$
$$x_2 = 1.$$

It is easy to show that this system has only one possible solution, $x_1 = 3$, $x_2 = 1$. The second equation gives 1 as the value of x_2. Substituting $x_2 = 1$ into the first equation gives $x_1 = 3$.

EXAMPLE 2   Consider the system

$$2x_1 + x_2 - x_3 = 5$$
$$x_2 + x_3 = 3$$
$$3x_3 = 6.$$

The last equation tells us that x_3 must equal 2. This value can then be substituted in the second equation, giving $x_2 = 1$. Now substitute the values for x_2 and x_3 in the first equation to obtain $x_1 = 3$. This system therefore has only one solution, $x_1 = 3$, $x_2 = 1$, $x_3 = 2$.

EXERCISE 1   Solve the following system.

$$2x_1 \qquad + x_3 - x_4 = 4$$
$$x_2 + x_3 + x_4 = 2$$
$$2x_3 - x_4 = 1$$
$$x_4 = 1$$

The preceding systems of equations are easy to solve because they have a special shape. Square linear systems whose coefficients a_{ij} are zero whenever $j < i$ are called **upper triangular** systems. All upper triangular systems such that $a_{ii} \neq 0$ for all i can be solved by repeating the substitution procedure used in the previous examples. This process is called **back substitution**. We solve for the variables from the last to the first (backward). We repeatedly substitute the values of known variables and solve for the one remaining variable in the equation. In the next example we show that the use of back substitution is not limited to triangular systems.

EXAMPLE 3   The nonsquare system

$$2x_1 + x_2 - x_3 + x_4 = 5$$
$$x_2 + x_3 - x_4 = 3 \qquad\qquad (1)$$
$$3x_3 + 6x_4 = 6$$

is not in triangular form. If we move the terms involving x_4 to the right-hand side of these equations, we can consider this system as a triangular system in the variables x_1, x_2, x_3.

$$2x_1 + x_2 - x_3 = 5 - x_4$$
$$x_2 + x_3 = 3 + x_4$$
$$3x_3 = 6 - 6x_4$$

We can now use back substitution to solve for x_1, x_2, and x_3 in terms of x_4.

$$x_1 = 3 - 3x_4$$
$$x_2 = 1 + 3x_4$$
$$x_3 = 2 - 2x_4$$

Have we found a solution to the original system? Yes! We have actually found an infinite number of solutions.

$$
\begin{aligned}
x_1 &= 3 - 3x_4 \\
x_2 &= 1 + 3x_4 \qquad \text{(where } x_4 \text{ is arbitrary)} \\
x_3 &= 2 - 2x_4 \\
x_4 &= x_4
\end{aligned}
\qquad (2)
$$

The variable x_4 is free to take on any value. No matter what value is chosen for x_4, it, together with the values for x_1, x_2, and x_3 determined from (2), will be a solution to (1). For example, when $x_4 = 7$, then $x_1 = -18$, $x_2 = 22$, $x_3 = -12$, $x_4 = 7$ is a particular solution to (1).

EXERCISE 2 Show that the values given by (2) are a solution of (1) for every choice of x_4.

Although the system in Example 3 had only one "extra" variable, the same method will work regardless of the number of extra variables.

EXERCISE 3 Solve for x_1, x_2, and x_3 in terms of x_4 and x_5.

$$
\begin{aligned}
x_1 + 2x_2 - x_3 + x_4 + x_5 &= 6 \\
x_2 + x_3 \qquad\quad - x_5 &= 2 \\
x_3 - x_4 + 2x_5 &= 3
\end{aligned}
$$

EXAMPLE 4 The system

$$
\begin{aligned}
x_1 + 2x_2 + 3x_3 + x_4 - x_5 &= 2 \\
x_3 + x_4 + x_5 &= -1 \\
x_4 - 2x_5 &= 4
\end{aligned}
\qquad (3)
$$

is a triangular system in the variables x_1, x_3, and x_4.

$$
\begin{aligned}
x_1 + 3x_3 + x_4 &= 2 - 2x_2 + x_5 \\
x_3 + x_4 &= -1 - x_5 \\
x_4 &= 4 + 2x_5
\end{aligned}
$$

We can use back substitution to solve for x_1, x_3, and x_4 in terms of x_2 and x_5.

$$
\begin{aligned}
x_1 &= 13 - 2x_2 + 8x_5 \\
x_3 &= -5 - 3x_5 \\
x_4 &= 4 + 2x_5
\end{aligned}
\qquad (4)
$$

circuit theory, and data analysis. Each example is carefully motivated, explained, and developed.

Every linear algebra text must contain exercises. They are vital because it is through working the exercises that the student actually confronts the subject and completes the learning process. Exercises will be found in the textual material to encourage the student to read the book in an active rather than passive manner. The problems at the end of each section begin with some fairly routine exercises. More substantial problems follow these. To assist the student, answers for computational exercises and for those problems marked with an asterisk are provided at the end of the book. All problems involving calculus are so indicated.

Since we have found that at this stage most students find set theoretical notation and the sigma notation more of a hinderence than a help, we have avoided their use in the text. We have also avoided using mathematical induction.

The book is organized so that it may be used for both a calculus-based and a non-calculus-based course. Only a very basic understanding of calculus is required for a calculus-based course. If calculus is not required, Chapter 4 and Sections 8.7 through 8.10 must be omitted. In the remainder of the book, the few problems and examples that require calculus are clearly marked.

Although we believe that there are strong pedagogical reasons for covering the topics in the book in the order in which they are presented, the instructor does have many options. For example, after covering Chapters 1, 2, and 3 there are essentially five options available: Chapter 4; Chapter 5; Chapter 6; Chapters 7 and 8; Chapter 9. We have noted the exact dependencies in the following chart.

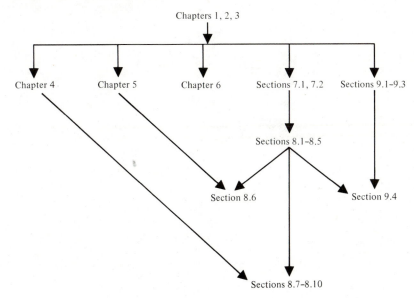

To our students at Knox College and to the students at Monmouth College who have suffered through using Xerox copies of this book in the various stages of its development, we wish to express our sincere thanks. Special thanks are due to those students, colleagues (especially George Converse and Lyle Welch at Monmouth College), and the reviewers who have made many valuable suggestions and criticisms. We also wish to express our gratitude to Betsy Kelly, Mavis Meadows, and Jonathan Young for their excellent typing of the manuscript. Finally, we wish to thank the entire staff at Macmillan for their effort.

D. M. S.
M. S.
F. H. Y.

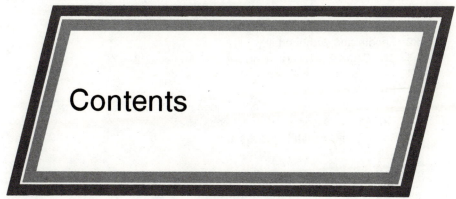

Contents

Chapter 3

Inconsistent Systems, Inner Products, and Projections   115

Chapter 4

Function Spaces   163

Chapter 5

Linear Transformations   181

Chapter 6

Abstract Vector Spaces   205

EXAMPLE 1   Consider the linear system

$$x_1 + x_2 = 4$$
$$x_2 = 1.$$

It is easy to show that this system has only one possible solution, $x_1 = 3$, $x_2 = 1$. The second equation gives 1 as the value of x_2. Substituting $x_2 = 1$ into the first equation gives $x_1 = 3$.

EXAMPLE 2   Consider the system

$$2x_1 + x_2 - x_3 = 5$$
$$x_2 + x_3 = 3$$
$$3x_3 = 6.$$

The last equation tells us that x_3 must equal 2. This value can then be substituted in the second equation, giving $x_2 = 1$. Now substitute the values for x_2 and x_3 in the first equation to obtain $x_1 = 3$. This system therefore has only one solution, $x_1 = 3$, $x_2 = 1$, $x_3 = 2$.

EXERCISE 1   Solve the following system.

$$2x_1 \qquad\;\; + x_3 - x_4 = 4$$
$$x_2 + x_3 + x_4 = 2$$
$$2x_3 - x_4 = 1$$
$$x_4 = 1$$

   The preceding systems of equations are easy to solve because they have a special shape. Square linear systems whose coefficients a_{ij} are zero whenever $j < i$ are called **upper triangular** systems. All upper triangular systems such that $a_{ii} \neq 0$ for all i can be solved by repeating the substitution procedure used in the previous examples. This process is called **back substitution**. We solve for the variables from the last to the first (backward). We repeatedly substitute the values of known variables and solve for the one remaining variable in the equation. In the next example we show that the use of back substitution is not limited to triangular systems.

EXAMPLE 3   The nonsquare system

$$2x_1 + x_2 - x_3 + x_4 = 5$$
$$x_2 + x_3 - x_4 = 3 \qquad\qquad (1)$$
$$3x_3 + 6x_4 = 6$$

is not in triangular form. If we move the terms involving x_4 to the right-hand side of these equations, we can consider this system as a triangular system in the variables x_1, x_2, x_3.

$$2x_1 + x_2 - x_3 = 5 - x_4$$
$$x_2 + x_3 = 3 + x_4$$
$$3x_3 = 6 - 6x_4$$

We can now use back substitution to solve for x_1, x_2, and x_3 in terms of x_4.

$$x_1 = 3 - 3x_4$$
$$x_2 = 1 + 3x_4$$
$$x_3 = 2 - 2x_4$$

Have we found a solution to the original system? Yes! We have actually found an infinite number of solutions.

$$
\begin{aligned}
x_1 &= 3 - 3x_4 \\
x_2 &= 1 + 3x_4 \quad \text{(where } x_4 \text{ is arbitrary)} \\
x_3 &= 2 - 2x_4 \\
x_4 &= x_4
\end{aligned}
\tag{2}
$$

The variable x_4 is free to take on any value. No matter what value is chosen for x_4, it, together with the values for x_1, x_2, and x_3 determined from (2), will be a solution to (1). For example, when $x_4 = 7$, then $x_1 = -18$, $x_2 = 22$, $x_3 = -12$, $x_4 = 7$ is a particular solution to (1).

EXERCISE 2   Show that the values given by (2) are a solution of (1) for every choice of x_4.

Although the system in Example 3 had only one "extra" variable, the same method will work regardless of the number of extra variables.

EXERCISE 3   Solve for x_1, x_2, and x_3 in terms of x_4 and x_5.

$$
\begin{aligned}
x_1 + 2x_2 - x_3 + x_4 + x_5 &= 6 \\
x_2 + x_3 \qquad\quad - x_5 &= 2 \\
x_3 - x_4 + 2x_5 &= 3
\end{aligned}
$$

EXAMPLE 4   The system

$$
\begin{aligned}
x_1 + 2x_2 + 3x_3 + x_4 - x_5 &= 2 \\
x_3 + x_4 + x_5 &= -1 \\
x_4 - 2x_5 &= 4
\end{aligned}
\tag{3}
$$

is a triangular system in the variables x_1, x_3, and x_4.

$$
\begin{aligned}
x_1 + 3x_3 + x_4 &= 2 - 2x_2 + x_5 \\
x_3 + x_4 &= -1 - x_5 \\
x_4 &= 4 + 2x_5
\end{aligned}
$$

We can use back substitution to solve for x_1, x_3, and x_4 in terms of x_2 and x_5.

$$
\begin{aligned}
x_1 &= 13 - 2x_2 + 8x_5 \\
x_3 &= -5 - 3x_5 \\
x_4 &= 4 + 2x_5
\end{aligned}
\tag{4}
$$

Whatever values are chosen for x_2 and x_5, they, together with the values for x_1, x_3, and x_4 determined from (4), will be a solution of the system (3).

EXERCISE 4 Show that for every choice of x_2 and x_5, they, together with the values given by (4), are a solution of (3).

As indicated in Examples 3 and 4, we can use back substitution to solve any system that can be converted to triangular form by moving some of the variables, the "extra" variables, to the right-hand side. When there are extra variables, they will be free to take on any value and hence will be called **free variables**. All of the other variables will be determined in terms of the free variables, and therefore will be called **determined variables**.

The systems in Examples 1 through 4 are examples of a general type of system that can always be solved by back substitution. Such systems have a "generalized triangular" shape. The first nonzero term in each equation is farther to the right than the first nonzero term in all previous equations. A system of this type is said to be in **echelon form** (or to be an **echelon system**). When a system is in echelon form the first nonzero coefficient in each equation is called a **pivot**. There is at most one pivot per variable and not more than one pivot per equation. Those variables that are associated with pivots are the determined variables and all the other variables are free variables. Any system that is in echelon form can be converted to a triangular system in its determined variables by moving all terms involving free variables to the right-hand sides of the equations.

EXERCISE 5 Consider the linear system

$$3x_1 + 2x_2 + x_3 - 2x_4 = 1$$
$$2x_3 - x_4 = 5$$
$$x_4 = 3.$$

(a) Identify the pivots.
(b) Identify the free variables.
(c) Identify the determined variables.
(d) Solve the system.

Now that we have a procedure which solves all echelon systems, we turn our attention to nonechelon systems.

EXAMPLE 5 Consider the system

$$x_1 + x_2 = 4$$
$$2x_1 + 3x_2 = 9. \tag{5}$$

This system is not in echelon form because the coefficient of x_1 in the second equation is 2. If we could eliminate this term (i.e., make its coefficient zero), then the system would be in echelon form and we could

BACK SUBSTITUTION

Given

A system of equations in echelon form.

Goal

To find all solutions of the system.

Procedure

1. Convert the system to a triangular system in the determined variables by moving all terms involving free variables to the right-hand sides of the equations.
2. Solve the last equation for the last determined variable.
3. Substitute this result in the next-to-last equation and solve.
4. Continue working toward the top, substituting all previous results in the next equation and solving for the next determined variable.

Note

If all variables are determined, this procedure gives the unique solution of the system. If there are free variables, then each determined variable is expressed in terms of the free variables. There are an infinite number of solutions in this case. No matter what values are chosen for the free variables, they, together with the resulting values for the determined variables, will be a solution of the system.

solve it using back substitution. We can eliminate this term as follows:

$$
\begin{array}{rl}
\text{second equation} & 2x_1 + 3x_2 = 9 \\
\text{subtract two times the first equation} & 2x_1 + 2x_2 = 8 \\
\hline
\text{new second equation} & x_2 = 1
\end{array}
$$

If we replace the second equation by the new equation found above, we have an echelon system that can be solved by back substitution.

$$
\begin{aligned}
x_1 + x_2 &= 4 \\
x_2 &= 1
\end{aligned}
\tag{6}
$$

The one big question remaining is the following: Do systems (5) and (6) have the same solutions?

EXERCISE 6   The solution of system (6) is $x_1 = 3$, $x_2 = 1$. Show that this is also a solution of system (5).

Example 5 suggests the following straightforward procedure for solving an arbitrary linear system.

1. Transform the system into echelon form.
2. Solve the echelon system.

If this procedure is to give correct answers, it is extremely important that the solutions of the system remain unchanged by the transformation to echelon form. We want all solutions of the original system to be solutions of the echelon system *and* we want all solutions of the echelon system to be solutions of the original system.

Definition   Two linear systems are called **equivalent** if they have the same solutions.

Using this terminology, we can restate our proposed solution procedure as follows:

1. Transform the system into an equivalent system that is in echelon form.
2. Solve the echelon system.

Before describing a procedure to accomplish step 1, we must discuss the basic operations that we can use to transform systems into equivalent systems.

The transformation in Example 5 was made by adding a multiple of one equation to another. The "old" second equation was replaced by a "new" equation. Specifically, it was replaced by the "old" second equation minus two times the first equation. Since we can use this operation to eliminate variables from equations, we call it the **elimination step**. It is one of three basic operations we use to transform systems into equivalent systems.

The Three Basic Operations

1. Subtract (or add) a multiple of one equation from another equation (the elimination step).
2. Multiply an equation by a nonzero constant.
3. Interchange two equations.

Each of these operations transforms a system into a new system. The first two operations actually change one of the equations. The last one changes only the order of the equations. Each of these operations is reversible; that is, undoing the operation involves an application of the same operation (possibly with different constants) (see Problem 4). The reader should note that we have already used two of these basic operations to perform the back substitution procedure. Substitution can be done by using the elimination step (see Problem 6). Solving for a variable involves multiplying an equation by a nonzero constant.

EXERCISE 7   Consider the system

$$3x_1 + 2x_2 = 17$$
$$6x_1 + 5x_2 = 32.$$

Use the elimination step to transform this system into an echelon system.

EXERCISE 8   Consider the system

$$2x_1 + 2x_2 - x_3 = 4$$
$$3x_3 = 6$$
$$x_2 - 2x_3 = 2.$$

(a) Use one of the basic operations to transform this system into an echelon system.

(b) Use one of the basic operations to transform the answer to part (a) into a system that has every pivot equal to 1.

Result 1   Each of the three basic operations transforms a system of equations into an equivalent system.

Proof   It is obvious that interchanging any pair of equations and multiplying an equation by a nonzero constant do not change the solutions of a system. The elimination step involves changing only one equation of the system. If we have values for the variables that make all the old equations true, then these same values will make $E_i + \alpha \cdot E_j$ true no matter which two equations (E_i and E_j) or constant α we choose. Thus all the solutions of the original system are solutions of the new system. The reversibility of the elimination operation implies that all solutions of the second system are also solutions of the first. This establishes the result.

In the next section we present a complete solution procedure. This procedure not only solves any system, but does so using only the three basic operations on systems.

PROBLEMS 1.3

1. Consider the system

$$x_1 + x_2 + 2x_3 = 12$$
$$2x_1 + 4x_2 + 3x_3 = 28.$$

(a) Use the elimination step to find an equivalent system with x_1 eliminated from the second equation.

(b) Use the elimination step to find an equivalent system with x_2 eliminated from the second equation.

(c) Use the elimination step to find an equivalent system with x_3 eliminated from the second equation.

(d) Which of the answers above are in echelon form?

*2. Identify the systems that are in echelon form. Explain why the others are not in echelon form.

(a)
$$2x_1 + x_2 \qquad\qquad = 2$$
$$x_1 \qquad\quad + x_4 = 3$$
$$x_2 + x_3 + x_4 = 1$$

(b)
$$3x_1 + 4x_2 = 5$$
$$x_2 = 3$$

(c)　　　$x_2 + 3x_3 = 13$
　　　$x_1 + x_2 - 2x_3 = 10$
　　　$x_2　　= -1$

(d)　　　　　$x_2 = 1$
　　　$2x_1 - 4x_2 = 0$

(e)　$x_1 + x_2 + 3x_3　　= 1$
　　　$-2x_3 + x_4 = 3$

(f)　$x_1 - x_2 = 20$
　　　$x_2 = -4$

(g)　$x_1 + x_2 + x_3 + x_4 + x_5 + x_6 = 3$
　　　$x_3 + x_4 + x_5 - x_6 = 4$
　　　　　　　$x_6 = 1$

(h)　$3x_1　　　= 2$
　　　$32x_2　　= 7$
　　　　$4x_3　= 5$
　　　　$5x_4 = 1$

3. Use back substitution to solve each of the following echelon systems.
　*(a)　$x_1 + 2x_2 = 6$
　　　　$3x_2 = 6$

　(b)　$2x_1 + x_2 = 5$
　　　　$-x_2 = 4$

　(c)　$3x_1 + 4x_2 = 7$
　　　　$3x_2 = 4$

　*(d)　$x_1 + 2x_2 + 3x_3 = 9$
　　　　$x_3 = 1$

　(e)　$2x_1 + x_2 + x_3 = 5$
　　　　$3x_2 + 2x_3 = 1$

　*(f)　$x_1 + 2x_2 + 3x_3 = 9$
　　　　$2x_2 + x_3 = 7$
　　　　$x_3 = 1$

　(g)　$2x_1 + x_2 + x_3 + x_4 = 6$
　　　　$2x_3 + x_4 = 8$

　*(h)　$2x_1 + 3x_2 + x_3 - x_4 + x_5 = 6$
　　　　$x_3 + x_5 = 4$
　　　　$x_4 - x_5 = 2$

　(i)　$2x_1 + x_2 + x_3 = 5$
　　　　$3x_2 + 2x_3 = 1$
　　　　$x_3 = 2$

　(j)　$x_1 + 2x_2 = 3$
　　　　$x_2 + 2x_3 = -2$
　　　　$x_3 + 2x_4 = 4$

　(k)　$x_1 + 2x_2 - 3x_3 + x_4 - x_5 = -2$
　　　　$x_2 - 2x_3 + x_4 - 2x_5 = -1$
　　　　$2x_3 - x_4 + x_5 = -3$
　　　　$x_4 - 2x_5 = 1$

　*(l)　$2x_1 + x_2 + x_3 + x_4 = 2$
　　　　$x_2 + x_3 + x_6 = 3$
　　　　$x_3 = -1$
　　　　$x_5 + x_6 = 4$

　*(m)　$x_1 + 2x_2 + 3x_3 + 4x_4 + 5x_5 + 6x_6 = 17$

4. (a)　Show that each of the three basic operations is reversible; that is, show that the changes made by any one of the operations can be undone by an operation of the same type.

(b) Show that the operation "multiply an equation by a constant" is not reversible. This explains why in the second operation a nonzero constant is specified.

5. The back substitution procedure described in this section finds an **upper echelon** system that is equivalent to the original system.
   (a) Describe what a system in **lower echelon** form would look like.
   (b) How would the back substitution procedure have to be changed in order for it to work on a lower echelon system?

6. (a) Substitute the value 4 for x_3 in the equation
$$2x_1 + 3x_2 + 2x_3 = 5.$$
   (b) Subtract two times the equation $x_3 = 4$ from the equation
$$2x_1 + 3x_2 + 2x_3 = 5.$$
   (c) Describe how substitution can be accomplished by using the elimination step.

1.4 SOLVING SYSTEMS—FORWARD ELIMINATION, GAUSSIAN ELIMINATION

In this section we develop a procedure (called forward elimination) that will transform any linear system into an equivalent echelon system. Then we show how forward elimination and back substitution can be combined to produce an algorithm (called Gaussian elimination) that solves all linear systems.

Forward Elimination

EXAMPLE 1 Consider the system

$$\begin{align*}
x_1 + 2x_2 + 3x_3 &= 6 \\
2x_1 + 5x_2 + x_3 &= 9 \\
x_1 + 4x_2 - 6x_3 &= 1.
\end{align*} \tag{1}$$

Let us show how to transform this system into an equivalent echelon system. First we must eliminate the terms involving x_1 in the second and third equations. To eliminate the x_1 in the second equation, we subtract twice the first equation from the second equation. To eliminate the x_1 in the third equation, we subtract the first equation from the third equation. (What we are doing is using the first equation to eliminate x_1 from all subsequent equations.) This gives us the following new system.

$$\begin{align*}
x_1 + 2x_2 + 3x_3 &= 6 \\
x_2 - 5x_3 &= -3 \\
2x_2 - 9x_3 &= -5
\end{align*}$$

Now we must eliminate the x_2 term from the third equation. This is done by subtracting two times the second equation from the third equation. (We are using the second equation to eliminate x_2 from the third equation.)

$$\begin{aligned} x_1 + 2x_2 + 3x_3 &= 6 \\ x_2 - 5x_3 &= -3 \\ x_3 &= 1 \end{aligned} \qquad (2)$$

This is an echelon system (in fact, a triangular system) which can be solved by back substitution.

EXAMPLE 2   Let us transform the following system into echelon form.

$$\begin{aligned} x_1 - 3x_2 + 3x_3 + x_4 + x_5 &= 34 \\ 2x_1 - 6x_2 - x_3 - 2x_4 - 5x_5 &= -8 \\ 3x_1 - 9x_2 - 5x_3 + x_4 - 11x_5 &= -20 \end{aligned}$$

After we eliminate x_1 we have the system

$$\begin{aligned} x_1 - 3x_2 + 3x_3 + x_4 + x_5 &= 34 \\ - 7x_3 - 4x_4 - 7x_5 &= -76 \\ - 14x_3 - 2x_4 - 14x_5 &= -122. \end{aligned}$$

Since the variable x_2 does not occur in the last two equations, it is a free variable. We must continue by moving to the next determined variable, in this case x_3. After we eliminate x_3 from the third equation, we have

$$\begin{aligned} x_1 - 3x_2 + 3x_3 + x_4 + x_5 &= 34 \\ -7x_3 - 4x_4 - 7x_5 &= -76 \\ 6x_4 &= 30. \end{aligned}$$

The system is now in echelon form. The free variables are x_2 and x_5. The determined variables are x_1, x_3, and x_4.

This systematic procedure for changing a system into echelon form is called **forward elimination**. First we use the initial equation to eliminate x_1 from all the other equations. Then we use the new second equation to eliminate the next determined variable from the third and subsequent equations. The procedure continues in the obvious way. After each series of eliminations is completed, we move down to the next equation and over to the next determined variable to begin the next series of eliminations.

EXAMPLE 3   Just to make sure that the forward elimination procedure is clearly understood, let us transform a larger system into echelon form.

$$\begin{aligned} 3x_1 - 4x_2 + x_3 + 2x_4 - 2x_5 + x_6 &= -19 \\ 3x_1 - 2x_2 + 2x_3 + x_4 + x_5 + x_6 &= -1 \\ 6x_1 - 6x_2 + 3x_3 + 4x_4 + x_5 + x_6 &= -9 \\ 3x_1 - 2x_2 + 2x_3 + 3x_4 + 6x_5 + x_6 &= 26 \end{aligned}$$

STEP 1.   Use the first equation to eliminate x_1 from all subsequent equations.

$$3x_1 - 4x_2 + x_3 + 2x_4 - 2x_5 + x_6 = -19$$
$$2x_2 + x_3 - x_4 + 3x_5 = 18$$
$$2x_2 + x_3 + 5x_5 - x_6 = 29$$
$$2x_2 + x_3 + x_4 + 8x_5 = 45$$

STEP 2.   Use the second equation to eliminate x_2 from all subsequent equations.

$$3x_1 - 4x_2 + x_3 + 2x_4 - 2x_5 + x_6 = -19$$
$$2x_2 + x_3 - x_4 + 3x_5 = 18$$
$$x_4 + 2x_5 - x_6 = 11$$
$$2x_4 + 5x_5 - x_6 = 27$$

STEP 3.   The next variable, x_3, is a free variable. Move over to the next determined variable, x_4, and use the third equation to eliminate x_4 from the last equation.

$$3x_1 - 4x_2 + x_3 + 2x_4 - 2x_5 + x_6 = -19$$
$$2x_2 + x_3 - x_4 + 3x_5 = 18$$
$$x_4 + 2x_5 - x_6 = 11$$
$$x_5 + x_6 = 5$$

The system is now in echelon form and, of course, can be solved using back substitution.

EXERCISE 1   Use forward elimination to transform the following system into echelon form.

$$3x_1 + x_2 + 2x_3 + x_4 + x_5 = 6$$
$$6x_1 + 2x_2 + 5x_3 + x_4 + 3x_5 = 12$$
$$3x_1 + x_2 + 4x_3 + x_4 - 2x_5 = 8$$

After one slight modification, the forward elimination algorithm will be complete.

Pivoting

Let us see what happens when we apply forward elimination to the following system.

$$x_1 + 2x_2 + x_3 + x_4 = 4$$
$$2x_1 + 4x_2 - x_3 + 2x_4 = 11 \qquad (3)$$
$$x_1 + x_2 + 2x_3 + 3x_4 = 1$$

The first step is to use the first equation to eliminate x_1 from all subsequent equations.

$$x_1 + 2x_2 + x_3 + x_4 = 4$$
$$- 3x_3 \qquad\quad = 3 \qquad\qquad (4)$$
$$-x_2 + x_3 + 2x_4 = -3$$

We now must move down to the second equation and over to the next determined variable. Since the variable x_2 is not a free variable (it was not eliminated from all of the remaining equations), we should attempt to use the second equation to eliminate x_2 from the third equation. However, the coefficient of x_2 in the second equation is zero and it is therefore impossible to use the second equation to eliminate x_2 from any other equation. Note that if the equations had initially been in a different order, we would have had no problem changing this system to echelon form. The initial order of the equations obviously does not affect the solutions of a system. As we saw in the last section, we can change the order of the equations without changing the solutions of the system. In addition, a change of order might enable us to continue the forward elimination procedure. For example, if we interchange the second and third equations in (4), the system is in echelon form (and the solution can be found using back substitution).

In forward elimination we repeatedly use one equation (the pivot equation) to eliminate a variable from all subsequent equations. As we have seen from the example, the coefficient of this variable must be nonzero. If this coefficient is zero, we must try to find a subsequent equation with a nonzero coefficient of the variable. If we can find such an equation, we then interchange this equation and the pivot equation. This process is called **pivoting** and the new equation becomes the pivot equation. We are now able to use the new pivot equation to eliminate the variable from all subsequent equations. If there is no subsequent equation with a nonzero coefficient for the variable, this variable is a free variable. We must move to the next variable and repeat the process.

From now on we assume that the forward elimination procedure includes pivoting. With the introduction of pivoting we are now using all three of the basic operations on equations that were introduced in Section 1.3.

We illustrate this procedure on the following system.

$$x_1 + x_2 + 3x_3 + x_4 = -1$$
$$2x_1 + 2x_2 - x_3 + 2x_4 = 5$$
$$2x_1 + 3x_2 + 4x_3 + 2x_4 = 3$$
$$2x_1 + 3x_2 + 4x_3 + 4x_4 = 3$$

STEP 1. Use the first equation to eliminate x_1 from all subsequent equations. (The first pivot is 1, the coefficient of x_1 in the first equation.) This gives us

$$x_1 + x_2 + 3x_3 + x_4 = -1$$
$$- 7x_3 \qquad\quad = 7$$
$$x_2 - 2x_3 \qquad\quad = 5$$
$$x_2 - 2x_3 + x_4 = 5$$

STEP 2.   Interchange the second and third equations, making the second pivot 1. Now use the new second equation to eliminate x_2 from all subsequent equations.

$$\begin{aligned} x_1 + x_2 + 3x_3 + x_4 &= -1 \\ x_2 - 2x_3 &= 5 \\ -7x_3 &= 7 \\ x_4 &= 0 \end{aligned}$$

The third pivot is -7 and the fourth pivot is 1. The system is now in echelon form.

EXERCISE 2   Use forward elimination to reduce the following system to echelon form. Indicate when each of the three basic operations is used. Identify the pivots.

$$\begin{aligned} x_1 + 2x_2 + x_3 &= 3 \\ 2x_1 + 4x_2 \quad\;\; &= 7 \\ x_2 + 2x_3 &= 4. \end{aligned}$$

FORWARD ELIMINATION

Given

An arbitrary linear system.

Goal

To reduce the system to echelon form.

Procedure

1. (Pivot) If necessary, interchange equations to make the coefficient of the first variable in the first equation nonzero.
2. (Eliminate) Use the first equation to eliminate the first variable from all subsequent equations.

Remark:

The first equation remains unchanged during the rest of the procedure and will be an equation in the final system. The remaining equations form a (smaller) system with one less equation and at least one less variable.

3a. (Repeat) If the smaller system referred to in the remark involves one or more variables *and* more than one equation, apply the full procedure to this smaller system.
3b. (Done) If the smaller system does not involve any variables *or* involves only one equation, stop. The original system has now been reduced to echelon form.

Inconsistent Systems

So far, the systems we have solved have had at least one solution. Let us see what forward elimination does to an inconsistent system.

EXAMPLE 4   Consider the inconsistent system

$$2x_1 + 3x_2 = 4$$
$$2x_1 + 3x_2 = 5.$$

Using the first equation to eliminate x_1 from the second equation, we obtain the equivalent system

$$2x_1 + 3x_2 = 4$$
$$0x_1 + 0x_2 = 1.$$

The new second equation is never true. No matter what values are substituted for x_1 and x_2, the resulting expression is false because 0 does not equal 1.

EXAMPLE 5   Consider the system

$$x_1 + 2x_2 = 5$$
$$-x_1 + x_2 = 1 \tag{5}$$
$$x_1 + x_2 = 6.$$

Eliminate x_1 from the second and third equations.

$$x_1 + 2x_2 = 5$$
$$3x_2 = 6$$
$$- x_2 = 1$$

Eliminate x_2 from the third equation.

$$x_1 + 2x_2 = 5$$
$$3x_2 = 6$$
$$0 = 3$$

The final equation above is the equation $0x_1 + 0x_2 = 3$. Since $0x_1 + 0x_2$ always equals 0, there cannot be any solution of this equation. It is false no matter what values are chosen for x_1 and x_2. Since any solution to (5) must be a solution to the false equation $0 = 3$, we conclude that (5) has no solutions.

Forward elimination announces that a system is inconsistent by generating a false equation $0 = a$, where $a \neq 0$. The forward elimination procedure will produce such a false equation if and only if the original system is inconsistent. (Why?)

EXERCISE 3 Show that the following system is inconsistent.

$$\begin{aligned} x_1 + x_2 &= 5 \\ 2x_1 + x_2 &= 6 \\ 4x_1 + 3x_2 &= 11 \end{aligned}$$

EXERCISE 4 Use forward elimination to show that there is no straight line passing through the three points $(0,0)$, $(1,1)$, $(2,3)$.

"Disappearing" Equations

Consider the following system.

$$\begin{aligned} x_1 + 2x_2 &= 5 \\ -x_1 + x_2 &= 1 \\ x_1 + x_2 &= 3 \end{aligned} \tag{6}$$

Eliminating x_1 from the second and third equations, we get

$$\begin{aligned} x_1 + 2x_2 &= 5 \\ 3x_2 &= 6 \\ -x_2 &= -2. \end{aligned}$$

Eliminating x_2 from the third equation reduces the system to

$$\begin{aligned} x_1 + 2x_2 &= 5 \\ 3x_2 &= 6 \\ 0 &= 0. \end{aligned} \tag{7}$$

We have now finished applying forward elimination to system (6). We are left with an echelon system that can be solved by back substitution; $x_1 = 1$, $x_2 = 2$. But we still need to discuss the significance of the equation $0 = 0$.

Actually, it is not at all surprising that the third equation "disappeared." The original third equation is two-thirds of the first equation minus one-third of the second equation. Since it can be expressed as a combination of the first two equations, the original third equation does not provide any new information; it is redundant. The equation $0 = 0$, the equation that "disappears," is the way forward elimination announces that one of the original equations is redundant.

There is another way to look at this situation. The equation $0 = 0$ in (7) is true for any choice of x_1 and x_2. This equation does not limit the solutions of system (7) in any way. Similarly, the third equation in (6) is just a combination of the first two equations and does not limit the solutions of that system. Thus it is very natural for the third equation to become $0 = 0$ in this situation.

EXERCISE 5   Find the solutions of

$$
\begin{aligned}
x_1 - x_2 \qquad\quad + 2x_4 &= 2 \\
2x_1 \qquad\; + x_3 + 3x_4 &= -1 \\
x_1 \qquad\; + 3x_3 \qquad\quad &= 4 \\
4x_1 - x_2 + 4x_3 + 5x_4 &= 5.
\end{aligned}
$$

The Full Algorithm

We have now developed the two-step solution procedure, which will solve any system. This procedure first applies forward elimination to convert the system to echelon form. If the echelon system is consistent, we use back substitution to find the solution(s). If the echelon system is inconsistent, there are no solutions. This full solution procedure is called **Gaussian elimination** (after K. F. Gauss, 1777–1855, distinguished German mathematician, physicist, and astronomer). This complete process can be accomplished by repeated use of the three basic operations that were introduced in Section 1.3.

GAUSSIAN ELIMINATION

Given
    An arbitrary linear system.

Goal
    To find all solutions of the system.

Procedure
1. Use forward elimination to reduce the system to echelon form.
2a. If there are any inconsistent equations, stop. The system has no solution.
2b. If there are no free variables, use back substitution to find the values of the variables. There will be one solution.
2c. If there are free variables, use back substitution to find the determined variables in terms of the free variables. There will be an infinite number of solutions.

Some Important Facts

In parts (a)–(f) of Theorem 1.1 we bring together the most significant facts about systems of equations. We refer to these facts often in subsequent chapters.

Theorem 1.1

(*a*) *A consistent system that has no free variables has only one solution.*

(*b*) *A consistent system that has one or more free variables has an infinite number of solutions.*

(*c*) *The number of free variables plus the number of determined variables equals the total number of variables.*

(*d*) *The number of determined variables is equal to the number of pivots.*

(*e*) *The number of determined variables is less than or equal to the number of equations.*

(*f*) *The system*

$$
\begin{aligned}
a_{11}x_1 + a_{12}x_2 + \cdots + a_{1n}x_n &= 0 \\
a_{21}x_1 + a_{22}x_2 + \cdots + a_{2n}x_n &= 0 \\
\vdots \qquad \vdots \qquad\qquad \vdots \quad\ \ \vdots & \\
a_{m1}x_1 + a_{m2}x_2 + \cdots + a_{mn}x_n &= 0
\end{aligned}
\tag{8}
$$

is consistent. If $m < n$, *then this system has an infinite number of solutions.*

Proof Parts (a)–(e) have been proved in our discussion of the Gaussian elimination procedure. We can always find one solution to (8) by setting $x_1 = x_2 = \cdots = x_n = 0$. If $m < n$, then at least one variable is free and thus there are an infinite number of solutions by part (b). This proves part (f).

A system like (8) which has every constant equal to 0 is called a **homogeneous system**. In the proof of part (f) of the theorem we showed that every homogeneous system has the solution $x_1 = x_2 = \cdots = x_n = 0$. This solution is called the **trivial solution** of the system. Systems that are not homogeneous are called **nonhomogeneous**, and solutions that are not the same as the trivial solution are called **nontrivial**. Using this language, we can restate part (f) as follows:

(*f′*) *A homogeneous system always has a trivial solution. If the homogeneous system has fewer equations than unknowns, then it has an infinite number of solutions; in particular, it always has a nontrivial solution.*

Gauss–Jordan Elimination

There is an alternative solution procedure, called **Gauss–Jordan elimination**, which is frequently used. In this procedure the pivot equation is first divided by the coefficient of the pivot variable. Then the pivot variable is eliminated from the prior equations as well as the subsequent ones. It can be shown that the Gauss–Jordan procedure requires 50% more multiplications than the Gaussian elimination procedure (see Chapter 9). If the only time-consuming operation were multiplication (as is indeed the case when computations are performed on a computer), then the Gauss–Jordan procedure would clearly be less efficient. But a significant portion of the work involved when doing calculations by hand is the writing of all the intermediate systems. Since Gauss–Jordan requires much less writing, it can be done at least as quickly as Gaussian elimination.

For most of the remainder of this book we will use the phrase "Gaussian elimination" as an abbreviation for either of these two complete solution procedures. We suggest that the student understand both procedures. Routine computations can be done by either method.

EXAMPLE 6   Let us solve the following system by Gauss–Jordan elimination.

$$x_1 + 3x_2 + 4x_3 = 8$$
$$2x_1 + 9x_2 + 6x_3 = 27$$
$$x_1 + 5x_2 + 6x_3 = 15$$

STEP 1.   Use the first equation to eliminate x_1 from all subsequent equations.

$$x_1 + 3x_2 + 4x_3 = 8$$
$$3x_2 - 2x_3 = 11$$
$$2x_2 + 2x_3 = 7$$

STEP 2.   Make the second pivot 1 and use the second equation to eliminate x_2 from *all* other equations.

$$x_1 \qquad + 6x_3 = -3$$
$$x_2 - \frac{2}{3}x_3 = \frac{11}{3}$$
$$\frac{10}{3}x_3 = -\frac{1}{3}$$

STEP 3.   Make the third pivot 1 and use the third equation to eliminate x_3 from *all* other equations.

$$x_1 \qquad = -\frac{12}{5}$$
$$x_2 \qquad = \frac{18}{5}$$
$$x_3 = -\frac{1}{10}$$

PROBLEMS 1.4

1. Use forward elimination to reduce the following systems to echelon form.

*(a) $2x_1 + 3x_2 = 6$
      $2x_1 + 4x_2 = 8$

(b) $x_1 + 6x_2 = 7$
      $3x_1 + x_2 = 4$

(c) $\begin{aligned} x_1 + x_2 &= 1 \\ x_1 + 2x_2 + x_3 &= 1 \\ x_2 + 2x_3 &= 1 \end{aligned}$

*(d) $\begin{aligned} x_1 + x_2 + x_3 + x_4 &= 1 \\ x_1 + 2x_2 + 3x_3 + 4x_4 &= 2 \\ x_1 + 2x_2 + 4x_3 + 5x_4 &= 3 \\ x_1 + 2x_2 + 4x_3 + 6x_4 &= 4 \end{aligned}$

2. Solve each of the following systems of equations using Gaussian elimination. Identify any free variables.

*(a) $\begin{aligned} x_1 + 3x_2 &= 2 \\ 2x_1 + 6x_2 &= 7 \end{aligned}$

(b) $\begin{aligned} 2x_1 + x_2 + 2x_3 &= 5 \\ 4x_1 + 5x_2 - 3x_3 &= 3 \end{aligned}$

*(c) $\begin{aligned} x_1 \quad\quad + 3x_3 &= 2 \\ 2x_1 - 3x_2 + 4x_3 &= 6 \\ x_1 - 7x_2 - 2x_3 &= 6 \end{aligned}$

(d) $\begin{aligned} x_1 + x_2 + x_3 &= 0 \\ 6x_1 + 6x_2 - 2x_3 &= -8 \\ 3x_2 - x_3 &= -3 \end{aligned}$

*(e) $\begin{aligned} x_1 + x_2 - x_3 + x_4 &= 7 \\ x_1 + 2x_2 \quad\quad + 2x_4 &= 15 \\ x_1 + x_2 \quad\quad\quad &= 17 \end{aligned}$

(f) $\begin{aligned} x_2 - x_3 &= 1 \\ x_1 - x_2 + x_3 &= 0 \\ 2x_1 + x_2 + 2x_3 &= 2 \end{aligned}$

*(g) $\begin{aligned} x_1 - x_2 - 3x_3 &= 2 \\ x_1 - 3x_2 - 13x_3 &= 14 \\ -3x_1 - 4x_2 + 4x_3 &= 0 \end{aligned}$

(h) $\begin{aligned} 2x_1 + 3x_2 + x_3 - x_4 &= 2 \\ 2x_1 - x_2 + x_3 - x_4 &= 4 \\ 4x_1 + 6x_2 + 2x_3 - 2x_4 &= 6 \end{aligned}$

*(i) $\begin{aligned} 2x_1 + 3x_2 + x_3 - x_4 &= 2 \\ 2x_1 - x_2 + x_3 - x_4 &= 4 \\ 4x_1 + 2x_2 + 2x_3 - 2x_4 &= 6 \end{aligned}$

(j) $\begin{aligned} x_1 + 2x_2 - 3x_3 &= 4 \\ x_1 - 4x_2 - 13x_3 &= 14 \\ -3x_1 - 6x_2 + 4x_3 &= 2 \end{aligned}$

*(k) $\begin{aligned} 4x_1 + 3x_2 + 2x_3 &= 0 \\ x_1 - 4x_2 + x_3 &= 1 \end{aligned}$

(l) $\begin{aligned} x_1 - 6x_2 \quad\quad + 2x_4 &= 0 \\ 2x_1 - 3x_2 \quad\quad - 4x_4 &= 0 \\ x_2 + 2x_3 + 3x_4 &= 0 \\ 3x_1 - 12x_2 + x_3 + x_4 &= 0 \end{aligned}$

*(m) $\begin{aligned} x_1 - 2x_2 + 7x_3 &= 1 \\ 2x_1 - 4x_2 + x_3 &= 0 \\ x_1 + x_2 + x_3 &= 1 \end{aligned}$

(n) $\begin{aligned} x_1 + x_2 + x_3 + x_4 + x_5 &= -2 \\ x_1 \quad\quad\quad\quad\quad + x_5 &= 4 \\ x_1 + x_2 \quad\quad\quad\quad &= 5 \end{aligned}$

*(o) $\begin{aligned} x_1 + 2x_2 + x_3 + 3x_4 &= 1 \\ x_1 + 2x_2 + 3x_3 - x_4 &= 5 \\ 2x_1 + 4x_2 + x_3 + x_4 &= 7 \end{aligned}$

(p) $x_1 + 2x_2 - x_3 - 2x_4 + 3x_5 - 3x_6 + 5x_7 = 0$

*(q) $3x_1 + x_3 - 2x_5 + 3x_7 = 0$

3. Solving a system by Gauss–Jordan elimination transforms the system into an equivalent system in special form, **reduced echelon form**.
   (a) Define reduced echelon form.
   (b) Give an algorithm for Gauss–Jordan elimination similar to the ones for Gaussian elimination and forward elimination that were given in the text.

4. Solve the following systems.

*(a) $\begin{aligned} x_1 + x_2 &= 2 \\ x_1 - x_2 &= 4 \end{aligned}$

(b) $\begin{aligned} x_1 - x_2 &= -1 \\ -x_1 + 2x_2 &= 4 \end{aligned}$

(c) $\begin{aligned} 2x_1 + x_2 &= 7 \\ 4x_1 - x_2 &= 17 \end{aligned}$

*(d) $x_1 + x_2 + x_3 = -2$ (e) $x_1 - x_2 + x_3 = 3$
  $-x_1 + x_2\ \ \ \ \ \ \ \ = 0$ $4x_1 - 3x_2 - x_3 = 6$
  $\ \ \ \ \ \ x_2 + x_3 = -1$ $3x_1 + x_2 + 2x_3 = 23$

(f) $-2x_1 - 2x_2 - 2x_3 = 0$ *(g) $x_1 + x_2 + x_3 + x_4 = 1$
      $\ \ \ \ \ \ - 2x_2 + 6x_3 = 0$ $2x_1 + x_2 - x_3 + 2x_4 = 9$
      $\ \ 2x_1 + 3x_2 + x_3 = 0$ $x_1 + 2x_2 + x_3 - x_4 = -6$
                                        $x_1 + x_2 - 2x_3 + x_4 = 7$

(h) $3x_1 + 7x_2 = 4$
      $2x_1 + 4x_2 = 7$

*(i) $x_1 + x_2\ = 1$
      $x_1 + 2x_2 + x_3\ \ \ \ \ \ \ \ \ \ \ \ \ \ \ \ \ \ \ = 3$
      $\ \ \ \ \ \ x_2 + 2x_3 + x_4\ \ \ \ \ \ \ \ \ \ \ \ = 3$
      $\ \ \ \ \ \ \ \ \ \ \ \ x_3 + 2x_4 + x_5\ \ \ \ \ \ \ = 1$
      $\ \ \ \ \ \ \ \ \ \ \ \ \ \ \ \ \ \ x_4 + 2x_5 + x_6 = -3$
      $\ x_5 + 2x_6 = -5$

1.5   VECTORS AND MATRICES

In this section we introduce vectors and matrices. Both of these concepts enable us to treat a collection of numbers as a single object. They also enable us to express a system of equations as one equation.

The system of equations (1) is determined by the coefficients of the variables (the a_{ij}'s), the constant terms (the b_i's), and the variables themselves (the x_i's). It is reasonable to try to think of each of these collections as a single object. Let us begin by looking at the variables.

$$
\begin{aligned}
a_{11}x_1 + a_{12}x_2 + \cdots + a_{1n}x_n &= b_1 \\
a_{21}x_1 + a_{22}x_2 + \cdots + a_{2n}x_n &= b_2 \\
\vdots \qquad\qquad \vdots \qquad\qquad \vdots \qquad \vdots & \\
a_{m1}x_1 + a_{m2}x_2 + \cdots + a_{mn}x_n &= b_m
\end{aligned}
\tag{1}
$$

One way to collect several quantities together is to write them in a list. If the order of the terms is important, we must be sure to use an ordered list. If we do this with the n variables, we will have what is called an **ordered n-tuple**

$$(x_1, x_2, \ldots, x_n).$$

We call an ordered n-tuple an **n-vector** or simply a **vector**. Given the n-vector $x = (x_1, x_2, \ldots, x_n)$ we call the number x_i the ith **coordinate** or the ith **component** of x. If x has every component equal to 0, we call x the **zero vector** and denote it by the symbol 0.

Although we have defined a vector as an ordered n-tuple, there are three distinct ways to visualize vectors:

1. As ordered n-tuples.
2. As points in Euclidean n-space.
3. As arrows.

The correspondence between points in Euclidean n-space and ordered n-tuples is well known, at least for $n = 2$ and 3. This correspondence is important to us because it will give us a way to look at vectors geometrically.

Arrows are directed line segments and as such they have both direction and length. For $n = 2$ and 3, there is a well-known and natural correspondence between the points in n-space and the arrows in n-space. Once a coordinate system has been determined, the arrow from the origin [the point $(0, 0, 0)$] to the point (x_1, x_2, x_3) has a direction and a magnitude (its length). Any direction–length combination determines precisely one such arrow. Arrows represent several important physical ideas; two of the most important are velocity and force. Every velocity (and every force) has both a direction and a magnitude. It thus determines a unique arrow (see Figure 1.2).

Since vectors can be thought of as arrows and arrows can be thought of as velocities, vectors should have all the properties of velocities. Velocities may be multiplied by a number. If the number is positive, the direction stays the same but the magnitude is changed. Velocities may also be added and subtracted using the parallelogram law (see Figure 1.3). We use these operations on velocities to define similar operations on vectors.

To add vectors we add the corresponding components. When $n = 2$ we have

$$(u_1, u_2) + (v_1, v_2) = (u_1 + v_1, u_2 + v_2),$$

and in general

$$(u_1, u_2, \ldots, u_n) + (v_1, v_2, \ldots, v_n) = (u_1 + v_1, u_2 + v_2, \ldots, u_n + v_n).$$

To multiply a vector by a number α (called a *scalar* in this context) each component is multiplied by the scalar. When $n = 2$ we have

$$\alpha(v_1, v_2) = (\alpha v_1, \alpha v_2),$$

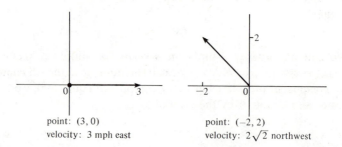

Figure 1.2

point: $(3, 0)$
velocity: 3 mph east

point: $(-2, 2)$
velocity: $2\sqrt{2}$ northwest

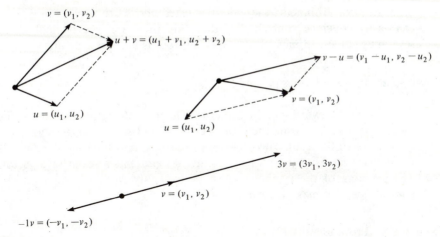

Figure 1.3

and in general

$$\alpha(v_1, v_2, \ldots, v_n) = (\alpha v_1, \alpha v_2, \ldots, \alpha v_n).$$

Note that when we add vectors, they must have the same number of components. The sum of two vectors is a vector and a scalar multiple of a vector is a vector. Multiplication of a vector by a scalar is called **scalar multiplication**.

Some simple geometric aspects of vector operations are illustrated in Figure 1.3. The sum of two vectors is the vector that is the diagonal of the parallelogram determined by the vectors. The difference between two vectors is a vector parallel to and the same length as the line segment joining them. Multiplication by a positive scalar stretches (or shrinks) the vector. Multiplication by -1 reverses the direction of the vector.

The following identities are a direct consequence of the definitions of vector operations.

Let α and β be scalars, and v, u, and w be vectors.

(1) $v + u = u + v$	(commutativity)
(2) $(v + u) + w = v + (u + w)$	(associativity)
(3) $v + 0 = v$	(zero vector)
(4) $v + (-1)v = v + (-v) = 0$	(negative)
(5) $(\alpha\beta)v = \alpha(\beta v)$	(quasi-associativity)
(6) $1v = v$	(identity scalar)
(7) $\alpha(v + u) = \alpha v + \alpha u$	(distributivity)
(8) $(\alpha + \beta)v = \alpha v + \beta v$	(distributivity)
(9) $0v = 0$	(zero scalar)

These identities should seem reasonable. Since vector operations are performed coordinatewise and the coordinates are numbers that satisfy the usual rules of arithmetic, the identities listed above are very natural. You can easily convince yourself that these identities are true by looking at several examples. In most cases a formal proof is not too difficult.

EXERCISE 1 Prove identities (1) and (5).

Identity (9) is worthy of special note. The 0 on the left-hand side of the identity is a scalar, while the 0 on the right-hand side is an n-vector—the zero vector $(0, 0, \ldots, 0)$. This ambiguous usage is unfortunately standard. It is usually clear from the context whether 0 represents a scalar or a vector. The meaning of the symbol 0 must be explicitly specified whenever the context does not make it clear.

EXERCISE 2 Prove identity (9).

Some treatments of vectors use an algebraic approach. They start with identities (1) through (8) above—the algebra of vectors. We have chosen the geometric approach because it is more concrete and more intuitive. It is important to recognize that vectors have algebraic and geometric properties simultaneously. This is why vectors are so important.

We have now shown how to look at the variables of system (1) as coordinates of a vector. Two collections of numbers remain, the coefficients and the constant terms. The constant terms can be disposed of quickly. We interpret the constant terms as coordinates of a vector—an m-vector since there are m of them.

We now consider the collection of coefficients. When we look at a system of equations and ignore everything except the coefficients, we see a two-dimensional array of numbers. A two-dimensional array of numbers is called a **matrix**. Every matrix can be thought of as a collection of rows (in our case, the coefficients from a certain equation) or as a collection of columns (the coefficients of a certain variable). We will use capital letters to designate matrices.

The matrix of coefficients of the system (1) is

$$A = \begin{bmatrix} a_{11} & a_{12} & a_{13} & \cdots & a_{1n} \\ a_{21} & a_{22} & a_{23} & \cdots & a_{2n} \\ a_{31} & a_{32} & a_{33} & \cdots & a_{3n} \\ \vdots & \vdots & \vdots & & \vdots \\ a_{m1} & a_{m2} & a_{m3} & \cdots & a_{mn} \end{bmatrix}.$$

The numbers $a_{k1}, a_{k2}, \ldots, a_{kn}$ are all in the same **row** of A, the kth row. Similarly, the numbers $a_{1k}, a_{2k}, \ldots, a_{mk}$ are the kth **column** of A. The number a_{ij} is the **entry** or **element** which is found in the ith row and the jth column of the matrix. The matrix A is an $m \times n$ (read m by n) matrix because it has m rows and n columns. We can consider each row of A as an

n-vector and each column as an m-vector. Matrices with the same number of columns as rows are called **square**.

EXAMPLE 1   Let A be the matrix $\begin{bmatrix} 3 & -1 & 0 \\ 2 & 7 & 1 \end{bmatrix}$. The rows of A can be thought of as 3-vectors and the columns as 2-vectors. If the element in the ith row and the jth column of A is called a_{ij}, then $a_{23} = 1$. Because A has 2 rows and 3 columns, A is a 2×3 (2 by 3) matrix.

The algebraic operations of addition and scalar multiplication are also defined for matrices. They are defined just as they are for vectors. Matrices are added by adding corresponding elements; that is, they are added coordinatewise. Like vectors, matrices must be the same size to be added. The sum of two matrices is a matrix of the same size. To multiply a matrix by a scalar we multiply every entry in the matrix by the scalar. The product of a scalar and a matrix is a matrix. A **zero matrix** (unfortunately designated by that much overworked symbol 0) is a matrix that has every entry equal to the number 0. Note that there are many different zero matrices—one for each possible shape.

If A, B, and C are matrices and α and β are scalars, then the following identities hold.

(1) $A + B = B + A$ \hspace{2em} (commutativity)
(2) $(A + B) + C = A + (B + C)$ \hspace{1em} (associativity)
(3) $A + 0 = A$ \hspace{2em} (zero matrix)
(4) $A + (-1)A = A + (-A) = 0$ \hspace{1em} (negative)
(5) $(\alpha\beta)A = \alpha(\beta A)$ \hspace{2em} (quasi-associativity)
(6) $1A = A$ \hspace{2em} (identity scalar)
(7) $\alpha(A + B) = \alpha A + \alpha B$ \hspace{1em} (distributivity)
(8) $(\alpha + \beta)A = \alpha A + \beta A$ \hspace{1em} (distributivity)
(9) $0A = 0$ \hspace{2em} (zero scalar)

These are just the ordinary rules of arithmetic applied to the entries of the matrices. (Compare this list with the list on page 35.)

EXAMPLE 2   If

$$A = \begin{bmatrix} 3 & -1 & 0 \\ 2 & 7 & 1 \end{bmatrix} \quad \text{and} \quad B = \begin{bmatrix} 5 & 4 & 1 \\ 2 & -3 & -4 \end{bmatrix},$$

then

$$A + B = \begin{bmatrix} 8 & 3 & 1 \\ 4 & 4 & -3 \end{bmatrix} \quad \text{and} \quad 5A = \begin{bmatrix} 15 & -5 & 0 \\ 10 & 35 & 5 \end{bmatrix}.$$

If $C = \begin{bmatrix} 1 & 2 \\ 3 & 4 \end{bmatrix}$, then $A + C$ is not defined because A and C are not the same shape (size).

EXERCISE 3   Let

$$A = \begin{bmatrix} 4 & 0 & 2 \\ 5 & -1 & 3 \end{bmatrix} \quad \text{and} \quad B = \begin{bmatrix} 2 & 4 & 0 \\ 1 & 1 & 3 \end{bmatrix}.$$

Compute $A - B$, $4A + 3B$, and $B + 2A$.

System (1) is completely determined by its coefficient matrix A, unknown vector x, and constant vector b. If we rewrite system (1) as $Ax = b$, we will have transformed the system of equations into a single equation, an equation involving a matrix and two vectors. To do this we must know what we mean by the product of a matrix and a vector. But that is easy—we define the product of the matrix A and the vector x so that it is the left-hand side of the system (1), interpreted as a vector. In other words, we define Ax so that it does equal b. Before giving the general formula we need to make a remark about how to write vectors in formulas involving matrices and to illustrate this remark with an example.

Whenever we multiply a matrix times a vector, we need to write the vector as a column vector. Column vectors will be the same as row vectors except that the entries will be written in a column. The reasons for this convention will be made clear in the next section.

Suppose that we start with the system

$$\begin{array}{rcl} 3x_1 + 2x_2 + 3x_3 &=& 5 \\ 2x_1 - x_2 + x_3 &=& 4 \\ x_1 + x_2 - x_3 &=& 6. \end{array} \tag{2}$$

Then $x = (x_1, x_2, x_3)$, $b = (5, 4, 6)$, and

$$A = \begin{bmatrix} 3 & 2 & 3 \\ 2 & -1 & 1 \\ 1 & 1 & -1 \end{bmatrix}.$$

If we interpret each side of (2) as a vector (and because a vector is being multiplied by a matrix, all vectors will be written as column vectors), we have the following vector equation:

$$\begin{bmatrix} 3x_1 + 2x_2 + 3x_3 \\ 2x_1 - x_2 + x_3 \\ x_1 + x_2 - x_3 \end{bmatrix} = \begin{bmatrix} 5 \\ 4 \\ 6 \end{bmatrix}. \tag{3}$$

The product Ax should be equal to the left-hand-side vector of (3). We thus have

$$Ax = \begin{bmatrix} 3 & 2 & 3 \\ 2 & -1 & 1 \\ 1 & 1 & -1 \end{bmatrix} \begin{bmatrix} x_1 \\ x_2 \\ x_3 \end{bmatrix} = \begin{bmatrix} 3x_1 + 2x_2 + 3x_3 \\ 2x_1 - x_2 + x_3 \\ x_1 + x_2 - x_3 \end{bmatrix}.$$

(Notice how we have remembered to write our vectors as columns again.) The second component of the product is obtained by using the second row of A, $(2, -1, 1)$, together with the vector x; multiplying corresponding components $(2x_1, -1x_2,$ and $1x_3)$; and then adding these products together $(2x_1 - x_2 + x_3)$. In general, the ith component of the product is obtained by using the ith row of A and the vector x. Corresponding components are multiplied and these products are then added together. This definition clearly requires the number of columns of the matrix to be equal to the number of components of the vector. It is also obvious from this definition that the product of an $m \times n$ matrix and an n-vector is an m-vector.

To summarize, we have

$$Ax = \begin{bmatrix} 3 & 2 & 3 \\ 2 & -1 & 1 \\ 1 & 1 & -1 \end{bmatrix} \begin{bmatrix} x_1 \\ x_2 \\ x_3 \end{bmatrix} = \begin{bmatrix} 3x_1 + 2x_2 + 3x_3 \\ 2x_1 - x_2 + x_3 \\ x_1 + x_2 - x_3 \end{bmatrix} = \begin{bmatrix} 5 \\ 4 \\ 6 \end{bmatrix} = b.$$

Using this, we are able to write the system (2) in the form

$$\begin{bmatrix} 3 & 2 & 3 \\ 2 & -1 & 1 \\ 1 & 1 & -1 \end{bmatrix} \begin{bmatrix} x_1 \\ x_2 \\ x_3 \end{bmatrix} = \begin{bmatrix} 5 \\ 4 \\ 6 \end{bmatrix}.$$

The general system (1) can be expressed as follows:

$$\begin{bmatrix} a_{11} & a_{12} & \cdots & a_{1n} \\ a_{21} & a_{22} & \cdots & a_{2n} \\ \vdots & \vdots & & \vdots \\ a_{m1} & a_{m2} & \cdots & a_{mn} \end{bmatrix} \begin{bmatrix} x_1 \\ x_2 \\ \vdots \\ x_n \end{bmatrix} = \begin{bmatrix} b_1 \\ b_2 \\ \vdots \\ b_m \end{bmatrix}.$$

We have now achieved our goal. The shorthand notation for linear systems enables us to write the system (1) as a single equation, an equation involving a matrix and two vectors.

EXAMPLE 3   Let

$$B = \begin{bmatrix} 2 & 4 & -2 & 3 \\ -1 & 0 & 2 & 3 \end{bmatrix} \quad \text{and} \quad x = (4, 1, -1, 3).$$

To compute Bx we first write x as a column vector,

$$x = \begin{bmatrix} 4 \\ 1 \\ -1 \\ 3 \end{bmatrix}.$$

The first component of Bx equals $2 \cdot 4 + 4 \cdot 1 + (-2)(-1) + 3 \cdot 3 = 23$. The second component equals $(-1)4 + 0 \cdot 1 + 2(-1) + 3 \cdot 3 = 3$. Thus

$$Bx = \begin{bmatrix} 2 & 4 & -2 & 3 \\ -1 & 0 & 2 & 3 \end{bmatrix} \begin{bmatrix} 4 \\ 1 \\ -1 \\ 3 \end{bmatrix} = \begin{bmatrix} 23 \\ 3 \end{bmatrix},$$

which can be rewritten in our usual notation as $(23, 3)$.

Exercise 4   Let

$$B = \begin{bmatrix} 1 & -3 \\ 2 & 5 \\ 1 & 4 \end{bmatrix} \quad \text{and} \quad x = (-1, 4).$$

Compute $B(3x)$, $(3B)x$, and $3(Bx)$.

Exercise 5   Write the following system as a single equation.

$$\begin{array}{rcl} 2x_1 + 3x_2 - x_3 &=& 1 \\ x_1 \phantom{{}+3x_2} + x_3 &=& 2 \\ 4x_1 - 3x_2 + x_3 &=& 7 \end{array}$$

Exercise 6   Write the system of equations associated with the following vector equation.

$$\begin{bmatrix} 3 & 1 & 4 & 2 \\ 5 & -1 & 2 & 1 \\ 8 & 1 & 2 & 1 \end{bmatrix} \begin{bmatrix} x_1 \\ x_2 \\ x_3 \\ x_4 \end{bmatrix} = \begin{bmatrix} 2 \\ 1 \\ -2 \end{bmatrix}$$

There is another extremely useful way to look at the product Ax. It will be used extensively in the rest of this book.

$$Ax = \begin{bmatrix} 3x_1 + 2x_2 + 3x_3 \\ 2x_1 - x_2 + x_3 \\ x_1 + x_2 - x_3 \end{bmatrix} = x_1 \begin{bmatrix} 3 \\ 2 \\ 1 \end{bmatrix} + x_2 \begin{bmatrix} 2 \\ -1 \\ 1 \end{bmatrix} + x_3 \begin{bmatrix} 3 \\ 1 \\ -1 \end{bmatrix}.$$

We see that we can interpret the product Ax as x_1 times the first column of A plus x_2 times the second column of A plus x_3 times the third column of A. If A is a larger matrix, then Ax equals the sum of all the (column) vectors obtained by multiplying the ith coordinate of x by the ith column of A. This way of looking at Ax is an important conceptual tool. It shows us that Ax is a sum of multiples of the columns of A.

Exercise 7   Let

$$A = \begin{bmatrix} 1 & 2 \\ -1 & 4 \end{bmatrix} \quad \text{and} \quad x = \begin{bmatrix} 1 \\ 2 \end{bmatrix}.$$

Show that Ax is the first column of A plus twice the second column of A.

If A and B are matrices, v and u vectors, and α a scalar, then the following identities are true.

(10) $A(v + u) = Av + Au$ (distributivity)
(11) $(A + B)v = Av + Bv$ (distributivity)
(12) $A(\alpha v) = \alpha(Av) = (\alpha A)v$ (quasi-associativity)
(13) $A0 = 0$ (zero vector)
(14) $0v = 0$ (zero matrix)

Of course, the indicated operations must be defined. This means that the matrices and vectors must be of the right size.

EXERCISE 8   Verify that (10), (11), and (12) are true when

$$A = \begin{bmatrix} 3 & -1 & 0 \\ 2 & 7 & 1 \end{bmatrix}, \qquad B = \begin{bmatrix} 5 & 4 & 1 \\ 2 & -3 & -4 \end{bmatrix},$$

$u = (1,3,4)$, $v = (-1,-2,3)$ and $\alpha = -3$.

Again, certain of these identities are worthy of special note. If A is an $m \times n$ matrix, then the 0 on the left-hand side of (13) is an n-vector while the 0 on the right-hand side is an m-vector. The 0 on the left-hand side of (14) is a zero matrix while the right-hand 0 is a zero vector. One can figure these things out almost all of the time.

EXERCISE 9   Explain why identities (13) and (14) are true.

EXAMPLE 4   The 2×2 matrix $I = \begin{bmatrix} 1 & 0 \\ 0 & 1 \end{bmatrix}$ is very interesting. $Iv = v$ for all 2-vectors v.

An $n \times n$ matrix A with $a_{ii} = 1$, for $i = 1, 2, \ldots, n$, and all other entries zero is called the $n \times n$ **identity matrix**. We will use the symbol I for identity matrices. Occasionally, we write I_n when it is necessary to indicate the size of the identity matrix. Iv always equals v. The elements a_{ii} in a matrix are called the **diagonal elements** or **diagonal entries**. The **main diagonal** of a matrix is all of its diagonal elements. A matrix is a **diagonal matrix** if every nonzero entry is on the diagonal, that is, if every element off the main diagonal is zero.

EXERCISE 10   Let

$$D = \begin{bmatrix} 3 & 0 & 0 \\ 0 & -1 & 0 \\ 0 & 0 & 2 \end{bmatrix} \qquad \text{and} \qquad v = (1, 2, -5).$$

(a) Compute Dv.
(b) In view of the answer to part (a), how would you multiply a diagonal matrix times a vector?

A word about notation. We normally use capital letters to designate matrices, lowercase letters to designate vectors, and Greek lowercase letters to designate numbers or scalars. Subscripted lowercase letters are either vectors or numbers, and the context usually indicates which. Letters at the beginning of the alphabet denote constant quantities, letters near the end usually denote variables or unknown quantities.

In the next section we use this matrix notation to solve systems of equations.

PROBLEMS 1.5

1. Let $u = (3, -1)$, $v = (2, 5)$, and $w = (0, 4)$. Compute $u + v$, $8u$, $-3w$, and $6u + 2v - w$.

*2. Let $u = (1, -2, 4)$, $v = (-3, 0, 9)$, and $w = (0, 0, 5)$. Compute $u - v$, $-3u + 5v$, and $u + v - 6w$.

3. Let $u = (4, -3, 2, 1)$ and $v = (4, 3, 2)$.
   (a) What is the third coordinate of u? the third coordinate of v?
   (b) Explain why we cannot compute $u + v$.

*4. If $u = (1, 3, 0, 6)$ and $v = (2, 1, -1, 4)$, find a 4-vector x such that:
   (a) $3x = 2u + v$ (b) $3u + 2x = u - 5v$

5. Let

$$A = \begin{bmatrix} -3 & 1 & 2 & 0 \\ 9 & 6 & -4 & 2 \end{bmatrix} \quad \text{and} \quad B = \begin{bmatrix} 1 & -1 & 1 & 5 \\ 0 & 3 & 0 & 2 \end{bmatrix}.$$

   Compute $A + B$, $-2A$, and $3A - B$.

*6. Repeat Problem 5 using the matrices

$$A = \begin{bmatrix} 0 & 0 & 0 \\ -1 & 2 & 3 \\ 0 & 1 & 6 \end{bmatrix} \quad \text{and} \quad B = \begin{bmatrix} 5 & 1 & 0 \\ 1 & 1 & 1 \\ -1 & 1 & 4 \end{bmatrix}.$$

*7. Let A and B be the matrices in Problem 5 and let $u = (0, 2, -1, 1)$ and $v = (3, 2, 1, -2)$. Compute Au, Av, Bu, $B(u + v)$, and $A(3v)$.

8. Let A and B be the matrices in Problem 6 and let $w = (4, -1, 2)$. Compute Aw, $(A + B)w$, and $-2Aw$.

9. Discuss the correspondence between the points on the plane and the ordered pairs of numbers. How does one determine a coordinate system for the plane? Once a coordinate system has been determined, is there one ordered pair for each point and one point for each ordered pair? What about arrows?

10. Let D be a diagonal matrix. Describe how a vector is changed when it is multiplied by D.

11. Prove that $Iv = v$.

12. Prove identities (2), (3), and (7) for vectors.

13. Prove identities (1), (3), (4), (8), and (9) for matrices.

14. Prove identities (11) and (12) for matrices.

1.6   MATRIX SOLUTION OF SYSTEMS

Now that we have a way to express every linear system as a matrix equation, we will show how this notation helps us to solve systems of equations.

Let us consider the system

$$2x_1 + x_2 - 2x_3 = 10$$
$$6x_1 + 4x_2 + 4x_3 = 2 \qquad\qquad (1)$$
$$10x_1 + 8x_2 + 6x_3 = 8.$$

The matrix of coefficients and the vector of constant terms completely determine this system. We associate with any system an **augmented matrix** consisting of the matrix of coefficients with an extra column added—a column that is the vector of constant terms. The augmented matrix for the system (1) is

$$\begin{bmatrix} 2 & 1 & -2 & \vdots & 10 \\ 6 & 4 & 4 & \vdots & 2 \\ 10 & 8 & 6 & \vdots & 8 \end{bmatrix}. \qquad\qquad (2)$$

Each row of this augmented matrix corresponds to an equation in the original system. The dotted line is to remind us of the equal signs in the original equations. Any operation performed on the equations of the system can also be performed on the rows of the augmented matrix. Let us look at an example.

Given the system

$$2x_1 + x_2 - 2x_3 = 10$$
$$6x_1 + 4x_2 + 4x_3 = 2$$
$$10x_1 + 8x_2 + 6x_3 = 8.$$

Use the first equation to eliminate x_1 from all subsequent equations.

$$2x_1 + x_2 - 2x_3 = 10$$
$$x_2 + 10x_3 = -28$$
$$3x_2 + 16x_3 = -42$$

Use the second equation to eliminate x_2 from all subsequent equations.

$$2x_1 + x_2 - 2x_3 = 10$$
$$x_2 + 10x_3 = -28$$
$$-14x_3 = 42$$

The system is now in triangular form. Solve for x_3 and substitute its value in all previous equations.

$$2x_1 + x_2 = 4$$
$$x_2 = 2$$
$$x_3 = -3$$

Given the augmented matrix

$$\begin{bmatrix} 2 & 1 & -2 & \vdots & 10 \\ 6 & 4 & 4 & \vdots & 2 \\ 10 & 8 & 6 & \vdots & 8 \end{bmatrix}.$$

Use the first row to make the first entry in each subsequent row zero.

$$\begin{bmatrix} 2 & 1 & -2 & \vdots & 10 \\ 0 & 1 & 10 & \vdots & -28 \\ 0 & 3 & 16 & \vdots & -42 \end{bmatrix}$$

Use the second row to make the second entry in each subsequent row zero.

$$\begin{bmatrix} 2 & 1 & -2 & \vdots & 10 \\ 0 & 1 & 10 & \vdots & -28 \\ 0 & 0 & -14 & \vdots & 42 \end{bmatrix}$$

The coefficient matrix is now triangular. Make the last pivot 1 and use the third row to make all previous entries in the third column zero.

$$\begin{bmatrix} 2 & 1 & 0 & \vdots & 4 \\ 0 & 1 & 0 & \vdots & 2 \\ 0 & 0 & 1 & \vdots & -3 \end{bmatrix}$$

Solve for x_2 and substitute its value in the first equation.

$$
\begin{aligned}
2x_1 \quad\quad\quad &= 2 \\
x_2 \quad\quad &= 2 \\
x_3 &= -3
\end{aligned}
$$

Make the second pivot 1 and use the second row to make all previous entries in the second column zero.

$$
\begin{bmatrix}
2 & 0 & 0 & \vdots & 2 \\
0 & 1 & 0 & \vdots & 2 \\
0 & 0 & 1 & \vdots & -3
\end{bmatrix}
$$

Solve for x_1

$$
\begin{aligned}
x_1 \quad\quad\quad &= 1 \\
x_2 \quad\quad &= 2 \\
x_3 &= -3
\end{aligned}
$$

Make the first pivot 1.

$$
\begin{bmatrix}
1 & 0 & 0 & \vdots & 1 \\
0 & 1 & 0 & \vdots & 2 \\
0 & 0 & 1 & \vdots & -3
\end{bmatrix}
$$

It should be clear that in this example we have been applying **row operations** to the augmented matrix in order to transform the coefficient matrix into the identity matrix. Since operating on the rows of a matrix is equivalent to operating on equations, the row operations we allow are identical with the basic operations on equations given in Section 1.3.

1. Subtract (or add) a multiple of one row from another row.
2. Multiply a row by a nonzero constant.
3. Interchange two rows.

Also notice that we have used a systematic solution procedure which is essentially the same as Gaussian elimination. The only difference is the order in which we perform the back substitutions. Once a value for a variable is determined, it is immediately substituted into all previous equations.

EXERCISE 1   Find the augmented matrix for the following system and use row operations on this augmented matrix to solve the system.

$$
\begin{aligned}
x_1 + 2x_2 - x_3 &= 1 \\
2x_1 + 5x_2 + x_3 &= 1 \\
4x_1 + 3x_2 + 2x_3 &= -12
\end{aligned}
$$

Many of the terms that apply to systems of equations are also applied to matrices. A square matrix is **upper triangular** when all entries below the main diagonal are zero. A matrix is in **echelon** form when the first nonzero entry in each row is farther to the right than the first nonzero terms in all preceding rows. The first nonzero entry in each row of an echelon matrix called a **pivot**. There is at most one pivot in each row and each column of an echelon matrix. Hence the number of pivots in an $m \times n$ echelon matrix must be less than or equal to both m and n. The columns that contain pivots are called **pivot columns**.

EXAMPLE 1   The matrix $\begin{bmatrix} 1 & 0 & 2 \\ 0 & 1 & 3 \end{bmatrix}$ is in echelon form. Both pivots are 1. The

matrix $\begin{bmatrix} -1 & 1 & 0 \\ 0 & 0 & 2 \end{bmatrix}$ is also in echelon form. Its pivots are -1 and 2. The

matrix $\begin{bmatrix} 2 & 1 \\ 0 & 0 \end{bmatrix}$ is in echelon form and its only pivot is 2. The matrix

$\begin{bmatrix} 2 & 1 \\ 1 & 0 \end{bmatrix}$ is not in echelon form.

EXERCISE 2   Use row operations to transform the following matrix to echelon form. Identify the pivots.

$$\begin{bmatrix} 2 & 3 & 4 & 1 \\ 4 & 6 & 2 & 8 \\ 2 & 3 & 10 & 4 \end{bmatrix}$$

We can use row operations on an augmented matrix to solve any system. However, we cannot always reduce the coefficient matrix to the identity matrix. This problem arises whenever there are any free variables or when the matrix has more rows than columns. Let us look at an example.

EXAMPLE 2   The system in Example 2 of Section 1.4 has the following augmented matrix.

$$\begin{bmatrix} 1 & -3 & 3 & 1 & 1 & : & 34 \\ 2 & -6 & -1 & -2 & -5 & : & -8 \\ 3 & -9 & -5 & 1 & -11 & : & -20 \end{bmatrix}$$

After clearing out the bottom of the first column we have

$$\begin{bmatrix} 1 & -3 & 3 & 1 & 1 & : & 34 \\ 0 & 0 & -7 & -4 & -7 & : & -76 \\ 0 & 0 & -14 & -2 & -14 & : & -122 \end{bmatrix}.$$

After clearing out the bottom of the third column, we have

$$\begin{bmatrix} 1 & -3 & 3 & 1 & 1 & : & 34 \\ 0 & 0 & -7 & -4 & -7 & : & -76 \\ 0 & 0 & 0 & 6 & 0 & : & 30 \end{bmatrix}.$$

Now we make the bottom pivot 1 and clear out the top of the fourth column.

$$\begin{bmatrix} 1 & -3 & 3 & 0 & 1 & : & 29 \\ 0 & 0 & -7 & 0 & -7 & : & -56 \\ 0 & 0 & 0 & 1 & 0 & : & 5 \end{bmatrix}$$

Finally, we make the second pivot 1 and clear out the top of its column.

$$\begin{bmatrix} 1 & -3 & 0 & 0 & -2 & : & 5 \\ 0 & 0 & 1 & 0 & 1 & : & 8 \\ 0 & 0 & 0 & 1 & 0 & : & 5 \end{bmatrix} \tag{3}$$

If we now interpret each row of the augmented matrix as an equation, we can find all solutions of the system. They are given by

$$x_1 = 5 + 3x_2 + 2x_5$$
$$x_2 = x_2$$
$$x_3 = 8 - x_5 \qquad \text{(where } x_2 \text{ and } x_5 \text{ are arbitrary).}$$
$$x_4 = 5$$
$$x_5 = x_5$$

Since every solution can be found from this set of equations, we say that these equations give the **general solution** of the system.

In this example we have reduced the coefficient matrix to a matrix with all pivots equal to 1 and with all pivot columns having only one nonzero term, the pivot itself. Echelon matrices with these two additional properties are called **reduced echelon matrices**. The Gauss–Jordan elimination procedure always reduces the coefficient matrix to reduced echelon form. Gaussian elimination usually leaves the pivots not equal to 1.

EXAMPLE 3   The matrix

$$A = \begin{bmatrix} 1 & 2 & 3 & 1 \\ 3 & 2 & 5 & 1 \\ 4 & 4 & 8 & 2 \end{bmatrix}$$

can be reduced to the echelon matrix

$$E = \begin{bmatrix} 1 & 2 & 3 & 1 \\ 0 & -4 & -4 & -2 \\ 0 & 0 & 0 & 0 \end{bmatrix}$$

by forward elimination. E can be further reduced to the matrix

$$R = \begin{bmatrix} 1 & 0 & 1 & 0 \\ 0 & 1 & 1 & 0.5 \\ 0 & 0 & 0 & 0 \end{bmatrix}$$

by back substitution. R is in reduced echelon form. The first two columns are the pivot columns. Both pivots are 1.

There is an alternative notation for the general solution of a linear system which is closely connected with the reduced echelon form of the augmented matrix. The general solution for Example 2 may be written in vector notation as follows:

$$\begin{bmatrix} x_1 \\ x_2 \\ x_3 \\ x_4 \\ x_5 \end{bmatrix} = \begin{bmatrix} 5 \\ 0 \\ 8 \\ 5 \\ 0 \end{bmatrix} + x_2 \begin{bmatrix} 3 \\ 1 \\ 0 \\ 0 \\ 0 \end{bmatrix} + x_5 \begin{bmatrix} 2 \\ 0 \\ -1 \\ 0 \\ 1 \end{bmatrix}. \tag{4}$$

This vector notation is actually quite natural. The augmented matrix (3) stands for a system of equations. If we add to that system the equations that express the arbitrary nature of the free variables ($x_2 = x_2$ and $x_5 = x_5$), we will not have changed the solutions of the system. The corresponding augmented matrix is

$$\begin{bmatrix} 1 & -3 & 0 & 0 & -2 & \vdots & 5 \\ 0 & 1 & 0 & 0 & 0 & \vdots & x_2 \\ 0 & 0 & 1 & 0 & 1 & \vdots & 8 \\ 0 & 0 & 0 & 1 & 0 & \vdots & 5 \\ 0 & 0 & 0 & 0 & 1 & \vdots & x_5 \end{bmatrix}.$$

(Notice that in the augmented matrix the equation $x_2 = x_2$ becomes the row $[0\ \ 1\ \ 0\ \ 0\ \ 0\!:\ \ x_2]$. The 1 indicates that the coefficient of x_2 on the left side of the equation is 1. The x_2 indicates that the right side of the equation is x_2. That is just the way it should be.) If we now reduce this augmented matrix by back substitution, we obtain

$$\begin{bmatrix} 1 & 0 & 0 & 0 & 0 & \vdots & 5 + 3x_2 + 2x_5 \\ 0 & 1 & 0 & 0 & 0 & \vdots & x_2 \\ 0 & 0 & 1 & 0 & 0 & \vdots & 8 - x_5 \\ 0 & 0 & 0 & 1 & 0 & \vdots & 5 \\ 0 & 0 & 0 & 0 & 1 & \vdots & x_5 \end{bmatrix}.$$

The solution vector on the right side can be written as

$$\begin{bmatrix} 5 + 3x_2 + 2x_5 \\ 0 + 1x_2 + 0x_5 \\ 8 + 0x_2 - 1x_5 \\ 5 + 0x_2 + 0x_5 \\ 0 + 0x_2 + 1x_5 \end{bmatrix},$$

which is identical with the solution given in (4).

Augmented matrices are useful when solving systems because we no longer have to write variables repeatedly. This can speed up computation considerably. But there is another way in which augmented matrices speed up computation even more. In certain applications it is necessary to solve several systems which all have the same coefficient matrix. In matrix notation, we have to solve $Ax = b_1$, $Ax = b_2, \ldots$, $Ax = b_k$ for k different vectors $b_1, b_2, \ldots, b_k$. To solve each of these systems we would use the augmented matrix $[A : b_i]$ and transform A to a reduced echelon matrix using row operations. But the transformation of A is identical for each of the k augmented matrices. We can reduce our work significantly by forming the augmented matrix

$$[A : b_1 \ \ b_2 \ \ \cdots \ \ b_k]$$

which has k extra columns added, one for each of the k different right-hand sides. We can then apply row operations to this matrix, solving the k different systems simultaneously with only *one* transformation of A to echelon form.

EXAMPLE 4 Let $A = \begin{bmatrix} 2 & 3 \\ 3 & 4 \end{bmatrix}$, $b_1 = \begin{bmatrix} 1 \\ 2 \end{bmatrix}$, $b_2 = \begin{bmatrix} 0 \\ 1 \end{bmatrix}$, and $b_3 = \begin{bmatrix} 1 \\ 0 \end{bmatrix}$. We wish to solve the three systems, $Ax = b_1$, $Ax = b_2$, $Ax = b_3$. First form the augmented matrix.

$$\begin{bmatrix} 2 & 3 \,\vdots\, 1 & 0 & 1 \\ 3 & 4 \,\vdots\, 2 & 1 & 0 \end{bmatrix}$$

Use the first row to clear out the rest of the first column.

$$\begin{bmatrix} 2 & 3 & \vdots & 1 & 0 & 1 \\ 0 & -\frac{1}{2} & \vdots & \frac{1}{2} & 1 & -\frac{3}{2} \end{bmatrix}$$

Solve for x_2.

$$\begin{bmatrix} 2 & 3 & \vdots & 1 & 0 & 1 \\ 0 & 1 & \vdots & -1 & -2 & 3 \end{bmatrix}$$

Back substitute.

$$\begin{bmatrix} 2 & 0 & \vdots & 4 & 6 & -8 \\ 0 & 1 & \vdots & -1 & -2 & 3 \end{bmatrix}$$

Solve for x_1.

$$\begin{bmatrix} 1 & 0 & \vdots & 2 & 3 & -4 \\ 0 & 1 & \vdots & -1 & -2 & 3 \end{bmatrix}$$

The three solutions can now be read from the appropriate columns. The solution to $Ax = \begin{bmatrix} 1 \\ 2 \end{bmatrix}$ is $x = \begin{bmatrix} 2 \\ -1 \end{bmatrix}$, the solution to $Ax = \begin{bmatrix} 0 \\ 1 \end{bmatrix}$ is $\begin{bmatrix} 3 \\ -2 \end{bmatrix}$, and the solution to $Ax = \begin{bmatrix} 1 \\ 0 \end{bmatrix}$ is $\begin{bmatrix} -4 \\ 3 \end{bmatrix}$.

In the next section we develop yet another method to solve the vector equation $Ax = b$.

PROBLEMS 1.6

*1. Which of the following matrices is in echelon form? If a matrix is in echelon form identify the pivots.

(a) $\begin{bmatrix} 2 & -1 & 1 \\ 0 & 3 & 1 \end{bmatrix}$
(b) $\begin{bmatrix} 1 & 0 & 2 \\ 0 & 3 & 1 \\ 0 & 5 & 0 \end{bmatrix}$
(c) $\begin{bmatrix} 1 & -1 & 0 & 1 & 0 \\ 0 & 0 & 1 & 2 & 0 \\ 0 & 0 & 0 & 0 & 1 \end{bmatrix}$

(d) $\begin{bmatrix} 1 & 0 & 3 & 0 \\ 0 & 2 & 1 & 0 \\ 0 & 0 & 0 & 0 \\ 0 & 0 & 0 & 0 \end{bmatrix}$
(e) $\begin{bmatrix} 1 & 3 & 0 & 1 \\ 0 & 2 & 0 & 2 \\ 0 & 0 & 0 & 1 \\ 0 & 0 & 2 & 0 \end{bmatrix}$
(f) $\begin{bmatrix} 1 & 0 & 1 & 0 & 2 \\ 0 & 1 & 1 & 1 & 3 \\ 0 & 0 & 0 & 0 & 0 \end{bmatrix}$

*2. Which of the matrices in Problem 1 are in reduced echelon form?

3. For each of the matrices A given in Problem 1, find all solutions of the linear system $Ax = 0$.

4. Rewrite the following systems of equations as vector equations. Then form the augmented matrix, find the general solution to the system, and express it in vector notation.

*(a)
$$-x_1 + x_2 + 2x_3 = 1$$
$$2x_1 - 3x_2 - x_3 = 5$$
$$-2x_1 + x_2 + 7x_3 = 9$$

(b)
$$x_1 - 3x_2 + x_3 + x_4 = 0$$
$$2x_1 - 6x_2 - x_3 - x_4 = 0$$
$$-x_1 + 3x_2 + x_3 + 2x_4 = 0$$

*(c)
$$3x_1 - x_2 + x_3 = 1$$
$$3x_1 + x_2 = 0$$
$$x_2 - 3x_3 = -1$$
$$6x_1 + x_2 - x_3 = 2$$

(d)
$$x_1 + 2x_2 + 3x_3 = 1$$
$$2x_1 + x_2 - x_3 = 0$$
$$2x_1 - x_2 = 0$$

*(e)
$$x_1 + x_2 + x_3 + x_4 + x_5 = 1$$
$$2x_3 - x_5 = 1$$
$$2x_1 + 2x_2 + x_3 - x_4 - x_5 = 1$$
$$x_1 + x_2 - x_3 - x_4 + x_5 = 1$$

(f)
$$3x_1 - x_2 = 0$$
$$2x_1 + x_2 = -2$$
$$7x_1 + x_2 = -4$$

5. In each part find the system of equations associated with the given vector equation. Then solve the system.

(a) $\begin{bmatrix} 2 & -4 & 0 \\ -1 & 2 & 3 \\ 1 & 1 & 2 \end{bmatrix} \begin{bmatrix} x_1 \\ x_2 \\ x_3 \end{bmatrix} = \begin{bmatrix} 2 \\ 0 \\ 1 \end{bmatrix}$ (b) $\begin{bmatrix} 2 & -1 \\ 1 & 2 \\ 3 & 1 \end{bmatrix} \begin{bmatrix} x_1 \\ x_2 \end{bmatrix} = \begin{bmatrix} 0 \\ 0 \\ 0 \end{bmatrix}$

6. In each part you are given all solutions of a linear system. Find the augmented reduced echelon matrix that led to this solution.

*(a) $\begin{bmatrix} x_1 \\ x_2 \\ x_3 \end{bmatrix} = \begin{bmatrix} -1 \\ 3 \\ 0 \end{bmatrix}$ ***(b)** $\begin{bmatrix} x_1 \\ x_2 \\ x_3 \end{bmatrix} = x_2 \begin{bmatrix} -2 \\ 1 \\ 0 \end{bmatrix} + \begin{bmatrix} 2 \\ 0 \\ 4 \end{bmatrix}$

*(c) $\begin{bmatrix} x_1 \\ x_2 \\ x_3 \\ x_4 \end{bmatrix} = x_2 \begin{bmatrix} -1 \\ 1 \\ 0 \\ 0 \end{bmatrix} + x_4 \begin{bmatrix} 3 \\ 0 \\ -1 \\ 1 \end{bmatrix} + \begin{bmatrix} 1 \\ 0 \\ 3 \\ 0 \end{bmatrix}$

7. What can you say about the reduced echelon form of the augmented matrix of an inconsistent linear system? In other words, how can you identify an inconsistent system after reducing its augmented matrix?

8. What can you say about the reduced echelon form of the augmented matrix of a linear system with a unique solution? In other words, how can you identify a system with a unique solution after reducing its augmented matrix?

9. What can you say about the reduced echelon form of the augmented matrix of a linear system with infinitely many solutions? In other words, how can you identify a system with an infinite number of solutions after reducing its augmented matrix?

*10. Use the augmented matrix technique to solve all of the following equations simultaneously, where the vector b takes on the following values: $(2, 0, 1)$,

$(4, 2, 1), (6, 0, 3), (6, 0, 0), (-2, 5, 4).$

$$\begin{bmatrix} 2 & -4 & 0 \\ -1 & 2 & 3 \\ 1 & 1 & 2 \end{bmatrix} \begin{bmatrix} x_1 \\ x_2 \\ x_3 \end{bmatrix} = b$$

1.7   MATRIX MULTIPLICATION

In previous sections we have seen how we can view a matrix as a collection of column vectors. We have also seen how to multiply a matrix times a column vector. Thus it is possible to define the product of two matrices to be the matrix that results from multiplying the first matrix times each of the columns of the second matrix, that is, by letting the jth column of the product AB be the product of A and the jth column of B.

Definition   Let A and B be two matrices such that the number of columns in A is equal to the number of rows in B. Then the product AB is the matrix whose jth column is the product of A and the vector that is the jth column of B. Letting b_j be the jth column of B, we can denote B by $[b_1 \quad b_2 \quad \cdots \quad b_n]$. Thus we have

$$AB = A[b_1 \quad b_2 \quad \cdots \quad b_n] = [Ab_1 \quad Ab_2 \quad \cdots \quad Ab_n].$$

EXERCISE 1   Let $A = \begin{bmatrix} 2 & 3 \\ -1 & 4 \end{bmatrix}$, $b_1 = \begin{bmatrix} 1 \\ 2 \end{bmatrix}$, $b_2 = \begin{bmatrix} 3 \\ 2 \end{bmatrix}$, and $b_3 = \begin{bmatrix} -1 \\ 0 \end{bmatrix}$. Compute Ab_1, Ab_2, Ab_3. Let $B = \begin{bmatrix} 1 & 3 & -1 \\ 2 & 2 & 0 \end{bmatrix}$. Compute AB.

Our definition of matrix multiplication tells us how to find the columns of the product matrix. But in practice we compute the product matrix one entry at a time. How do we compute a specific entry in the product matrix? In other words, if A, B, and C are matrices with $C = AB$, how do we actually compute c_{ij}, the entry in the ith row and jth column of C?

If b_j is the jth column of B, then c_{ij} is the ith component of Ab_j. Thus we only use the jth column of B to compute c_{ij}. But we know that the ith entry in Ab_j is obtained by using the ith row of A. Thus computing c_{ij} will involve only the ith row of A and the jth column of B. It is not hard to show that the following formula is valid:

$$c_{ij} = a_{i1}b_{1j} + a_{i2}b_{2j} + a_{i3}b_{3j} + \cdots + a_{ir}b_{rj},$$

where r equals the number of columns of A and also the number of rows of B.

$$\begin{bmatrix} a_{11} & \cdots & a_{1r} \\ a_{i1} & \cdots & a_{ir} \\ a_{m1} & \cdots & a_{mr} \end{bmatrix} \begin{bmatrix} b_{11} \cdots & b_{1j} & \cdots & b_{1n} \\ & \vdots & \\ b_{r1} \cdots & b_{rj} & \cdots & b_{rn} \end{bmatrix} = \begin{bmatrix} \\ & c_{ij} \\ \\ \end{bmatrix}$$

The restriction on the sizes of A and B in the definition of matrix multiplication is a natural one. If the jth column of AB equals Ab_j, then we must be able to compute Ab_j. This means that the number of columns of A must equal the number of entries in b_j. The latter number is obviously the same as the number of rows in B. One should also note that AB is a matrix whose size can be determined from the sizes of A and B. If A is an $m \times r$ matrix and B is an $r \times n$ matrix, then AB is an $m \times n$ matrix. In words, the number of rows in the product is the number of rows in the first matrix; the number of columns in the product is the number of columns in the second matrix (see Problem 7).

EXAMPLE 1   Given

$$A = \begin{bmatrix} 0 & 1 & 2 \\ 3 & 1 & 4 \end{bmatrix} \quad \text{and} \quad B = \begin{bmatrix} 1 & 2 & 1 \\ 3 & 1 & -2 \\ 4 & 2 & 4 \end{bmatrix},$$

AB is a 2×3 matrix. The entry in the first row and second column of AB will be obtained by taking the first row of A, $(0, 1, 2)$, the second column of B, $(2, 1, 2)$, multiplying corresponding components, and adding $(0 \cdot 2 + 1 \cdot 1 + 2 \cdot 2 = 5)$. The complete product is $AB = \begin{bmatrix} 11 & 5 & 6 \\ 22 & 15 & 17 \end{bmatrix}$.

Most people multiply matrices with the aid of their index fingers. The left index finger moves across a row of the first matrix and the right index finger moves down a column of the second matrix. If the left index finger moves across row i and the right index finger moves down column j, and one adds up the products of the corresponding entries, the result is the entry in the ith row and jth column of the product.

EXERCISE 2   Compute AB, where

$$A = \begin{bmatrix} 1 & 2 \\ 1 & 4 \\ 2 & 1 \end{bmatrix} \quad \text{and} \quad B = \begin{bmatrix} 1 & 2 \\ -3 & 4 \end{bmatrix}.$$

If A, B, and C are matrices and v is a vector, we have the following identities.

(15) $(AB)C = A(BC)$ (associativity)
(16) $A(B + C) = AB + AC$ (distributivity)
(17) $(A + B)C = AC + BC$ (distributivity)
(18) $A(\alpha B) = (\alpha A)B = \alpha(AB)$ (quasi-associativity)
(19) $A(Bv) = (AB)v$ (quasi-associativity)

These identities are reasonable. Proofs (although sometimes laborious and dull) are not impossible. It is again true that all objects must have the correct shape so that the indicated operations are defined.

One very familiar identity is missing from our list—the commutative rule, $AB = BA$. It is missing because it is *not true*. Although there are some pairs of matrices that satisfy $AB = BA$, they are very exceptional.

EXAMPLE 2   Let $A = \begin{bmatrix} 1 & 2 \\ 3 & 4 \end{bmatrix}$ and $B = \begin{bmatrix} -1 & 0 \\ 2 & 4 \end{bmatrix}$. Then $AB = \begin{bmatrix} 3 & 8 \\ 5 & 16 \end{bmatrix}$ and $BA = \begin{bmatrix} -1 & -2 \\ 14 & 20 \end{bmatrix}$. $AB \neq BA$.

EXERCISE 3   Show that if AB and BA are defined and equal, then A and B are square matrices of the same size.

EXERCISE 4   Let $I = \begin{bmatrix} 1 & 0 \\ 0 & 1 \end{bmatrix}$ and $B = \begin{bmatrix} 2 & 3 & 4 \\ 5 & 6 & -2 \end{bmatrix}$. Compute IB.

The matrix I in Exercise 4 is the 2×2 identity matrix. The identity matrices and the zero matrices have some special properties.

(20) $AI = A$,   (21) $IA = A$,
(22) $0A = 0$,   (23) $A0 = 0$.

You are probably wondering why identities (20) and (21) are written as two separate identities. If A is an $m \times n$ matrix with $m \neq n$, then the formulas involve two different identity matrices. In (20), $I = I_n$ and in (21), $I = I_m$.

EXERCISE 5   If A is an $m \times n$ matrix, what are the possible sizes of the zero matrices in (22)?

The discussion above is not the only one that can be used to motivate the definition of matrix multiplication. Problems 9 and 10 give some other motivation for this definition.

We can easily solve a linear equation $ax = b$ when a and b are numbers. We multiply both sides of the equation by the reciprocal of a, that is, by the number $a^{-1} = 1/a$. Is it possible to extend the idea of a reciprocal to matrices so that linear systems can be solved in a similar way? Given the linear system $Ax = b$, does there exist a matrix A^{-1} such that we can multiply both sides of the equation by A^{-1}, obtaining $x = A^{-1}b$ as the solution? Unfortunately, the answer is frequently *no*. But there are some special cases where the answer is yes.

The **inverse** of a matrix A is a matrix B such that

$$AB = I = BA.$$

The inverse of A (when it exists) will be denoted by the symbol A^{-1}. If AB and BA are both defined and equal, then both A and B must be $n \times n$ matrices for some n. Thus only square matrices can have inverses. If a square matrix has an inverse, it is called **nonsingular** or **invertible**. Square matrices that do not have inverses are called **singular**.

If we are to use inverses to solve equations, then we must develop a method to find them (when they exist). To find A^{-1} is to solve the equation $AB = I$ for B. But this equation can be viewed as a collection of equations all of which have the same coefficient matrix. The ith column of B satisfies the equation $Ax = e_i$, where e_i is the ith column of the identity matrix, that is, the vector whose ith component is 1 and whose other components are 0. If every one of these systems has a solution, then A will have an inverse. If any one of these systems is inconsistent, then A^{-1} will not exist. (This process only finds a matrix B such that $AB = I$. We must show that this matrix B which we have found also satisfies $BA = I$. We do this in Chapter 2.)

EXAMPLE 3   Let

$$A = \begin{bmatrix} 1 & 1 & 1 \\ 2 & 3 & 3 \\ 3 & 4 & 5 \end{bmatrix}.$$

To find the inverse of A form the augmented matrix $[A : I]$ and use row operations to reduce A to the identity matrix. The augmented matrix will then be $[I : A^{-1}]$.

$$\begin{bmatrix} 1 & 1 & 1 & : & 1 & 0 & 0 \\ 2 & 3 & 3 & : & 0 & 1 & 0 \\ 3 & 4 & 5 & : & 0 & 0 & 1 \end{bmatrix}$$

Clearing out the first column below the pivot, we obtain

$$\begin{bmatrix} 1 & 1 & 1 & : & 1 & 0 & 0 \\ 0 & 1 & 1 & : & -2 & 1 & 0 \\ 0 & 1 & 2 & : & -3 & 0 & 1 \end{bmatrix}.$$

Clearing out the second column above and below the pivot, we obtain

$$\begin{bmatrix} 1 & 0 & 0 & : & 3 & -1 & 0 \\ 0 & 1 & 1 & : & -2 & 1 & 0 \\ 0 & 0 & 1 & : & -1 & -1 & 1 \end{bmatrix}.$$

Clearing out the third column above the pivot, we have

$$\begin{bmatrix} 1 & 0 & 0 & : & 3 & -1 & 0 \\ 0 & 1 & 0 & : & -1 & 2 & -1 \\ 0 & 0 & 1 & : & -1 & -1 & 1 \end{bmatrix}.$$

A^{-1} is on the right-hand side of the final augmented matrix.

EXERCISE 6   Verify that $A^{-1}A = AA^{-1} = I$ for the matrices A and A^{-1} in Example 3.

EXAMPLE 4   Let

$$A = \begin{bmatrix} 1 & 1 & 1 \\ 1 & 2 & 1 \\ 2 & 3 & 2 \end{bmatrix}.$$

To compute A^{-1}, we apply forward elimination to the augmented matrix

$$\left[\begin{array}{ccc:ccc} 1 & 1 & 1 & 1 & 0 & 0 \\ 1 & 2 & 1 & 0 & 1 & 0 \\ 2 & 3 & 2 & 0 & 0 & 1 \end{array}\right].$$

We obtain

$$\left[\begin{array}{ccc:ccc} 1 & 1 & 1 & 1 & 0 & 0 \\ 0 & 1 & 0 & -1 & 1 & 0 \\ 0 & 0 & 0 & -1 & -1 & 1 \end{array}\right].$$

Since at least one of the equations $Ax = e_i$ is inconsistent (in fact, all of them are), A does not have an inverse.

EXERCISE 7 Compute A^{-1} if $A = \begin{bmatrix} 2 & 3 \\ 3 & 4 \end{bmatrix}$.

One important computational note is appropriate here. The computation of the inverse of an $n \times n$ matrix A involves solving a system with n *different* right-hand sides. Thus it is foolish to find the inverse of A if it will only be used to solve *one* system. We would be trading a problem that involved only one right-hand side for one involving n right-hand sides and a final multiplication.

We conclude this section with a definition that will prove useful later. If A is an $m \times n$ matrix, the **transpose** of A, denoted by A^T, is the $n \times m$ matrix obtained from A by interchanging its rows and columns. The entry in the ith row and jth column of A^T is the number a_{ji}, the number found in the jth row and ith column of A.

EXAMPLE 5 If

$$A = \begin{bmatrix} 1 & 3 & -4 \\ 2 & 1 & 0 \\ 3 & 1 & 4 \end{bmatrix}, \quad \text{then} \quad A^T = \begin{bmatrix} 1 & 2 & 3 \\ 3 & 1 & 1 \\ -4 & 0 & 4 \end{bmatrix}.$$

If $A = [1 \quad 4]$, then $A^T = \begin{bmatrix} 1 \\ 4 \end{bmatrix}$.

The following identities involving transposes are true.

(1) $(A^T)^T = A$ (2) $(\alpha A)^T = \alpha A^T$
(3) $(A + B)^T = A^T + B^T$ (4) $(AB)^T = B^T A^T$

The student is asked in Problem 16 to explain these rules.

EXERCISE 8 Compute A^T, B^T, $(AB)^T$, and $B^T A^T$ when

$$A = \begin{bmatrix} 1 & 2 & 3 \\ 4 & -1 & 2 \end{bmatrix} \quad \text{and} \quad B = \begin{bmatrix} 1 & 4 \\ 2 & -3 \\ 3 & -1 \end{bmatrix}.$$

PROBLEMS 1.7

*1. Let

$$A = \begin{bmatrix} 2 & 1 & 3 \\ 4 & -1 & 2 \end{bmatrix}, \quad B = \begin{bmatrix} 2 & 3 \\ 0 & -1 \\ 1 & -2 \end{bmatrix}, \quad \text{and} \quad v = \begin{bmatrix} 1 \\ 2 \end{bmatrix}.$$

Compute AB, $(AB)v$, Bv, and $A(Bv)$.

2. Let A and B be as in Problem 1 and let

$$v = \begin{bmatrix} 2 \\ 0 \\ 1 \end{bmatrix}.$$

Compute BA, $(BA)v$, Av, and $B(Av)$.

*3. Let $A = [1 \quad 2 \quad -1]$ and

$$B = \begin{bmatrix} 2 \\ 0 \\ 1 \end{bmatrix}.$$

Compute AB and BA.

4. Let

$$A = \begin{bmatrix} 5 & 2 \\ -3 & 1 \end{bmatrix}, \quad B = \begin{bmatrix} 2 & -1 & 3 \\ 0 & 4 & 1 \end{bmatrix}, \quad C = \begin{bmatrix} 1 & 3 \\ 2 & -4 \end{bmatrix},$$

$$D = \begin{bmatrix} -2 & 1 & 3 \\ -1 & 0 & 2 \\ 4 & 1 & 4 \end{bmatrix}, \quad E = [2 \quad 1 \quad 3], \quad F = \begin{bmatrix} -1 \\ 3 \\ 1 \end{bmatrix}.$$

Compute the following matrices:
*(a) AB, AC, BD, BF, EF, and FE.
   (b) EE^T, F^TD, $(A + C)B$, $D(E^T + F)$, and $3C^TB$.

5. Show that $A^3 [= AAA] = 0$, where

$$A = \begin{bmatrix} 1 & 1 & 3 \\ 5 & 2 & 6 \\ -2 & -1 & -3 \end{bmatrix}.$$

6. Show that $A^2 = A$, where

$$A = \begin{bmatrix} 2 & -2 & -4 \\ -1 & 3 & 4 \\ 1 & -2 & -3 \end{bmatrix}.$$

7. Explain why the product of an $m \times r$ matrix and an $r \times n$ matrix is an $m \times n$ matrix.

8. Verify identity (18) by showing that the entries in the ith row and the jth column of $A(\alpha B)$, $(\alpha A)B$, and $\alpha(AB)$ are all equal. (Assume that A is an $m \times r$ matrix and B is an $r \times n$ matrix.)

9. Given the $m \times r$ matrix $A = [a_{ij}]$, the $r \times n$ matrix $B = [b_{ij}]$, the $m \times n$ matrix $C = [c_{ij}]$, and the n-vector v.
   (a) Show that the ith component of Cv is
$$c_{i1}v_1 + c_{i2}v_2 + \cdots + c_{in}v_n.$$

(b) Show that the ith component of Bv is

$$b_{i1}v_1 + b_{i2}v_2 + \cdots + b_{in}v_n.$$

(c) Determine the ith component of $A(Bv)$.
(d) What is the coefficient of v_j in the ith component of $A(Bv)$?
(e) If Cv is to be equal to $A(Bv)$, what must c_{ij} equal?
(f) If identity (19) is to be true, the C defined above must be equal to AB. Is it?

10. Let

$$w_1 = a_{11}y_1 + a_{12}y_2$$
$$w_2 = a_{21}y_1 + a_{22}y_2$$

and

$$y_1 = b_{11}x_1 + b_{12}x_2$$
$$y_2 = b_{21}x_1 + b_{22}x_2.$$

(a) Find c_{11}, c_{12}, c_{21}, and c_{22} such that
$$w_1 = c_{11}x_1 + c_{12}x_2$$
$$w_2 = c_{21}x_1 + c_{22}x_2.$$

(b) Interpret all of these systems as matrix-vector equations. What is the relationship between the three coefficient matrices?
(c) Investigate this problem when there are three y's (y_1, y_2, and y_3) and four x's (x_1, x_2, x_3, and x_4).

11. Let

$$A = \begin{bmatrix} 1 & 1 & -1 \\ 2 & 0 & 3 \\ 4 & 3 & -2 \end{bmatrix} \quad \text{and} \quad B = \begin{bmatrix} -9 & 1 & 3 \\ 16 & 2 & -5 \\ 6 & 1 & -2 \end{bmatrix}.$$

Compute AB and BA.

*12. Show that the matrix

$$A = \begin{bmatrix} 1 & 1 & 1 \\ 0 & 2 & 3 \\ 3 & 5 & 2 \end{bmatrix}$$

is nonsingular. Then use A^{-1} to solve the system $Ax = b$, where $b = (1, -1, 3)$.

13. If possible, find the inverses of the following matrices.

*(a) $\begin{bmatrix} 1 & 3 \\ -2 & 6 \end{bmatrix}$
*(b) $\begin{bmatrix} 2 & -6 \\ -4 & 12 \end{bmatrix}$
(c) $\begin{bmatrix} 1 & -1 & 1 \\ 4 & 5 & 6 \\ 5 & 7 & 9 \end{bmatrix}$

*(d) $\begin{bmatrix} 1 & -1 & 2 & 3 \\ 0 & 2 & 1 & 2 \\ 0 & 0 & -1 & 3 \\ 0 & 0 & 0 & 1 \end{bmatrix}$
(e) $\begin{bmatrix} 1 & 2 & 3 & 1 \\ 1 & 3 & 3 & 2 \\ 2 & 4 & 3 & 3 \\ 1 & 1 & 1 & 1 \end{bmatrix}$
(f) $\begin{bmatrix} 1 & -1 & 1 \\ 2 & 0 & 3 \\ 3 & -1 & 2 \end{bmatrix}$

14. (a) Show that if A, B, and C are matrices and A is nonsingular, then $AB = AC$ implies that $B = C$.
(b) Find matrices A, B, and C such that $A \neq 0$, $AB = AC$, and $B \neq C$.

15. Suppose that A is an $n \times n$ matrix such that $(I - A)(I + A) = 0$. Show that A is nonsingular and that $A^{-1} = A$.

16. Explain why each of the rules involving transposes is true.

17. Define the **trace** of the $n \times n$ matrix A to be the sum of the diagonal entries. If the trace of A is denoted $\operatorname{tr} A$ and $A = [a_{ij}]$, then $\operatorname{tr} A = a_{11} + a_{22} + \cdots + a_{nn}$. Show that:
   (a) $\operatorname{tr}(\alpha A) = \alpha \cdot \operatorname{tr} A$
   (b) $\operatorname{tr}(A + B) = \operatorname{tr} A + \operatorname{tr} B$
   (c) $\operatorname{tr}(AB) = \operatorname{tr}(BA)$

18. Let D be the diagonal matrix
$$\begin{bmatrix} 3 & & \\ & 1 & \\ & & -1 \end{bmatrix}.$$
   (a) Compute D^2, D^3, and D^4.
   (b) On the basis of the answers to part (a), describe D^k for $k > 0$.
   (c) Describe the relationship between D^n and D for an arbitrary diagonal matrix D.

19. A square matrix A is called **symmetric** if $A^T = A$ and **skew-symmetric** if $A^T = -A$.
   (a) Prove that every diagonal element of a skew-symmetric matrix is zero.
   (b) Describe what a symmetric matrix looks like. Do the same for a skew-symmetric matrix.
   (c) Find all $n \times n$ matrices that are both symmetric and skew-symmetric.
   (d) For any square matrix A, show that $A^T A$ is symmetric.
   (e) For any square matrix A, prove that $\frac{1}{2}(A + A^T)$ is a symmetric matrix. Why does A have to be square?
   (f) For any square matrix A, prove that $\frac{1}{2}(A - A^T)$ is a skew-symmetric matrix.
   (g) Show that any square matrix A can be expressed as a sum $A = S + K$, where S is a symmetric matrix and K is a skew-symmetric matrix.
   (h) Show that the representation above is unique; that is, show that if $A = S + K = S' + K'$, where S, S' are symmetric and K, K' are skew-symmetric, then $S = S'$ and $K = K'$.

1.8   DIFFERENCE EQUATIONS AND MARKOV PROCESSES

In Section 1.2 we presented several applications of systems of equations. Our goal in this section is to present additional applications of linear systems, in particular some applications that illustrate the power and utility of matrix notation. One other common theme is found in the applications presented here. Each of them involves an attempt to predict the future. In all cases the predictions are based on certain (verifiable) information about the present together with certain (we hope true) assumptions about the future.

Predicting Population (One Species)

When one studies a particular animal species (e.g., bacteria, fish, deer, human beings) one frequently wishes to make reasonable predictions about the future population of that species. All changes in the population of an isolated species are caused by births and deaths. It is not too difficult to determine the present birth rate and the present death rate of the species. But it is clearly impossible to determine the rates of births and deaths in the future. Therefore, if we wish to make predictions about the future population we are forced to make some reasonable assumptions about these rates in the future. We will assume that the future rates of births and deaths will be the same as the present rates. (This may or may not be true. If it is true, our predictions will be correct. If not, our predictions may be in error.) Let us express these ideas in appropriate symbols.

Choose some basic time interval as the unit of time. This may be 1 year, 1 minute, 3 days, and so on, depending on the species being studied. Let $p^{(0)}$ be the initial population and $p^{(k)}$ be the population after k units of time have passed. Let i be the rate of change (increase if positive, decrease if negative) of the population over one unit of time. (We can interpret i as the birth rate minus the death rate. It is normally positive unless the species is dying out.) Then we have the following **difference equation**:

$$p^{(k+1)} - p^{(k)} = ip^{(k)} \tag{1}$$

or, equivalently,

$$p^{(k+1)} = (1 + i)p^{(k)}. \tag{2}$$

The two sides of equation (1) are the two different ways to compute the change in population during the $(k + 1)$st time interval. Equations (1) and (2) hold when k is a nonnegative integer.

It is possible to obtain a simple formula for $p^{(k)}$. We know that $p^{(k)} = (1 + i)p^{(k-1)}$. But we also know that $p^{(k-1)} = (1 + i)p^{(k-2)}$ because (2) holds for every positive integer k. These two equalities can be combined to give $p^{(k)} = (1 + i)^2 p^{(k-2)}$. Continuing this process, we eventually conclude that

$$p^{(k)} = (1 + i)^k p^{(0)}. \tag{3}$$

We have now accomplished our goal. Given the present population $p^{(0)}$ of a species and the (assumed constant) rate of change i of the species, we can predict the population $p^{(k)}$ after k time intervals have elapsed.

EXERCISE 1　If the initial population of bacteria is 100 and the rate of increase is 25% per day, what is a reasonable prediction for the bacteria population after 4 days? (This problem can be done by multiplication. If we wished to predict the population after 40 days, we would not use multiplication to determine 1.25^{40}. Instead, we would use logarithmic and exponential functions to calculate the answer.)

Predicting Population (More Than One Species)

The ideas discussed above can also be used to predict the future populations of two or more species. If the species are independent (i.e., if they do not interact in any way), then there is nothing new to be done. But the possibility of interaction (e.g., competition, predition, symbiosis, scavenging) creates the need for a more complex model of population growth. Again, let us express our ideas in appropriate symbols for the case of two different species.

Let $p_1^{(k)}$ and $p_2^{(k)}$ be the population of species 1 and 2 after k time units have passed. Under the same assumptions made above, we would expect $p_1^{(k+1)} - p_1^{(k)} = ip_1^{(k)}$ for some rate of change i. But we are considering the case of interacting species, and the presence of the second species must also influence the population of the first. We will assume that this added influence is proportional to the population of the second species. (Of course, this assumption may not always be valid. When it is not, our predictions may be in error.) We are thus assuming that future changes in the population of the first species satisfy the equation

$$p_1^{(k+1)} - p_1^{(k)} = ip_1^{(k)} + bp_2^{(k)}$$

for some numbers i and b. A similar equation holds for the population of the second species. Changing notation so that a_{ij} represents the rate of change of species i due to the presence of species j, we have the following equations:

$$\begin{aligned} p_1^{(k+1)} - p_1^{(k)} &= a_{11}p_1^{(k)} + a_{12}p_2^{(k)} \\ p_2^{(k+1)} - p_2^{(k)} &= a_{21}p_1^{(k)} + a_{22}p_2^{(k)}. \end{aligned} \tag{4}$$

If $p^{(k)}$ is the vector $(p_1^{(k)}, p_2^{(k)})$ and A is the matrix $[a_{ij}]$, then (4) can be written as the **difference equation**

$$p^{(k+1)} - p^{(k)} = Ap^{(k)}$$

or, equivalently,

$$p^{(k+1)} = (I + A)p^{(k)}, \tag{5}$$

where I is the identity matrix.

This matrix equation is the matrix analogue of equation (2) and can be solved in a similar fashion, giving

$$p^{(k)} = (I + A)^k p^{(0)}. \tag{6}$$

We will refer to $p^{(k)}$ as the population vector and A as the growth matrix.

EXAMPLE 1 Suppose that we have two species, the first of which is increasing and is not influenced by the second, and the second of which scavenges on the dead of the first. If the scavenging animal cannot live except on the dead of the other species, then the entries in the growth matrix A would be as follows:

$a_{11} > 0$, the first species is increasing;
$a_{12} = 0$, the second species does not influence the first;
$a_{21} > 0$, the second scavenges on the first;
$a_{22} < 0$, the second species will die out if it cannot scavenge upon the first.

EXERCISE 2 Suppose that we have two competing species. What are the signs of the entries in A?

EXAMPLE 2 Suppose that the growth matrix A for two species is $\begin{bmatrix} 0.1 & 0.01 \\ 0.05 & 0.1 \end{bmatrix}$.
If the initial population vector is $\begin{bmatrix} 10{,}000 \\ 100{,}000 \end{bmatrix}$, then the prediction of future population is given by

$$p^{(k)} = \begin{bmatrix} 1.1 & 0.01 \\ 0.05 & 1.1 \end{bmatrix}^k \begin{bmatrix} 10{,}000 \\ 100{,}000 \end{bmatrix}.$$

In particular, $p^{(1)}$, the population after one time interval, equals $\begin{bmatrix} 12{,}000 \\ 110{,}500 \end{bmatrix}$
and $p^{(2)}$ equals $\begin{bmatrix} 14{,}305 \\ 122{,}150 \end{bmatrix}$.

EXERCISE 3 Suppose that the growth matrix A in a two-species model is $\begin{bmatrix} 1.05 & 0.01 \\ 0.02 & 0.06 \end{bmatrix}$. What is the population of each species after three time intervals if the initial population vector is $(100, 1000)$?

Exercise 3 shows that equation (6) is not easy to use when one wishes to compute answers to specific problems. Raising a matrix to the tenth power is neither easy nor enjoyable. Even cubing a 6×6 matrix is tedious. We must develop a procedure to quickly raise a matrix to a power. This topic is discussed in Chapter 8.

Predicting the Age Distribution of a Population

There are situations where we are interested in the numbers of animals alive in certain age groups. For example, to determine appropriate Social Security tax levels, there must be some predictions of the age distribution of the U.S. population. Let us develop a model (the **Leslie model**) to make such predictions.

Let the basic unit of time be 5 years. We wish to estimate the size of each age group of the population a certain number of time units in the future. Define variables as follows:

$p_1^{(k)}$ number of people 0–4 years old after k intervals,

$p_2^{(k)}$ number of people 5–9 years old after k intervals,

$p_3^{(k)}$ number of people 10–14 years old after k intervals,

$\vdots$

$p_{20}^{(k)}$ number of people 95–99 years old after k intervals.

(We assume that the number of persons 100 and over is negligible.) Let us relate the population at a certain time to the population 5 years later. A certain fraction of people in the ith age group will survive to become members of the $(i + 1)$st age group. In addition, the people in the ith age group have been responsible for births in proportion to their numbers. Let s_i be the fraction of the ith age group that survives after 5 years and b_i be the birth rate for the ith age group. If we assume that future births and deaths will continue in the same proportion, then we have the following relations:

$$p_{i+1}^{(k+1)} = s_i p_i^{(k)}, \qquad i = 1,\ldots,19$$
$$p_1^{(k+1)} = b_1 p_1^{(k)} + b_2 p_2^{(k)} + \cdots + b_{20} p_{20}^{(k)}. \tag{7}$$

The values for $b_1,\ldots,b_{20}$ and $s_1,\ldots,s_{19}$ will of course be obtained from studies of the present behavior of the population.

The system of equations (7) can be rewritten as a matrix equation. If A is the matrix

$$
\begin{bmatrix}
b_1 & b_2 & b_3 & \cdots & b_{19} & b_{20} \\
s_1 & 0 & 0 & \cdots & 0 & 0 \\
0 & s_2 & 0 & \cdots & 0 & 0 \\
0 & 0 & s_3 & \cdots & 0 & 0 \\
& & \vdots & & \vdots & \\
0 & 0 & 0 & \cdots & s_{19} & 0
\end{bmatrix}
$$

and $p^{(k)}$ the vector $(p_1^{(k)}, p_2^{(k)},\ldots, p_{20}^{(k)})$, then (7) becomes

$$p^{(k+1)} = Ap^{(k)}.$$

This matrix equation has a solution

$$p^{(k)} = A^k p^{(0)}.$$

Again we see that a complete solution of our problem is obtained by computing powers of a matrix.

In the Leslie model for predicting population distributions, the basic time unit and the time span of each population group must be equal. By choosing age groups that span a time interval equal to our basic unit of time, we have succeeded in obtaining a special kind of matrix A. Matrices of this form are called **Leslie matrices** after P. H. Leslie, who was one of the first to investigate this method for predicting population distributions.

This model was developed to study the age distribution of a population. But there are other questions that can be investigated once we know the Leslie matrix, A, for the population. Is there a distribution of the population that is stable under the birth and death processes described by the matrix A? (In symbols, does there exist a vector p such that $Ap = \lambda p$ for some constant λ?) What is the natural rate of growth of this population? How can we determine this growth rate from the matrix A? Is this population likely to grow in a stable manner, approaching some fixed age distribution, or will there be wild fluctuations in the age distribution? Some of these questions are considered in Chapter 8.

Markov Processes

Suppose that we are studying the geographical distribution of the U.S. population. What we are interested in are the population shifts between different regions. Let us divide the United States into two regions: the Sun Belt and everywhere else. Suppose that the population movement between these two regions has been studied over several years and the following figures determined: percentage of Sun Belt population moving out, 15% per year; percentage of other population moving into Sun Belt, 20% per year. (These figures were invented by the authors and have no connection with reality.) Let the Sun Belt be region 1 and everywhere else be region 2. In addition, let $f_1^{(k)}$ ($f_2^{(k)}$) be the fraction of the population living in region 1 (region 2) after k years. [Note that $f_1^{(k)} + f_2^{(k)} = 1$ for all $k \geq 0$.] Then the following equation holds for all $k \geq 0$:

$$\begin{bmatrix} f_1^{(k+1)} \\ f_2^{(k+1)} \end{bmatrix} = \begin{bmatrix} 0.85 & 0.20 \\ 0.15 & 0.80 \end{bmatrix} \begin{bmatrix} f_1^{(k)} \\ f_2^{(k)} \end{bmatrix}$$

or, letting A be the coefficient matrix and $f^{(k)}$ the distribution vector,

$$f^{(k+1)} = Af^{(k)}.$$

The process described by this difference equation is called a **Markov process** (after the Russian mathematician A. A. Markov, 1856–1922, who first studied processes of this type). In a Markov process there are a certain number of states (two in our example). The entries in the coefficient (or **transition**) matrix indicate how the distribution among the states changes over one time interval; it gives information about the transitions between states. In particular, the entry in the ith row and the jth column of the transition matrix is the fraction of objects in state j which move to state i (0.85 of the Sun Belt people stay while 0.15 move out). In general, a Markov process is described by a difference equation

$$f^{(k+1)} = Af^{(k)}, \tag{8}$$

where the transition matrix A has columns that add up to 1 and the **state vectors** $f^{(k)}$ all have coordinates that sum to 1.

Suppose that the initial distribution of the population is 15% in the Sun Belt and 85% in the other region. What can we say about the future distribution of the population? What predictions can we make? Of course, these predictions will only be as good as our assumptions—specifically, our assumption that the future movement of population will continue as it has in the past.

By this time it should be obvious that equation (8) has a solution given by

$$f^{(k)} = A^k f^{(0)}.$$

So in our example we can predict the population distribution k years from now by computing

$$\begin{bmatrix} 0.85 & 0.20 \\ 0.15 & 0.80 \end{bmatrix}^k \begin{bmatrix} 0.15 \\ 0.85 \end{bmatrix}.$$

Again we have the problem of raising a matrix to a power (see Chapter 8).

EXERCISE 4 . Compute $f^{(2)}$ for the Sun Belt example.

There are other questions that can be asked. What will be the long-range behavior of the population distribution? Will it fluctuate wildly or approach some equilibrium? Is there an equilibrium population distribution; that is, is there a state vector f such that $f = Af$? Is there more than one such vector?

EXERCISE 5 Show that $f = (\frac{4}{7}, \frac{3}{7})$ satisfies $f = Af$, where A is the transition matrix for the Sun Belt example.

PROBLEMS 1.8

1. Consider a population divided into three age groups with the following Leslie matrix.

$$\begin{bmatrix} 0 & 2 & 0 \\ 0.72 & 0 & 0 \\ 0 & 0.5 & 0 \end{bmatrix}$$

(a) Show that if the initial population distribution is $(200, 120, 50)$, the population in each age group increases by $\frac{1}{5}$ after one time interval has passed.

(b) In part (a) it was shown that the relative proportion of animals in each age group remained the same after one time interval has passed. Will this continue indefinitely?

2. Suppose that the Leslie matrix for a population satisfies $Av = \lambda v$ for some vector v.
   (a) Show that $A^2v = \lambda^2 v$, $A^3v = \lambda^3 v$, and $A^4v = \lambda^4 v$.
   (b) If the initial population distribution for this population is given by v, describe the future population distribution.
   *(c) What is the natural growth rate of this population?
   *(d) Are there any stipulations that you should make as to which values of λ are acceptable or meaningful?

*3. A certain fish population has been studied. Of the eggs that hatch each year, $\frac{1}{100}$ survive. Of the 1-year-old fish, $\frac{1}{5}$ survive. Of the 2-year-old fish, $\frac{1}{2}$ survive. All of the 3-year-old fish spawn and then die. The number of eggs that hatch averages 1100 per 3-year-old fish. What is the Leslie matrix for this population if the basic unit of time is 1 year?

4. A certain fish population has the following Leslie matrix. (The basic time interval is 1 year.)

$$A = \begin{bmatrix} 0 & 0 & 110 \\ 0.05 & 0 & 0 \\ 0 & 0.2 & 0 \end{bmatrix}$$

If the initial population is 1000 in each age group, what will be the population of each age group after 1 year? 2 years? 3 years? 4 years?
    Compute A^3. How does this help you predict the future distribution of the fish population? Describe the way this fish population changes from year to year.

*5. (a) How many multiplications must be performed to cube a 6×6 matrix?
    (b) How many multiplications must be performed to square an $n \times n$ matrix?

6. Suppose that $p^{(k+1)} = (I + A)p^{(k)}$ for all $k \geq 0$. Show that $p^{(k)} = (I + A)^k p^{(0)}$.

7. Apply the Markov process idea to the distribution of sales among three supermarkets. What is the interpretation of the coefficients in the transition matrix? What questions are of interest, and what is their mathematical formulation?

Chapter 2

Fundamental Concepts of Linear Algebra

In Chapter 1 we introduced vectors and matrices to help solve linear systems. But solving linear systems is not the only purpose of linear algebra. In this chapter we develop some of the basic ideas that make linear algebra a useful conceptual tool. These ideas are fundamental to linear algebra, are closely related to the material of Chapter 1, and will increase our understanding of linear systems.

2.1 VECTOR SPACES

In Chapter 1 we saw how to use Gaussian elimination to find the solutions of a system of linear equations. If the system has n variables, each solution is an n-vector $x = (x_1, x_2, \ldots, x_n)$ which can be thought of as a point in **Euclidean n-space**, the collection of all n-vectors. We will denote this space by R^n. For $n = 1$, 2, and 3, these spaces can be represented geometrically. R^1 is just the ordinary number line and each 1-vector (x_1) corresponds to a unique point on the number line. The space R^2 of all 2-vectors is represented by the usual xy-plane. The components of a 2-vector (x_1, x_2) are the coordinates of the corresponding point in the plane. R^3 is represented by points in ordinary three-dimensional space. The components of a 3-vector $x = (x_1, x_2, x_3)$ are the coordinates of a point in 3-space.

Within the space R^n we can add vectors and we can multiply vectors by scalars. Each of these operations has an obvious but important property. If

65

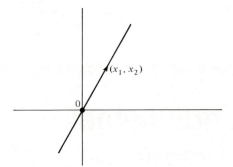

Figure 2.1

we add two vectors in R^n, the resulting vector is again a vector in R^n. If we multiply a vector in R^n by a scalar, the result is again in R^n. Because of this we say that R^n is **closed** under vector addition and scalar multiplication. Performing these algebraic operations on vectors in R^n cannot lead us outside R^n. We discuss next some examples of other collections of vectors that are closed under vector addition and scalar multiplication.

EXAMPLE 1   The solutions of the linear equation $2x_1 - x_2 = 0$ are the vectors on the straight line in R^2 passing through the origin with slope 2 (see Figure 2.1). Adding two vectors on this line or multiplying a vector on this line by a scalar yields another vector on the same line. Indeed, if $u = (u_1, u_2)$ and $v = (v_1, v_2)$ are two solutions of the equation $2x_1 - x_2 = 0$, then $2u_1 - u_2 = 0$ and $2v_1 - v_2 = 0$ and therefore

$$2(u_1 + v_1) - (u_2 + v_2) = 2u_1 - u_2 + 2v_1 - v_2 = 0 + 0 = 0.$$

Hence $u + v = (u_1 + v_1, u_2 + v_2)$ is a solution of $2x_1 - x_2 = 0$. Similarly, if α is any scalar, then $\alpha u = (\alpha u_1, \alpha u_2)$ is a solution of $2x_1 - x_2 = 0$ because

$$2\alpha u_1 - \alpha u_2 = \alpha(2u_1 - u_2) = \alpha \cdot 0 = 0.$$

Therefore, the solutions of the linear equation $2x_1 - x_2 = 0$ form a set of 2-vectors that is closed under vector addition and scalar multiplication.

EXAMPLE 2   Let us consider a plane in R^3 which contains the origin (see Figure 2.2). It follows from the definition of vector addition that when we add two vectors in such a plane, the resulting vector is also in the plane. Similarly, multiplying a vector in such a plane by a scalar gives us a vector that is also in the plane. Thus any plane that contains the origin is closed under vector addition and scalar multiplication.

EXERCISE 1   The general equation for a plane in R^3 containing the origin is $ax_1 + bx_2 + cx_3 = 0$. Verify that if $u = (u_1, u_2, u_3)$ and $v = (v_1, v_2, v_3)$ are solutions of this equation and α is a scalar, then $u + v$ and αu are also solutions of the equation.

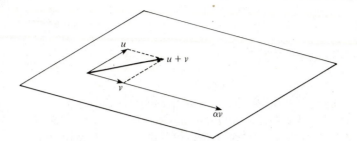

Figure 2.2

EXAMPLE 3 Consider a homogeneous linear system $Ax = 0$, where A is an $m \times n$ matrix. This system has at least one solution, the n-vector $x = 0$. However, there may be other solutions. Let V be the collection of all solutions. Then V is a collection of n-vectors. If u and v are two n-vectors in V, then $Au = 0$ and $Av = 0$ and hence

$$A(u + v) = Au + Av = 0 + 0 = 0.$$

Thus $u + v$ is in V. This shows that the solution set V is closed under vector addition. If α is any scalar, then αu is also in V because $A(\alpha u) = \alpha(Au) = \alpha 0 = 0$. Thus the set of solutions of a homogeneous linear system is also closed under scalar multiplication.

Collections of n-vectors that are closed under the operations of vector addition and scalar multiplication are so important that we give them a special name.

Definition   A collection V of n-vectors is called a **vector space** if V has the following two properties:

1. (Closure under vector addition) If we add any two vectors u and v belonging to V, then the sum $u + v$ again belongs to V.
2. (Closure under scalar multiplication) If we multiply any vector v in V by any scalar α, then the product αv again belongs to V.

Note that every vector space V consisting of n-vectors contains the zero n-vector. To see this, choose any vector v in V. (Here and everywhere else in this book we assume that a vector space contains at least one vector, that is, is not empty.) Then by property 2 with $\alpha = 0$, $0v = 0$ must be in V. Also, V contains with every vector v its negative $-v$ (property 2 with $\alpha = -1$).

The Euclidean n-spaces are examples of vector spaces. Every plane in R^3 containing the origin is also a vector space (see Example 2). Since these planes are subsets of R^3, it is appropriate to call them subspaces of R^3.

Definition   A vector space V is called a **subspace** of a vector space W if every vector in V also belongs to W.

Figure 2.3

Thus a subspace of a vector space W is simply a subset of W which is closed under vector addition and scalar multiplication. The smallest subspace of a vector space is the collection consisting of only the zero vector. We denote this subspace by $\{0\}$. It is closed under vector addition and scalar multiplication because $0 + 0 = 0$ and $\alpha 0 = 0$ for any α. The subspace $\{0\}$ is called the **trivial** subspace of W. Any other subspace is called **nontrivial**.

It is easy to determine the subspaces of R^3. Certainly, R^3 is a subspace of itself. In Example 2 we saw that a plane in R^3 containing the origin is a vector space and hence a subspace of R^3. Every straight line in R^3 passing through the origin is a subspace of R^3 because sums and multiples of vectors on such a line are again on the same line (see Figure 2.3 and Exercise 2). Finally, there is the trivial subspace $\{0\}$ consisting of the origin alone. We have found the following subspaces of R^3:

1. R^3 itself.
2. Planes containing the origin.
3. Straight lines containing the origin.
4. $\{0\}$.

We will see later that these are the only subspaces of R^3.

EXERCISE 2   The general equation of a straight line in R^3 passing through the origin is $(x_1, x_2, x_3) = t(a_1, a_2, a_3)$, where (a_1, a_2, a_3) is a fixed nonzero vector and t is an arbitrary scalar. Verify that if $u = (u_1, u_2, u_3)$ and $v = (v_1, v_2, v_3)$ are vectors on this line and α is any scalar, then $u + v$ and αu are on this line.

EXERCISE 3   Verify that each of the following is a subspace of R^2, the xy-plane.

(a) $\{0\}$.
(b) Straight lines containing the origin.
(c) R^2 itself.

It is important to realize that straight lines which do not pass through the origin are not vector spaces. They are not closed under vector addition or under scalar multiplication (see Figure 2.4). Similarly, planes or straight lines in R^3 that do not contain the origin are not vector spaces. In general, a set of vectors that does not contain the zero vector cannot be a vector space because a vector space must contain the zero vector.

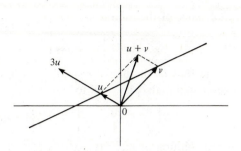

Figure 2.4

EXAMPLE 4 Consider the set V of all 2-vectors whose components are integers. This set is closed under vector addition, but fails to be closed under scalar multiplication. For instance, the vector $(1, 2)$ belongs to V but $0.5(1, 2) = (0.5, 1)$ is not in V. Thus V is not a vector space.

EXAMPLE 5 The set W of all 2-vectors that lie in the first or third quadrant of the xy-plane is not a vector space. If x is a vector in W, then every multiple αx of x is also in W. Hence W is closed under scalar multiplication. However, W fails to be closed under vector addition. The vectors $(2, 3)$ and $(-1, -5)$ are in W but their sum $(1, -2)$ is not.

Examples 4 and 5 show that a collection of vectors can have one of the two defining properties for a vector space without having the other.

In Example 3 we saw that the set V of all solutions of a homogeneous system $Ax = 0$ is closed under vector addition and scalar multiplication. Hence V is a vector space. If the matrix A has n columns, every solution of $Ax = 0$ is an n-vector and therefore V is a subspace of R^n.

Definition Let A be an $m \times n$ matrix. Then the subspace of R^n consisting of all solutions of the homogeneous linear system $Ax = 0$ is called the **null space** of the matrix A. We denote the null space of A by $N(A)$.

For instance, the null space of the matrix $A = [2 \quad -1]$ is the straight line in R^2 passing through the origin with slope 2 (see Example 1).

EXAMPLE 6 The null space of the matrix $A = [1 \quad 2 \quad 1]$ consists of all solutions of the equation $x_1 + 2x_2 + x_3 = 0$. This equation is the equation of a plane in R^3 passing through the origin (see Exercise 1). Thus the null space $N(A)$ is a plane in R^3 passing through the origin. The general solution of $Ax = 0$ is $x = x_2(-2, 1, 0) + x_3(-1, 0, 1)$, where x_2 and x_3 are free variables. Hence the null space of A consists of all sums of multiples of the vectors $(-2, 1, 0)$ and $(-1, 0, 1)$.

EXAMPLE 7 The null space of the matrix

$$A = \begin{bmatrix} 1 & 2 & 1 \\ 1 & 3 & 2 \end{bmatrix}$$

consists of all solutions of the linear system $Ax = 0$. Gaussian elimination reduces the augmented matrix $[A : 0]$ to

$$\begin{bmatrix} 1 & 0 & -1 & \vdots & 0 \\ 0 & 1 & 1 & \vdots & 0 \end{bmatrix}.$$

Hence the general solution of $Ax = 0$ is $x = x_3(1, -1, 1)$, where x_3 is a free variable. Thus the null space of A consists of all multiples of the vector $(1, -1, 1)$. It is a straight line in R^3 passing through the origin (see Exercise 2).

EXAMPLE 8 The reader can check that the only solution of the system $Ax = 0$, where $A = \begin{bmatrix} 1 & 2 \\ 3 & 4 \end{bmatrix}$, is the trivial solution $x = (0, 0)$. Thus $N(A) = \{0\}$.

Although the collection of all solutions to a homogeneous linear system $Ax = 0$ is a vector space, the collection of all solutions of a nonhomogeneous linear system $Ax = b$, $b \neq 0$, is never a vector space. The reason for this is very simple. Since $A0 = 0 \neq b$, the zero vector is not a solution of $Ax = b$. In fact, as we saw in Chapter 1, a nonhomogeneous linear system may be inconsistent and therefore have no solutions at all.

However, the set of all vectors b such that the system $Ax = b$ is consistent is a vector space. To see this, let A be an $m \times n$ matrix and let us call an m-vector b *attainable* if $Ax = b$ has a solution. Suppose that b and c are two attainable vectors. Thus $Ax = b$ and $Ax = c$ have solutions, say, $x = u$ and $x = v$, respectively. Then $x = u + v$ is a solution of $Ax = b + c$, because $A(u + v) = Au + Av = b + c$. Hence $b + c$ is an attainable vector and we see that the set of attainable vectors is closed under vector addition. This set is also closed under scalar multiplication. If α is any scalar, then $x = \alpha u$ is a solution of $Ax = \alpha b$ since $A(\alpha u) = \alpha Au = \alpha b$. Hence αb is an attainable vector. This proves that the set of all attainable vectors is a vector space.

Definition Let A be an $m \times n$ matrix. The subspace of R^m consisting of all vectors b for which the system $Ax = b$ is consistent is called the **column space** of the matrix A. We denote the column space of A by $C(A)$.

Thus a vector b lies in the column space of a matrix A if and only if the system $Ax = b$ has at least one solution.

EXAMPLE 9 To describe the column space of the matrix

$$A = \begin{bmatrix} 2 & 1 \\ 0 & 1 \\ 2 & 1 \end{bmatrix}$$

we use forward elimination to reduce the augmented matrix

$$\begin{bmatrix} 2 & 1 & \vdots & b_1 \\ 0 & 1 & \vdots & b_2 \\ 2 & 1 & \vdots & b_3 \end{bmatrix} \quad \text{to} \quad \begin{bmatrix} 2 & 1 & \vdots & b_1 \\ 0 & 1 & \vdots & b_2 \\ 0 & 0 & \vdots & b_3 - b_1 \end{bmatrix}.$$

It follows that the system $Ax = b$ is consistent if and only if $b_3 - b_1 = 0$. Hence the column space of A consists of all vectors $b = (b_1, b_2, b_3)$ whose first and third components are equal.

EXAMPLE 10 Consider the matrix

$$A = \begin{bmatrix} 1 & 2 & -1 \\ 2 & 4 & -2 \\ -4 & -8 & 4 \end{bmatrix}.$$

Gaussian elimination reduces the augmented matrix

$$\begin{bmatrix} 1 & 2 & -1 & \vdots & b_1 \\ 2 & 4 & -2 & \vdots & b_2 \\ -4 & -8 & 4 & \vdots & b_3 \end{bmatrix} \quad \text{to} \quad \begin{bmatrix} 1 & 2 & -1 & \vdots & b_1 \\ 0 & 0 & 0 & \vdots & b_2 - 2b_1 \\ 0 & 0 & 0 & \vdots & b_3 + 4b_1 \end{bmatrix}.$$

Therefore, the system $Ax = b$ is consistent if and only if the components of the vector $b = (b_1, b_2, b_3)$ satisfy the two equations

$$b_2 - 2b_1 = 0$$
$$b_3 + 4b_1 = 0.$$

Thus the column space of A is the set of all 3-vectors whose components satisfy both of these equations.

Examples 9 and 10 illustrate a general method for describing the column space of an $m \times n$ matrix A. Form the augmented matrix $[A : b]$, where b is an arbitrary m-vector, and reduce it to $[E : b']$, where E is an echelon matrix. The components of b' are then given in terms of the components of b. Each zero row of the echelon matrix E defines an equation which the components of b must satisfy if there is to be a solution to $Ax = b$. These equations are called **constraints**. If b satisfies every constraint, then the system $Ax = b$ is consistent. Thus the column space of A consists of all m-vectors whose components satisfy the constraints. If the echelon matrix E has no zero rows, then $Ax = b$ is consistent for every m-vector b and hence the column space of A is all of R^m, that is, $C(A) = R^m$.

PROBLEMS 2.1

1. Prove that the following collections of vectors are vector spaces.
   (a) The collection of all 3-vectors x whose components satisfy $x_1 = 5x_3$ and $x_1 + x_2 - x_3 = 0$.

(b) The collection of all 4-vectors x whose components satisfy $x_1 + x_2 + x_3 + x_4 = 0$.

(c) The collection of all 5-vectors whose second and fourth components are equal.

*2. Determine which of the following collections of vectors are vector spaces.

(a) The collection of all 3-vectors whose first component is nonnegative.

(b) The collection of all 4-vectors x whose components satisfy $x_1^2 + x_2^2 + x_3^2 + x_4^2 = 0$.

(c) The collection of all 4-vectors x whose components satisfy $x_1 = x_2 = x_3 = x_4$.

(d) The collection of all 5-vectors x whose components satisfy $x_1 + x_2 + x_3 \le x_4 + x_5$.

(e) The collection consisting of the vectors $(0,0,0,0)$, $(1,1,1,1)$, and $(2,2,2,2)$.

(f) The collection of all 2-vectors x whose components satisfy $x_1^2 + x_2^2 = 1$.

*3. The following collections of vectors are not vector spaces. In each case determine which of the two defining properties of a vector space are not satisfied.

(a) The collection of all 5-vectors x with first component 1.

(b) The collection of all 4-vectors x whose components satisfy $x_1 x_2 x_3 x_4 = 0$.

(c) The collection of all 3-vectors x whose components satisfy $x_1 + x_2 + x_3 = 2$.

(d) The collection of all 3-vectors x whose second component is an integer.

4. Suppose that v and w are solutions of a nonhomogeneous system $Ax = b$, $b \ne 0$. Show that $v + w$ and αv, where $\alpha \ne 1$, are not solutions of $Ax = b$.

*5. Give a geometric description of the null spaces of the following matrices.

(a) $\begin{bmatrix} 3 & 4 \\ -6 & -8 \end{bmatrix}$ (b) $\begin{bmatrix} 1 & 2 \\ 3 & 4 \end{bmatrix}$ (c) $\begin{bmatrix} 0 & 0 \\ 0 & 0 \end{bmatrix}$

(d) $\begin{bmatrix} 1 & -2 & 2 \\ 4 & -6 & 2 \end{bmatrix}$ (e) $\begin{bmatrix} 1 & -2 & 2 \\ 3 & -6 & 6 \end{bmatrix}$ (f) $\begin{bmatrix} 1 & 0 & 1 \\ 2 & 1 & 0 \\ 3 & 1 & 1 \end{bmatrix}$

*6. For each of the following matrices, describe its column space by finding the constraints a vector b must satisfy in order to be in the column space.

(a) $\begin{bmatrix} 3 & -1 \\ 9 & -3 \\ 6 & -2 \end{bmatrix}$ (b) $\begin{bmatrix} 1 & 2 & 0 \\ 3 & 1 & 1 \\ 5 & 5 & 1 \end{bmatrix}$ (c) $\begin{bmatrix} 1 & 2 & 0 \\ 1 & 1 & 2 \\ 0 & 2 & 3 \end{bmatrix}$

*7. Let A be an $n \times n$ matrix. Show that the collection V of all n-vectors x that satisfy $Ax = x$ is a subspace of R^n.

8. Suppose that U and V are subspaces of R^n. Then the set of all vectors that are in both U and V is called the *intersection* of U and V. We denote the intersection of U and V by $U \cap V$.

(a) Show that $U \cap V$ is a subspace of R^n.

(b) Suppose that U and V are two distinct planes in R^3 passing through the origin. Describe their intersection.

(c) Describe $U \cap V$ if U and V are two distinct straight lines passing through the origin.

2.2 LINEAR COMBINATIONS

In Section 2.1 we saw that a linear system $Ax = b$ is consistent if and only if the vector b lies in the column space of the matrix A. In addition, we saw that the column space can be described by constraint equations. In this section we give another description of the column space of a matrix, which also explains why it is called the column space.

Consider the linear system

$$\begin{bmatrix} 2 & 1 \\ 0 & 1 \\ 2 & 1 \end{bmatrix} \begin{bmatrix} x_1 \\ x_2 \end{bmatrix} = \begin{bmatrix} b_1 \\ b_2 \\ b_3 \end{bmatrix}. \tag{1}$$

Since

$$\begin{bmatrix} 2 & 1 \\ 0 & 1 \\ 2 & 1 \end{bmatrix} \begin{bmatrix} x_1 \\ x_2 \end{bmatrix} = \begin{bmatrix} 2x_1 + x_2 \\ 0x_1 + x_2 \\ 2x_1 + x_2 \end{bmatrix} = x_1 \begin{bmatrix} 2 \\ 0 \\ 2 \end{bmatrix} + x_2 \begin{bmatrix} 1 \\ 1 \\ 1 \end{bmatrix},$$

we see that system (1) has a solution if and only if the equation

$$x_1 \begin{bmatrix} 2 \\ 0 \\ 2 \end{bmatrix} + x_2 \begin{bmatrix} 1 \\ 1 \\ 1 \end{bmatrix} = \begin{bmatrix} b_1 \\ b_2 \\ b_3 \end{bmatrix}$$

has a solution. Therefore, the vector $b = (b_1, b_2, b_3)$ is in the column space of the coefficient matrix A in (1) if and only if b is equal to a sum of multiples of the columns of A.

In general, if A is an $m \times n$ matrix and b is an m-vector, then the linear system $Ax = b$ can be rewritten in the form

$$x_1 v_1 + x_2 v_2 + \cdots + x_n v_n = b$$

where $v_1, v_2, \ldots, v_n$ are the columns of A. It follows that the vector b is in the column space of A if and only if b can be expressed as a sum of multiples of the columns of A.

Definition A vector b is called a **linear combination** of the vectors $v_1, v_2, \ldots, v_n$ if b can be expressed in the form

$$b = \alpha_1 v_1 + \alpha_2 v_2 + \cdots + \alpha_n v_n,$$

where the α_i's are scalars.

In this terminology, a vector b is in the column space of a matrix A if and only if b is a linear combination of the columns of A. This of course means that the column space of A consists precisely of all possible linear combinations of the columns of A. This is why we called $C(A)$ the "column space" of A.

EXAMPLE 1   Let us determine whether the vector $b = (-3, 12, 12)$ is a linear combination of the vectors $v_1 = (-1, 3, 1)$, $v_2 = (0, 2, 4)$, and $v_3 = (1, 0, 2)$. It follows from the discussion above that the vector b is a linear combination of v_1, v_2, v_3 if and only if b is in the column space of the matrix

$$\begin{bmatrix} -1 & 0 & 1 \\ 3 & 2 & 0 \\ 1 & 4 & 2 \end{bmatrix}.$$

Gaussian elimination reduces the augmented matrix

$$\begin{bmatrix} -1 & 0 & 1 & \vdots & -3 \\ 3 & 2 & 0 & \vdots & 12 \\ 1 & 4 & 2 & \vdots & 12 \end{bmatrix} \quad \text{to} \quad \begin{bmatrix} 1 & 0 & 0 & \vdots & 2 \\ 0 & 1 & 0 & \vdots & 3 \\ 0 & 0 & 1 & \vdots & -1 \end{bmatrix}.$$

The solution is $x_1 = 2$, $x_2 = 3$, $x_3 = -1$. It follows that $b = 2v_1 + 3v_2 - v_3$ and hence b is a linear combination of v_1, v_2, and v_3.

EXAMPLE 2   The system

$$\begin{bmatrix} 3 & -6 \\ 2 & -4 \end{bmatrix} \begin{bmatrix} x_1 \\ x_2 \end{bmatrix} = \begin{bmatrix} 1 \\ 5 \end{bmatrix}$$

has no solutions. Thus the vector $(1, 5)$ is not a linear combination of the vectors $(3, 2)$ and $(-6, -4)$. Equivalently, the vector $(1, 5)$ does not lie in the column space of the coefficient matrix of the system.

In general, given vectors b and $v_1, v_2, \ldots, v_n$ in R^m, we can determine whether b is a linear combination of the v_i's as follows. The matrix form of the equation

$$x_1 v_1 + x_2 v_2 + \cdots + x_n v_n = b$$

is $Ax = b$, where the columns of the matrix A are the vectors $v_1, v_2, \ldots, v_n$. Then b is a linear combination of the v_i's if and only if the system $Ax = b$ has a solution.

EXERCISE 1   Determine whether b is a linear combination of the vectors $v_1, v_2, \ldots$, where:

(a) $b = (3, 1)$, $v_1 = (2, 4)$, $v_2 = (-1, 3)$, $v_3 = (5, 9)$
(b) $b = (0, 1, 4)$, $v_1 = (1, 3, -5)$, $v_2 = (2, 9, 4)$

We now have two ways to describe the column space of a matrix A. In Section 2.1 we saw that a vector b is in the column space of A if and only if b satisfies the constraint equations obtained from the zero rows of the echelon form $[E : b']$ of the augmented matrix $[A : b]$. From our work in this section we know that b is in the column space of A if and only if b is a linear combination of the columns of A. These two descriptions of $C(A)$ illustrate two general methods for describing a vector space V. The first is to impose constraints that the components of any vector in V must satisfy (see

Problem 1 in Section 2.1). The second is to specify a set of vectors $v_1, v_2, \ldots, v_n$ in V such that any other vector in V can be expressed as a linear combination of these vectors. In this situation we say that the vectors $v_1, v_2, \ldots, v_n$ span V.

Definition   Suppose that $v_1, v_2, \ldots, v_n$ are vectors in a vector space V. We say that these vectors **span** V if V consists of all linear combinations of $v_1, v_2, \ldots, v_n$, that is, if every vector v in V can be expressed in the form

$$v = \alpha_1 v_1 + \alpha_2 v_2 + \cdots + \alpha_n v_n,$$

where the α_i's are scalars.

In this terminology, the columns of a matrix span the column space of the matrix. Next we give several examples of vectors that span a given vector space.

EXAMPLE 3   The vectors $e_1 = (1, 0)$ and $e_2 = (0, 1)$ span R^2 since every 2-vector b is a linear combination of e_1 and e_2.

$$b = (b_1, b_2) = b_1(1, 0) + b_2(0, 1) = b_1 e_1 + b_2 e_2$$

Similarly, the vectors $e_1 = (1, 0, 0)$, $e_2 = (0, 1, 0)$, and $e_3 = (0, 0, 1)$ span R^3 because for any vector b in R^3 we have

$$\begin{aligned} b &= (b_1, b_2, b_3) \\ &= b_1(1, 0, 0) + b_2(0, 1, 0) + b_3(0, 0, 1) \\ &= b_1 e_1 + b_2 e_2 + b_3 e_3. \end{aligned}$$

In general, we define n special vectors $e_1, e_2, \ldots, e_n$ in R^n by letting e_i be the n-vector with ith component equal to 1 and all other components equal to 0.

$$\begin{aligned} e_1 &= (1, 0, 0, \ldots, 0) \\ e_2 &= (0, 1, 0, \ldots, 0) \\ e_3 &= (0, 0, 1, \ldots, 0) \\ &\;\vdots \\ e_n &= (0, 0, 0, \ldots, 1) \end{aligned}$$

Notice that the vector e_i is simply the ith column (as well as the ith row) of I_n, the $n \times n$ identity matrix. If $b = (b_1, b_2, \ldots, b_n)$, then

$$b = b_1 e_1 + b_2 e_2 + \cdots + b_n e_n.$$

Thus the vectors $e_1, e_2, \ldots, e_n$ span R^n.

EXAMPLE 4   The null space of the matrix $A = \begin{bmatrix} 1 & 2 & -1 \end{bmatrix}$ consists of all solutions of the equation $x_1 + 2x_2 - x_3 = 0$. The general solution is given by $x_1 = -2x_2 + x_3$, where x_2 and x_3 are free variables. In vector form the general solution is

$$(x_1, x_2, x_3) = x_2(-2, 1, 0) + x_3(1, 0, 1)$$

and we see that the solutions of $Ax = 0$ are precisely the linear combinations of the vectors $u = (-2, 1, 0)$ and $v = (1, 0, 1)$. Hence these two vectors span the null space of the matrix A. Note that u and v are the solutions of $Ax = 0$ obtained by setting $x_2 = 1$, $x_3 = 0$ and $x_2 = 0$, $x_3 = 1$, respectively.

EXAMPLE 5   The general solution of the homogeneous system $Ax = 0$, where

$$A = \begin{bmatrix} 1 & -2 & 0 & 3 & -8 \\ 0 & 0 & 1 & 4 & -5 \\ 0 & 0 & 0 & 0 & 0 \end{bmatrix},$$

is $x_1 = 2x_2 - 3x_4 + 8x_5$, $x_3 = -4x_4 + 5x_5$. The free variables are x_2, x_4, and x_5. For each of these free variables, there is a solution with this free variable equal to 1 and all other free variables equal to 0. In this way we obtain three solutions of $Ax = 0$.

$$
\begin{aligned}
u &= (2, 1, 0, 0, 0) & & x_2 = 1, x_4 = 0, x_5 = 0 \\
v &= (-3, 0, -4, 1, 0) & & x_2 = 0, x_4 = 1, x_5 = 0 \\
w &= (8, 0, 5, 0, 1) & & x_2 = 0, x_4 = 0, x_5 = 1
\end{aligned}
$$

Using these three solutions, we find that the vector form of the general solution of $Ax = 0$ is $x = x_2 u + x_4 v + x_5 w$. Hence the vectors u, v, and w span the null space of the matrix A.

Examples 4 and 5 illustrate a general procedure for finding a set of vectors that span the null space of a matrix A. First reduce the homogeneous system $Ax = 0$ to echelon form, $Ex = 0$. If there are no free variables, then $x = 0$ is the only solution and the null space of A is trivial. If there are free variables, then to each free variable x_i there corresponds a nonzero solution vector v_i obtained by setting x_i equal to 1 and all other free variables equal to 0. Every solution of $Ax = 0$ is a linear combination of these nonzero vectors v_i. Hence these vectors span the null space of A.

EXERCISE 2   For each of the following matrices, find a set of vectors that span its null space.

$$
\text{(a)} \begin{bmatrix} 1 & 0 & 1 & 2 \\ 2 & 1 & 2 & 3 \\ 3 & 1 & 3 & 5 \end{bmatrix} \qquad \text{(b)} \begin{bmatrix} 1 & 2 & -1 & 0 \\ 2 & 4 & 1 & 3 \\ 1 & 2 & 3 & 4 \end{bmatrix}
$$

In the discussion above we started with a vector space V and found a set of vectors $v_1, v_2, \ldots, v_n$ that spanned V. What happens if we start with a set of vectors $v_1, v_2, \ldots, v_n$ in R^m? Is there a vector space V which these vectors span? Yes, V is the column space of the matrix A which has $v_1, v_2, \ldots, v_n$ as its columns. V is the set of all possible linear combinations of $v_1, v_2, \ldots, v_n$ and is called the **subspace of R^m spanned** by $v_1, v_2, \ldots, v_n$. We can also verify directly from the definition that the collection of all linear combinations of $v_1, v_2, \ldots, v_n$ is a vector space. If u and w are linear combinations of

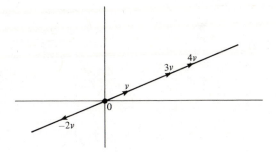

Figure 2.5

the v_i's, say

$$u = \delta_1 v_1 + \delta_2 v_2 + \cdots + \delta_n v_n,$$
$$w = \beta_1 v_1 + \beta_2 v_2 + \cdots + \beta_n v_n,$$

and α is any scalar, then $u + w$ and αu are also linear combinations of the v_i's because

$$u + w = (\delta_1 + \beta_1) v_1 + (\delta_2 + \beta_2) v_2 + \cdots + (\delta_n + \beta_n) v_n$$

and

$$\alpha u = (\alpha \delta_1) v_1 + (\alpha \delta_2) v_2 + \cdots + (\alpha \delta_n) v_n.$$

Since the set of linear combinations of the v_i's is closed under vector addition and scalar multiplication, it is a vector space.

EXAMPLE 6   Consider a single vector $v \neq 0$. The subspace spanned by v consists of all the multiples of v. If v is in R^2 or R^3, these multiples form a straight line through the origin (see Figure 2.5). Although we cannot easily visualize the whole vector space R^m if $m > 3$, we can still think of the subspace spanned by a single nonzero m-vector as a straight line through the origin.

EXAMPLE 7   Let u and v be two nonzero m-vectors. If one of these vectors is a multiple of the other, say $u = \delta v$, then u and v lie on the same straight line through the origin. Such vectors are called **collinear**. Every linear combination of u and v is a multiple of v because

$$\alpha u + \beta v = \alpha \delta v + \beta v = (\alpha \delta + \beta) v.$$

Hence the subspace of R^m spanned by u and v is just the line spanned by v. If u and v are not collinear, the subspace of R^m spanned by u and v can be thought of as a plane containing the origin (see Figure 2.6).

The column space $C(A)$ of an $m \times n$ matrix A is the subspace of R^m spanned by the columns of A. It follows that $C(A) = R^m$ if and only if the columns of A span R^m. Thus $C(A) = R^m$ if and only if for every vector b in R^m the linear system $Ax = b$ has at least one solution. If E is the echelon form of A, then $Ax = b$ has a solution for every m-vector b if and only if E

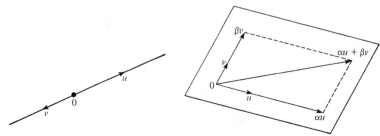

Figure 2.6

has no zero rows, that is, if and only if there are no constraint equations. Thus to determine whether a set $v_1, \ldots, v_n$ of m-vectors span R^m, we proceed as follows. Form the matrix A with columns $v_1, v_2, \ldots, v_n$ and reduce A to an echelon matrix E. These vectors span R^m if and only if E has no zero rows.

EXAMPLE 8   Let us show that the vectors $v_1 = (1, 0, 0)$, $v_2 = (2, 2, 4)$, and $v_3 = (-1, 0, 1)$ span R^3. The matrix

$$A = \begin{bmatrix} 1 & 2 & -1 \\ 0 & 2 & 0 \\ 0 & 4 & 1 \end{bmatrix}$$

with columns v_1, v_2, and v_3 can be reduced to the echelon matrix

$$E = \begin{bmatrix} 1 & 2 & -1 \\ 0 & 2 & 0 \\ 0 & 0 & 1 \end{bmatrix}.$$

Therefore, the column space of A is R^3. Equivalently, the vectors v_1, v_2, and v_3 span R^3.

Let A be an $m \times n$ matrix. Besides the null space and the column space of A, there is a third vector space associated with the matrix A. It is the space of all linear combinations of the rows of A. This space is a subspace of R^n because the rows of A are n-vectors.

Definition   If A is an $m \times n$ matrix, then the subspace of R^n spanned by the rows of A is called the **row space** of A. We denote the row space of A by $R(A)$.

For example, the row space of the matrix

$$A = \begin{bmatrix} 1 & 2 & -1 \\ 2 & -2 & 3 \\ 7 & 2 & 3 \end{bmatrix} \tag{2}$$

is the subspace of R^3 consisting of all linear combinations of the three rows $(1, 2, -1)$, $(2, -2, 3)$, and $(7, 2, 3)$. To see the significance of the row space in the theory of linear equations, consider the homogeneous system

$$\begin{bmatrix} 1 & 2 & -1 \\ 2 & -2 & 3 \\ 7 & 2 & 3 \end{bmatrix} \begin{bmatrix} x_1 \\ x_2 \\ x_3 \end{bmatrix} = \begin{bmatrix} 0 \\ 0 \\ 0 \end{bmatrix}. \tag{3}$$

Since every vector in the row space of A is a linear combination of the rows of A, every vector in the row space corresponds to a linear combination of the equations in the system (3). For example, if we denote the rows of A by v_1, v_2, and v_3, then $v = 2v_1 - v_2 + v_3 = (7, 8, -2)$ is a vector in the row space of A corresponding to the equation

$$7x_1 + 8x_2 - 2x_3 = 0.$$

This equation is of course twice the first equation minus the second equation plus the third equation. Hence if we replace any row of A with the vector v, we do not change the row space of A and we do not change the null space of A.

We will now show that if A is any matrix, row operations on A do not change the row space of A. To see why, suppose that the matrix B is obtained from A by an application of one of the three row operations. The rows of B are linear combinations of the rows of A and hence contained in the row space of A. Therefore, $R(B)$ is a subspace of $R(A)$. Since the row operations are reversible, we can obtain A from B by appropriate row operations. Hence $R(A)$ is also a subspace of $R(B)$. Therefore, the two row spaces are identical; that is, $R(A) = R(B)$. In particular, if we reduce a matrix A to an echelon matrix E, the row space does not change and consequently $R(A) = R(E)$. Since the nonzero rows of E clearly span $R(E)$, they also span $R(A)$. This proves the following result.

Result 1   If a matrix A has been reduced to an echelon matrix E by row operations, then $R(A) = R(E)$ and hence the nonzero rows of E span the row space of A.

For example, the matrix A in (2) can be reduced to the echelon matrix

$$E = \begin{bmatrix} 1 & 2 & -1 \\ 0 & -6 & 5 \\ 0 & 0 & 0 \end{bmatrix}. \tag{4}$$

Thus the row space of A is spanned by the vectors $(1, 2, -1)$ and $(0, -6, 5)$.

EXERCISE 3   Express each row of the matrix A in (2) as a linear combination of the nonzero rows of the echelon matrix E in (4). Also express each row of E as a linear combination of the rows of A.

PROBLEMS 2.2

*1. In each part determine whether the vector u is a linear combination of the v_i's.
   (a) $u = (3, -5)$, $v_1 = (1, 2)$, $v_2 = (-2, 6)$
   (b) $u = (1, 1, 1)$, $v_1 = (1, 0, 1)$, $v_2 = (0, 3, 5)$
   (c) $u = (2, -2, \frac{1}{6}, \frac{1}{6})$, $v_1 = (1, -1, 0, 0)$, $v_2 = (2, 0, 1, 1)$, $v_3 = (0, 3, 1, 1)$
   (d) $u = (0, 1, 0, 1, 0)$, $v_1 = (1, 2, 2, 1, 1)$, $v_2 = (\frac{2}{3}, 1, \frac{4}{3}, 1, \frac{2}{3})$

*2. Which of the following sets of vectors span R^3?
   (a) $(2, 0, 0)$, $(1, 1, 0)$, $(-1, 3, 4)$
   (b) $(-3, 0, 0)$, $(1, 0, 1)$, $(2, 0, -3)$
   (c) $(5, 0, 0)$, $(-1, 2, 0)$
   (d) $(1, 0, 2)$, $(1, 1, 1)$, $(0, 2, 1)$, $(1, 1, 2)$

*3. Which of the following sets of vectors span R^4?
   (a) $(1, 0, 0, 0)$, $(0, 3, 0, 0)$, $(2, 1, 2, 0)$, $(1, 1, 1, 0)$
   (b) $(-2, 0, 0, 0)$, $(1, 1, 0, 0)$, $(-1, 2, 1, 1)$
   (c) $(1, 0, 0, 0)$, $(1, 1, 0, 0)$, $(1, 1, 1, 0)$, $(1, 1, 1, 1)$
   (d) $(0, 1, 1, 0)$, $(1, 1, 1, 1)$, $(-1, 1, -1, 1)$, $(1, 2, 3, 4)$

*4. For each of the following matrices, find a set of vectors that span the null space of the matrix.

   (a) $\begin{bmatrix} 2 & -3 \\ 4 & -6 \end{bmatrix}$

   (b) $\begin{bmatrix} 1 & 1 & -1 & 3 \\ 2 & 2 & 1 & 0 \end{bmatrix}$

   (c) $\begin{bmatrix} 1 & 2 & 0 & 1 & 1 \\ 3 & 6 & 1 & 2 & 1 \end{bmatrix}$

   (d) $\begin{bmatrix} 0 & 1 & 0 & 1 & 0 \\ 1 & 2 & 0 & 1 & 0 \\ 0 & 1 & 0 & 1 & 2 \end{bmatrix}$

5. Suppose that the vector v is a linear combination of the vectors $v_1, v_2, \ldots, v_n$. Explain why the subspaces spanned by $v_1, v_2, \ldots, v_n$ and $v_1, v_2, \ldots, v_n, v$ are identical.

*6. Let V be the vector space spanned by vectors v_1, v_2, and v_3. Show that the vectors $u_1 = v_1$, $u_2 = v_1 + v_2$, and $u_3 = v_1 + v_2 + v_3$ also span V.

7. Suppose that the vectors $w_1, w_2, \ldots, w_m$ are linear combinations of the vectors $v_1, v_2, \ldots, v_n$. Explain why the vector space W spanned by the w_i's is a subspace of the vector space V spanned by the v_i's.

8. Suppose that U and V are subspaces of R^n. Then the *sum* of U and V, denoted by $U + V$, is defined as the set of all vectors that can be expressed as a sum of a vector in U and a vector in V. Thus a vector w belongs to $U + V$ if and only if there are vectors u in U and v in V such that $w = u + v$.
   (a) Show that $U + V$ is a subspace of R^n.
   (b) Show that both U and V are subspaces of $U + V$.
   (c) If U is spanned by the vectors $u_1, u_2, \ldots, u_r$ and V is spanned by the vectors $v_1, v_2, \ldots, v_s$, show that $U + V$ is spanned by $u_1, u_2, \ldots, u_r, v_1, v_2, \ldots, v_s$.
   (d) Describe $U + V$ if U and V are two distinct straight lines in R^3 passing through the origin.
   (e) Describe $U + V$ if U is a plane and V is a straight line in R^3 both of which pass through the origin.

2.3   LINEAR INDEPENDENCE

In Section 1.4 we saw that if a system of equations contains redundant equations, then some of these equations disappear during the Gaussian elimination process. For example, reducing the homogeneous linear system

$$\begin{bmatrix} 1 & 2 & -1 \\ 2 & -2 & 3 \\ 7 & 2 & 3 \end{bmatrix} \begin{bmatrix} x_1 \\ x_2 \\ x_3 \end{bmatrix} = \begin{bmatrix} 0 \\ 0 \\ 0 \end{bmatrix} \tag{1}$$

to echelon form yields

$$\begin{bmatrix} 1 & 2 & -1 \\ 0 & -6 & 5 \\ 0 & 0 & 0 \end{bmatrix} \begin{bmatrix} x_1 \\ x_2 \\ x_3 \end{bmatrix} = \begin{bmatrix} 0 \\ 0 \\ 0 \end{bmatrix}. \tag{2}$$

The third equation $7x_1 + 2x_2 + 3x_3 = 0$ in (1) becomes $0 = 0$ in (2) because it is redundant. In system (1), three times the first equation plus two times the second equation gives the third equation. Thus the third equation is just a combination of the first two equations and does not provide any new information about the variables. In other words, the equations in (1) are not independent of each other. In terms of the coefficient matrix

$$A = \begin{bmatrix} 1 & 2 & -1 \\ 2 & -2 & 3 \\ 7 & 2 & 3 \end{bmatrix} \tag{3}$$

this means that the third row v_3 of A is a linear combination of its first two rows v_1 and v_2; that is, $v_3 = 3v_1 + 2v_2$.

Definition   A set of vectors $v_1, v_2, \ldots, v_n$ is called **linearly dependent** (or simply **dependent**) if it is possible to express one of the vectors, say v_i, as a linear combination of the remaining ones,

$$v_i = \alpha_1 v_1 + \cdots + \alpha_{i-1} v_{i-1} + \alpha_{i+1} v_{i+1} + \cdots + \alpha_n v_n.$$

Referring to our previous discussion, the rows of the matrix A in (3) are dependent because the third row is a linear combination of the first two rows. It follows from this that every vector in the row space of A is a linear combination of the first two rows of A. For if $v = \alpha_1 v_1 + \alpha_2 v_2 + \alpha_3 v_3$ is a vector in the row space of A, then

$$v = \alpha_1 v_1 + \alpha_2 v_2 + \alpha_3 v_3 = \alpha_1 v_1 + \alpha_2 v_2 + \alpha_3 (3v_1 + 2v_2)$$
$$= (\alpha_1 + 3\alpha_3) v_1 + (\alpha_2 + 2\alpha_3) v_2.$$

Hence the row space of A is spanned by v_1 and v_2. This shows that the vector v_3 provides no essential information about the row space and therefore is redundant.

In general, whenever a set of vectors is linearly dependent, at least one of the vectors is redundant and may be removed from the set without changing the space spanned by the original set. That is, if the vectors $v_1, v_2, \ldots, v_n$ span a vector space V and are linearly dependent, then one of them, say v_i, is a linear combination of the remaining ones and the vectors $v_1, \ldots, v_{i-1}, v_{i+1}, \ldots, v_n$ also span V.

A set of vectors that does not contain any redundant vectors is called linearly independent.

Definition   A set of vectors $v_1, v_2, \ldots, v_n$ is called **linearly independent** (or simply **independent**) if it is not linearly dependent. That is, the vectors are independent if it is impossible to express any one of the vectors as a linear combination of the remaining ones.

If a set of vectors $v_1, v_2, \ldots, v_n$ is independent, then each one of the vectors provides essential information about the space V spanned by the vectors. If we delete one of the vectors, the remaining ones do not span V. They span a smaller space.

For instance, the rows v_1, v_2, and v_3 of the matrix

$$\begin{bmatrix} 2 & 0 & 0 \\ 0 & 3 & 0 \\ 0 & 0 & 1 \end{bmatrix}$$

are independent. The first row cannot be expressed as a linear combination of the last two because any linear combination of v_2 and v_3 has its first component equal to zero and hence cannot be equal to v_1. Any linear combination of v_1 and v_3 has its second component equal to zero and hence cannot be equal to v_2. Similarly, v_3 cannot be expressed as a linear combination of v_1 and v_2.

Let us examine the geometric meaning of linear dependence and linear independence in the case of two vectors v_1 and v_2. These two vectors are linearly dependent if and only if one of them is a multiple of the other. In other words, v_1 and v_2 are dependent if and only if they are collinear. If they are not collinear, then they are independent and span a plane (see Figure 2.7).

EXAMPLE 1   Three vectors in R^3, no two of which are collinear, are linearly dependent if and only if they lie in the same plane through the origin. If one of the vectors is a linear combination of the remaining two vectors, then this vector lies in the plane spanned by the others and hence all three vectors lie

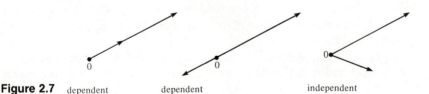

Figure 2.7   dependent   dependent   independent

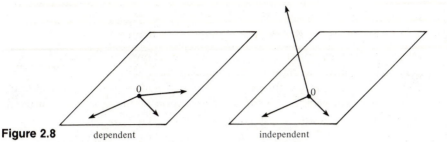

Figure 2.8   dependent   independent

in the same plane. Conversely, if they lie in the same plane, then one of the vectors is a linear combination of the others because this plane can be spanned by two of the vectors. If we have three vectors in R^n, no two of which are collinear, then the same argument shows that they are dependent if and only if they lie in the same plane through the origin (see Figure 2.8).

EXERCISE 1   Show that three vectors in R^3 are linearly dependent if and only if there exists a plane through the origin which contains all three of the vectors. Because of Example 1 you only have to consider the case where at least two of the vectors are collinear.

We have defined a set of vectors to be linearly dependent if one of the vectors is a linear combination of the remaining ones. We now give a different but equivalent definition of linear dependence. This definition leads to a practical procedure for determining whether a given set of vectors is dependent or independent. Suppose that $v_1, v_2, \ldots, v_n$ is a set of vectors. It is of course possible to express the zero vector as a linear combination of these vectors,

$$0 = 0v_1 + 0v_2 + \cdots + 0v_n.$$

Let us assume that we can express the zero vector as a linear combination of $v_1, v_2, \ldots, v_n$ in a nontrivial way, that is,

$$0 = \alpha_1 v_1 + \alpha_2 v_2 + \cdots + \alpha_n v_n,$$

where at least one of the scalars, say α_i, is not equal to zero. Then we can solve for the vector v_i,

$$v_i = \beta_1 v_1 + \cdots + \beta_{i-1} v_{i-1} + \beta_{i+1} v_{i+1} + \cdots + \beta_n v_n,$$

where $\beta_j = -\alpha_j/\alpha_i, j \neq i$. This shows that v_i is a linear combination of the remaining vectors and hence the vectors $v_1, v_2, \ldots, v_n$ are linearly dependent. Conversely, let us assume that these vectors are linearly dependent and show that the zero vector can be expressed as a linear combination of $v_1, v_2, \ldots, v_n$ in a nontrivial way. Since the vectors are dependent, one of the vectors, say v_i, is a linear combination of the remaining vectors,

$$v_i = \gamma_1 v_1 + \cdots + \gamma_{i-1} v_{i-1} + \gamma_{i+1} v_{i+1} + \cdots + \gamma_n v_n.$$

Subtracting v_i from both sides of this equation gives

$$0 = \alpha_1 v_1 + \alpha_2 v_2 + \cdots + \alpha_n v_n,$$

where $\alpha_j = \gamma_j$ for $j \neq i$ and $\alpha_i = -1 \neq 0$. This shows that the zero vector is a nontrivial linear combination of the vectors $v_1, v_2, \ldots, v_n$. Thus the vectors $v_1, v_2, \ldots, v_n$ are linearly dependent if and only if it is possible to express the zero vector as a linear combination of the v_i's in a nontrivial way, that is, if and only if the equation

$$x_1 v_1 + x_2 v_2 + \cdots + x_n v_n = 0 \tag{4}$$

has a nontrivial solution (a solution with at least one x_i not equal to zero). Equivalently, the vectors $v_1, v_2, \ldots, v_n$ are linearly independent if and only if the only solution of equation (4) is the trivial solution $x_1 = x_2 = \cdots = x_n = 0$.

This discussion proves the following result, which gives us alternative definitions for both linear dependence and linear independence.

Result 1   (a) A set of vectors $v_1, v_2, \ldots, v_n$ is linearly dependent if and only if it is possible to express the zero vector as a linear combination

$$\alpha_1 v_1 + \alpha_2 v_2 + \cdots + \alpha_n v_n = 0,$$

where at least one of the α_i's is not equal to zero.
(b) A set of vectors $v_1, v_2, \ldots, v_n$ is linearly independent if and only if the zero vector cannot be expressed as a nontrivial linear combination of the v_i's, that is, whenever $\alpha_1 v_1 + \alpha_2 v_2 + \cdots + \alpha_n v_n = 0$, it follows that $\alpha_1 = \alpha_2 = \cdots = \alpha_n = 0$.

Now suppose that $v_1, v_2, \ldots, v_n$ are vectors in R^m. Then the matrix form of the equation (4) is $Ax = 0$, where A is the $m \times n$ matrix with columns $v_1, v_2, \ldots, v_n$. Thus the v_i's (the columns of A) are dependent if and only if the homogeneous system $Ax = 0$ has a nontrivial solution. This proves:

Result 2   A homogeneous linear system $Ax = 0$ has a nontrivial solution if and only if the columns of A are linearly dependent. Equivalently, the system $Ax = 0$ does not have a nontrivial solution if and only if the columns of A are linearly independent.

This result leads to a computational procedure for determining whether a given set of vectors $v_1, v_2, \ldots, v_n$ in R^m is dependent or independent. Let A be the matrix with columns $v_1, v_2, \ldots, v_n$. Then use row operations to reduce A to an echelon matrix E. The systems $Ax = 0$ and $Ex = 0$ are equivalent. Now there are two possibilities. If every column of E is a pivot column, then the system $Ax = 0$ has no free variables and consequently $x = 0$ is the only solution. In this case, the vectors $v_1, v_2, \ldots, v_n$ (the columns of A) are independent. If E has a column that does not contain a pivot, then the system $Ax = 0$ has at least one free variable. Assigning a nonzero value to this free variable yields a nontrivial solution of $Ax = 0$. In this case the vectors $v_1, v_2, \ldots, v_n$ are dependent. Note that to determine the pivot columns it is not necessary for E to be in reduced echelon form.

EXAMPLE 2   To determine whether the vectors $v_1 = (2, -2, 4)$, $v_2 = (3, 0, 1)$, and $v_3 = (-1, 1, -2)$ are linearly independent or linearly dependent, we reduce the matrix

$$A = \begin{bmatrix} 2 & 3 & -1 \\ -2 & 0 & 1 \\ 4 & 1 & -2 \end{bmatrix} \quad \text{to} \quad E = \begin{bmatrix} 2 & 3 & -1 \\ 0 & 3 & 0 \\ 0 & 0 & 0 \end{bmatrix}.$$

Since the third column of E does not contain a pivot, the vectors v_1, v_2, and v_3 are dependent. If we solve $Ax = 0$, we obtain $x_1 = 0.5x_3$, $x_2 = 0$, where x_3 is free. Setting, for instance, $x_3 = 1$ yields $0.5v_1 + 0v_2 + 1v_3 = 0$.

EXAMPLE 3   The vectors $v_1 = (1, -2)$ and $v_2 = (3, 1)$ are linearly independent because row operations reduce the matrix

$$A = \begin{bmatrix} 1 & 3 \\ -2 & 1 \end{bmatrix} \quad \text{to} \quad E = \begin{bmatrix} 1 & 3 \\ 0 & 7 \end{bmatrix}$$

and both columns of E are pivot columns. The only solution of $Ax = 0$ is $x = 0$.

EXAMPLE 4   The n-vectors $e_1, e_2, \ldots, e_n$ defined in Example 3 in Section 2.2 are linearly independent because the matrix with columns $e_1, e_2, \ldots, e_n$ is the $n \times n$ identity matrix I and every column of I is a pivot column.

EXAMPLE 5   A set consisting of one nonzero vector v is linearly independent because the only solution of $x_1 v = 0$ is $x_1 = 0$. However, the set consisting of only the zero vector is linearly dependent; $x_1 = 1$ is a nontrivial solution of $x_1 0 = 0$. In general, any set of vectors containing the zero vector is dependent. (Why?)

EXERCISE 2   Which of the following sets of vectors are linearly independent?

(a) $(-3, 1)$, $(4, -2)$
(b) $(-3, 1)$, $(4, -2)$, $(7, 2)$
(c) $(-1, 2, 0, 2)$, $(5, 0, 1, 1)$, $(8, -6, 1, -5)$
(d) $(0, -3, 1)$, $(2, 4, 1)$, $(-2, 8, 5)$

The columns of the triangular matrix

$$E = \begin{bmatrix} 2 & 3 & 4 \\ 0 & 6 & 1 \\ 0 & 0 & 5 \end{bmatrix}$$

are independent because all three columns of E are pivot columns. In general, the pivot columns of any echelon matrix form an independent set of vectors. To see why, consider the echelon matrix

$$E = \begin{bmatrix} 2 & 1 & 3 & 4 \\ 0 & 0 & 5 & 2 \\ 0 & 0 & 0 & 4 \\ 0 & 0 & 0 & 0 \end{bmatrix}.$$

The first, third, and fourth columns of E are the pivot columns. If we let D be the matrix obtained by deleting the columns of E that do not contain pivots,

$$D = \begin{bmatrix} 2 & 3 & 4 \\ 0 & 5 & 2 \\ 0 & 0 & 4 \\ 0 & 0 & 0 \end{bmatrix},$$

then every column of D is a pivot column. It follows that the columns of D, that is, the pivot columns of E, are independent. It is also true that the nonzero rows of E are independent. This follows from the form of the echelon matrix E. The form that we are referring to is the progression of the pivots from left to right as we move down the rows. Suppose that we had

$$\alpha_1(2, 1, 3, 4) + \alpha_2(0, 0, 5, 2) + \alpha_3(0, 0, 0, 4) = (0, 0, 0, 0).$$

Looking at the first components, we obtain $2\alpha_1 = 0$, hence $\alpha_1 = 0$. This gives

$$\alpha_2(0, 0, 5, 2) + \alpha_3(0, 0, 0, 4) = (0, 0, 0, 0).$$

Now the third components give $5\alpha_2 = 0$, hence $\alpha_2 = 0$. Therefore,

$$\alpha_3(0, 0, 0, 4) = (0, 0, 0, 0),$$

which gives $\alpha_3 = 0$. This shows that the zero vector cannot be expressed as a linear combination of the rows of E except with all coefficients equal to zero. Clearly, this reasoning can be extended to any echelon matrix. This proves the following result.

Result 3   If E is an echelon matrix, then:

(a) The pivot columns of E are linearly independent.
(b) The nonzero rows of E are linearly independent.

We saw in Section 2.2 that the solution vectors of a homogeneous system $Ax = 0$ obtained by setting in turn each free variable equal to 1 and all other free variables equal to 0 span the null space of the matrix A. These solution vectors are also linearly independent. For example, consider the homogeneous system

$$\begin{bmatrix} 2 & -4 & 2 \\ -1 & 2 & -1 \\ 3 & -6 & 3 \end{bmatrix} \begin{bmatrix} x_1 \\ x_2 \\ x_3 \end{bmatrix} = \begin{bmatrix} 0 \\ 0 \\ 0 \end{bmatrix}.$$

Gaussian elimination reduces this system to

$$\begin{bmatrix} 1 & -2 & 1 \\ 0 & 0 & 0 \\ 0 & 0 & 0 \end{bmatrix} \begin{bmatrix} x_1 \\ x_2 \\ x_3 \end{bmatrix} = \begin{bmatrix} 0 \\ 0 \\ 0 \end{bmatrix}.$$

The free variables are x_2 and x_3. Setting $x_2 = 1$, $x_3 = 0$ yields the solution vector $(2, 1, 0)$, while setting $x_2 = 0$, $x_3 = 1$ yields $(-1, 0, 1)$. The general solution of the system is

$$x = (x_1, x_2, x_3) = x_2(2, 1, 0) + x_3(-1, 0, 1) \tag{5}$$

and the vectors $(2, 1, 0)$ and $(-1, 0, 1)$ span the null space of the coefficient matrix. Since the second and third components of the linear combination in (5) are x_2 and x_3, respectively, it follows that the only solution of

$$(0, 0, 0) = x_2(2, 1, 0) + x_3(-1, 0, 1)$$

is $x_2 = x_3 = 0$ (all free variables equal to zero). Therefore, the solution vectors $(2, 1, 0)$ and $(-1, 0, 1)$ are independent. The general result is:

Result 4   Let A be a matrix. Then the solution vectors of the homogeneous system $Ax = 0$ obtained by setting in turn each free variable equal to one and all other free variables equal to zero are linearly independent and span the null space of A.

It follows immediately from this result that the only solution of $Ax = 0$ having all free variables equal to zero is the zero solution. If there are no free variables, then $Ax = 0$ has no nontrivial solution and $N(A) = \{0\}$.

We conclude this section with an important theorem about homogeneous systems. Suppose that A is an $m \times n$ matrix and we reduce A to an echelon matrix E. Then each nonzero row of E contains precisely one pivot, and no two pivots are in the same column. Consequently, the number of pivot columns of E is equal to the number of nonzero rows of E. Hence the number of pivot columns is less than or equal to m, the total number of rows of E. The total number of columns of E is n. It follows that if $m < n$, then not all columns of E can be pivot columns. Therefore, there will be at least one free variable, say x_i, in the system $Ax = 0$. Assigning a nonzero value to x_i, for instance $x_i = 1$, yields a nontrivial solution of $Ax = 0$. This proves the following theorem [which the reader should compare with Theorem 1.1(f)].

Theorem 2.1   *If A is an $m \times n$ matrix and $m < n$, then the homogeneous system $Ax = 0$ has nontrivial solutions.*

In other words, a homogeneous system of linear equations with more unknowns than equations must have a nontrivial solution. It follows from this theorem that any set of more than m vectors in R^m is dependent. If

$v_1, v_2, \ldots, v_n$ are vectors in R^m and $m < n$, then the system $Ax = 0$, where the columns of A are the v_i's, has a nontrivial solution.

PROBLEMS 2.3

*1. Which of the following sets of vectors are linearly independent?
  (a) $(-1, 3)$, $(5, 9)$
  (b) $(6, 8, 1)$, $(0, 0, 0)$, $(-1, 1, -1)$
  (c) $(2, 1)$, $(3, 1)$, $(4, 1)$
  (d) $(4, 3, 0)$, $(6, 1, 1)$, $(0, 0, 1)$, $(1, 1, 1)$, $(1, 2, 3)$
  (e) $(1, 2, 0)$, $(2, 1, 1)$, $(3, 1, 0)$
  (f) $(1, 1, 1, 1, 1)$, $(0, 1, 0, 1, 0)$, $(-1, 2, 1, 3, 1)$

2. Show that the vectors $(1, 0, 3, 1)$, $(-1, 1, 0, 1)$, $(2, 3, 0, 0)$, and $(1, 1, 6, 3)$ are linearly dependent. Can each one of these vectors be expressed as a linear combination of the others?

*3. Find an independent set of vectors that span the null space of the given matrix.

(a) $\begin{bmatrix} 1 & 0 & -1 & -2 \\ 0 & 1 & 3 & 4 \end{bmatrix}$ (b) $\begin{bmatrix} 1 & 2 & 3 \\ 1 & 2 & 5 \end{bmatrix}$ (c) $\begin{bmatrix} 0 & 1 & 2 \\ 0 & 2 & 4 \\ 0 & 3 & 6 \end{bmatrix}$

4. If one of the vectors $v_1, v_2, \ldots, v_n$ is the zero vector, show that the vectors are linearly dependent.

*5. If two of the vectors $v_1, v_2, \ldots, v_n$ are equal, show that the vectors are linearly dependent.

*6. Show that if the vectors v_1, v_2, and v_3 are independent, then the vectors $u_1 = 2v_1$, $u_2 = v_1 + v_2$, and $u_3 = -v_1 + v_3$ are also independent.

7. Show that if the vectors $v_1, v_2, \ldots, v_m, v_{m+1}, \ldots, v_n$ are independent, then the vectors $v_1, v_2, \ldots, v_m$ are independent. More generally, explain why any subset of a linearly independent set of vectors must be linearly independent.

8. Suppose that $v_1, v_2, \ldots, v_k$ is an independent set of vectors. Show that if the vector v is not a linear combination of $v_1, v_2, \ldots, v_k$, then $v_1, v_2, \ldots, v_k, v$ is an independent set of vectors.

*9. Suppose that the columns of the matrix A are independent and the vectors $v_1, v_2, \ldots, v_n$ are independent. Show that the vectors $Av_1, Av_2, \ldots, Av_n$ are independent.

10. Suppose that A is an $m \times n$ matrix and $b_1, b_2, \ldots, b_k$ is an independent set of vectors in R^m. Suppose that v_i is a solution of $Ax = b_i$, $i = 1, \ldots, k$. Show that $v_1, v_2, \ldots, v_k$ is an independent set of vectors in R^n.

2.4   BASIS AND DIMENSION

In this section we introduce the concepts of a basis for a vector space and the dimension of a vector space. Our intuitive concept of dimension is reflected in words such as length, width and height. We think of lines as one-dimensional, planes as two-dimensional, and ordinary space as three-dimensional. We will now make these intuitive ideas precise.

We saw in Section 2.2 that if we reduce a matrix A to an echelon matrix E, then the nonzero rows of E span the row space of A. This spanning set contains no redundant information about $R(A)$ because the nonzero rows of E are linearly independent (Result 3 in Section 2.3). This combination of linear independence and spanning is so important that sets of vectors having these two properties are given a special name.

Definition   A set of vectors $v_1, v_2, \ldots, v_n$ in a vector space V is called a **basis** for V if these vectors are linearly independent and span V.

Thus a basis for a vector space V is a set of vectors that contains all necessary information about V (the vectors span V) and no redundant information about V (the vectors are independent). For example, the nonzero rows of an echelon matrix obtained from a matrix A form a basis for the row space of A. In the next section we develop a systematic procedure to find bases for the other two spaces associated with A, the column space and the null space of A.

Before giving several other examples of bases for vector spaces, let us make one important point. If the vectors $v_1, v_2, \ldots, v_n$ form a basis for V, then any given vector v in V can be expressed as a linear combination of the basis vectors in one and only one way. Since the v_i's span V, v is a linear combination of the v_i's, say

$$v = \alpha_1 v_1 + \alpha_2 v_2 + \cdots + \alpha_n v_n.$$

If $v = \beta_1 v_1 + \beta_2 v_2 + \cdots + \beta_n v_n$ is another expression of v as a linear combination of the basis vectors, then

$$0 = v - v = (\alpha_1 - \beta_1)v_1 + (\alpha_2 - \beta_2)v_2 + \cdots + (\alpha_n - \beta_n)v_n.$$

Since the v_i's are independent, we conclude that $\alpha_i - \beta_i = 0$ and hence $\alpha_i = \beta_i$ for all i. This proves:

Result 1   Suppose that the vectors $v_1, v_2, \ldots, v_n$ are a basis for the vector space V. If v is a vector in V, then v can be expressed as a linear combination of the basis vectors $v_1, v_2, \ldots, v_n$ in precisely one way.

EXAMPLE 1   Using row operations, we can reduce the matrix

$$A = \begin{bmatrix} 1 & 2 & 0 & 1 \\ 2 & 5 & 1 & 4 \\ 3 & 8 & 2 & 7 \end{bmatrix} \quad \text{to} \quad E = \begin{bmatrix} 1 & 2 & 0 & 1 \\ 0 & 1 & 1 & 2 \\ 0 & 0 & 0 & 0 \end{bmatrix}.$$

The nonzero rows of the echelon matrix E are linearly independent and span the row space of A. Hence the vectors $(1, 2, 0, 1)$ and $(0, 1, 1, 2)$ form a basis for $R(A)$. Every vector in the row space of A can be expressed as a linear combination of these two basis vectors in exactly one way.

EXAMPLE 2 The vectors $e_1, e_2, \ldots, e_n$ in R^n defined by

$$
\begin{aligned}
e_1 &= (1, 0, 0, \ldots, 0), \\
e_2 &= (0, 1, 0, \ldots, 0), \\
e_3 &= (0, 0, 1, \ldots, 0), \\
&\ \ \vdots \\
e_n &= (0, 0, 0, \ldots, 1)
\end{aligned}
$$

are linearly independent and span R^n (Example 4 in Section 2.3 and Example 3 in Section 2.2). Hence they form a basis for R^n. This particular basis is called the **standard basis** for R^n. Every n-vector $b = (b_1, b_2, \ldots, b_n)$ can be expressed as a linear combination of the e_i's in exactly one way,

$$
b = b_1 e_1 + b_2 e_2 + \cdots + b_n e_n.
$$

EXAMPLE 3 The vectors $e_1 = (1, 0)$ and $e_2 = (0, 1)$ are a basis for R^2. Geometrically, it should be clear that any pair of noncollinear vectors in R^2 also form a basis for R^2. For instance, to prove algebraically that the vectors $v_1 = (1, 2)$ and $v_2 = (-2, -3)$ form a basis for R^2, we consider the equation

$$
x_1 \begin{bmatrix} 1 \\ 2 \end{bmatrix} + x_2 \begin{bmatrix} -2 \\ -3 \end{bmatrix} = \begin{bmatrix} b_1 \\ b_2 \end{bmatrix},
$$

where b is an arbitrary 2-vector. Gaussian elimination reduces the augmented matrix

$$
\begin{bmatrix} 1 & -2 & \vdots & b_1 \\ 2 & -3 & \vdots & b_2 \end{bmatrix} \quad \text{to} \quad \begin{bmatrix} 1 & 0 & \vdots & -3b_1 + 2b_2 \\ 0 & 1 & \vdots & b_2 - 2b_1 \end{bmatrix}.
$$

The unique solution is $x_1 = -3b_1 + 2b_2$, $x_2 = b_2 - 2b_1$. It follows that every vector b in R^2 is a linear combination of v_1 and v_2; that is, v_1 and v_2 span R^2. Furthermore, setting $b_1 = b_2 = 0$, we see that the only solution of $x_1 v_1 + x_2 v_2 = 0$ is $x_1 = x_2 = 0$. Hence v_1 and v_2 are linearly independent. Since we have exhibited two different bases for R^2, this example shows that a vector space can have more than one basis.

EXERCISE 1 Verify that the two vectors $v_1 = (5, 2)$ and $v_2 = (-3, 1)$ form a basis for R^2.

EXAMPLE 4 Let us show that the vectors $v_1 = (1, 1, 0)$, $v_2 = (2, 3, 1)$, and $v_3 = (1, 1, 1)$ form a basis for R^3. Let $b = (b_1, b_2, b_3)$ be an arbitrary vector in R^3. Consider the equation $x_1 v_1 + x_2 v_2 + x_3 v_3 = b$. Gaussian elimination reduces the augmented matrix

$$\left[\begin{array}{ccc:c} 1 & 2 & 1 & b_1 \\ 1 & 3 & 1 & b_2 \\ 0 & 1 & 1 & b_3 \end{array}\right] \quad \text{to} \quad \left[\begin{array}{ccc:c} 1 & 0 & 0 & 2b_1 - b_2 - b_3 \\ 0 & 1 & 0 & b_2 - b_1 \\ 0 & 0 & 1 & b_3 - b_2 + b_1 \end{array}\right].$$

The unique solution is

$$x_1 = 2b_1 - b_2 - b_3, x_2 = b_2 - b_1, x_3 = b_3 - b_2 + b_1.$$

Thus every vector in R^3 is a linear combination of v_1, v_2, and v_3. Setting $b = 0$, we conclude that the only solution of $x_1 v_1 + x_2 v_2 + x_3 v_3 = 0$ is $x_1 = x_2 = x_3 = 0$. Hence v_1, v_2, and v_3 are linearly independent.

Although a vector space can have more than one basis (see Examples 3 and 4), one of the most important theorems in linear algebra states that any two bases for a given vector space have the same number of vectors. To prove this, we need a preliminary result.

Result 2   Let V be a vector space that is spanned by the m vectors $w_1, w_2, \ldots, w_m$. If $v_1, v_2, \ldots, v_n$ are independent vectors in V, then $n \le m$. That is, if V is spanned by a set of m vectors, then no set of more than m vectors in V can be independent.

Proof   Let A be the matrix with columns $w_1, w_2, \ldots, w_m$. Then V is the column space of A. Since the v_i's are in V, it follows that each of the n systems

$$Ax = v_j, \qquad j = 1, \ldots, n$$

is consistent. Let $b_j, j = 1, \ldots, n$, be vectors in R^m such that $Ab_j = v_j$. Then $AB = C$, where B is the matrix with columns $b_1, b_2, \ldots, b_n$ and C is the matrix with columns $v_1, v_2, \ldots, v_n$. Note that B is an $m \times n$ matrix. Just suppose for a moment that $m < n$. Then $Bx = 0$ has a nontrivial solution and hence there must be a nonzero vector u such that $Bu = 0$. This implies that $Cu = ABu = 0$ and thus the system $Cx = 0$ has a nontrivial solution. But this is impossible because the columns of C are the independent vectors $v_1, v_2, \ldots, v_n$. Therefore, $m < n$ cannot be true and we conclude that $n \le m$.

Now suppose that $v_1, v_2, \ldots, v_n$ and $w_1, w_2, \ldots, w_m$ are two bases for a vector space V. Since the v_i's are linearly independent and the w_i's span V, it follows that $n \le m$. Similarly, since the w_i's are linearly independent and the v_i's span V, we also have $m \le n$. Therefore, $m = n$, and we have proved the following theorem.

Theorem 2.2   *Any two bases for a vector space have the same number of vectors.*

This theorem allows us to give a precise definition of the dimension of a vector space.

Definition   We say that a vector space V has **dimension n** (or that V is **n-dimensional**) if V has a basis consisting of n vectors. The dimension of V is denoted by $\dim V$.

Thus, in order to determine the dimension of a vector space V, we have to find a basis for V. Then the number n of vectors in this basis is the dimension of V and all bases for V consist of n vectors. For example, the standard basis for R^n (see Example 2) consists of n vectors. Therefore, the dimension of R^n is n (as it should be) and every basis for R^n consists of n vectors. The reader may wonder whether every vector space has a basis and therefore a well-defined dimension. It is true that all nontrivial subspaces of the Euclidean spaces R^n have bases (see Problem 8). The trivial subspace $\{0\}$ contains no independent vectors, and therefore it cannot have a basis (or its "basis" consists of no vectors). For this reason we say that the space $\{0\}$ has dimension 0. Vector spaces that have bases consisting of finitely many vectors are called **finite-dimensional**. (Of course, the trivial space $\{0\}$ is also called finite-dimensional.)

Any set of k linearly independent vectors $u_1, u_2, \ldots, u_k$ in a vector space V spans a k-dimensional subspace U of V because the vectors $u_1, u_2, \ldots, u_k$ form a basis for U. In particular, a single nonzero vector u in V spans a one-dimensional subspace (a straight line through the origin) and two independent vectors u_1, u_2 span a two-dimensional subspace (a plane through the origin).

Theorem 2.3   *Suppose that V is a vector space of dimension n. Then no set of more than n vectors in V can be linearly independent and V cannot be spanned by fewer than n vectors.*

Proof   Suppose that V has dimension n. Then V is spanned by a basis of n vectors and hence no set of more than n vectors can be independent (Result 2). Also, the n vectors in a basis for V are independent and therefore (again by Result 2) any set of vectors spanning V must consist of at least n vectors.

Thus the dimension of a vector space is the maximum number of linearly independent vectors in V and also the minimum number of vectors needed to span V. For instance, R^n has dimension n and consequently it is impossible to span R^n with fewer than n vectors and no set of more than n vectors in R^n can be linearly independent.

To check whether a given set $v_1, v_2, \ldots, v_n$ is a basis for a vector space V, we have to determine whether these vectors are independent *and* span V. However, if we already know that the dimension of V is n, it suffices to check whether the vectors $v_1, v_2, \ldots, v_n$ are independent *or* span V.

Theorem 2.4   *Suppose that V is a vector of space of dimension n and $v_1, v_2, \ldots, v_n$ is a set of n vectors in V.*

(a) If the vectors are linearly independent, then they form a basis for V.
(b) If the vectors span V, then they form a basis for V.

$$
\begin{bmatrix}
1 & 2 & 1 & \vdots & b_1 \\
1 & 3 & 1 & \vdots & b_2 \\
0 & 1 & 1 & \vdots & b_3
\end{bmatrix}
\quad \text{to} \quad
\begin{bmatrix}
1 & 0 & 0 & \vdots & 2b_1 - b_2 - b_3 \\
0 & 1 & 0 & \vdots & b_2 - b_1 \\
0 & 0 & 1 & \vdots & b_3 - b_2 + b_1
\end{bmatrix}.
$$

In other words, n independent vectors in an n-dimensional vector space automatically span the space. Similarly, n vectors spanning an n-dimensional space are automatically independent.

Proof   Suppose that the vectors $v_1, v_2, \ldots, v_n$ are independent and that v is any vector in V. By Theorem 2.3 the $n + 1$ vectors $v_1, v_2, \ldots, v_n, v$ must be dependent and hence there exist scalars $\alpha_1, \alpha_2, \ldots, \alpha_n, \alpha_{n+1}$, not all equal to zero, such that

$$\alpha_1 v_1 + \alpha_2 v_2 + \cdots + \alpha_n v_n + \alpha_{n+1} v = 0. \tag{1}$$

Now α_{n+1} cannot be equal to zero because otherwise (1) would imply that $v_1, v_2, \ldots, v_n$ are dependent. Hence we can solve (1) for v,

$$v = \beta_1 v_1 + \beta_2 v_2 + \cdots + \beta_n v_n,$$

where $\beta_i = -\alpha_i / \alpha_{n+1}$. Since v was an arbitrary vector in V, the vectors $v_1, v_2, \ldots, v_n$ span V. This proves part (a) of the theorem.

   To prove part (b), suppose that $v_1, v_2, \ldots, v_n$ span V. If these vectors were dependent, one of them would be a linear combination of the remaining $n - 1$ vectors and these remaining vectors would span V. This is impossible by Theorem 2.3. Thus $v_1, v_2, \ldots, v_n$ are independent, which proves part (b).

EXAMPLE 5   The three vectors $v_1 = (3, 0, 0)$, $v_2 = (1, 2, 0)$, and $v_3 = (5, 1, 4)$ are independent because all columns of the matrix

$$A = \begin{bmatrix} 3 & 1 & 5 \\ 0 & 2 & 1 \\ 0 & 0 & 4 \end{bmatrix}$$

are pivot columns. Since $\dim R^3 = 3$, it follows from Theorem 2.4 that v_1, v_2, v_3 is a basis for R^3. In general, any independent set of n vectors in R^n is a basis for R^n.

   Ideas similar to those used in the proof of Theorem 2.4 can be used to show that any linearly independent set of vectors in a vector space V can be expanded to a basis for V and that any set of vectors spanning V can be contracted to a basis for V. To be more precise, if $v_1, v_2, \ldots, v_k$ is an independent set of vectors in an n-dimensional vector space V and $k < n$, it is possible to find $n - k$ vectors $v_{k+1}, \ldots, v_n$ such that $v_1, v_2, \ldots, v_n$ is a basis for V. Similarly, if the vectors $w_1, w_2, \ldots, w_r$ span V and $n < r$, then it is possible to obtain a basis for V by discarding $r - n$ of the spanning vectors (see Problem 7). In the next section we see how to determine which vectors should be added to an independent set or discarded from a spanning set in order to obtain a basis for V.

   It is easy to describe the relationship between the dimension of a vector space V and the dimension of a subspace of V. If U is a subspace of V, then $\dim U \le \dim V$, and $\dim U = \dim V$ if and only if $U = V$. To see this, suppose that $\dim U = m$ and $\dim V = n$. Then U has a basis $u_1, u_2, \ldots, u_m$ consisting of m vectors. These vectors are independent and in V, and therefore $m \leq n$ by Theorem 2.3. If $m = n$, then (by Theorem 2.4) $u_1, u_2, \ldots, u_n$ is a basis for V and hence $U = V$.

We conclude this section with some remarks about dimension. All one-dimensional subspaces of Euclidean n-space are straight lines through the origin. Although two such lines need not be identical, they have the same geometric properties. If V is a line in R^2 and W is a line in R^3 (or in R^7) and if we disregard how they are positioned in the bigger spaces, then the two lines are geometrically identical. Similarly, any two-dimensional subspace of Euclidean n-space looks like a plane and hence all two-dimensional vector spaces are essentially the same. More generally, any two n-dimensional vector spaces are essentially the same. Mathematicians often state this by saying that all vector spaces of a given dimension are **isomorphic**, a term derived from Greek which means having the same form or shape. For example, every three-dimensional subspace of R^4 is isomorphic to R^3 and can be visualized as ordinary space. This is extremely helpful when one attempts to visualize the geometry of R^4 (or of any higher-dimensional Euclidean space).

PROBLEMS 2.4

*1. In each part determine whether the given vectors form a basis for R^2.
    (a) $v_1 = (1, -1)$, $v_2 = (3, 0)$
    (b) $v_1 = (2, -3)$, $v_2 = (-6, 9)$
    (c) $v_1 = (1, 1)$, $v_2 = (0, 2)$, $v_3 = (2, 3)$
    (d) $v_1 = (1, 1)$, $v_2 = (0, 8)$

*2. In each part determine whether the given vectors form a basis for R^3.
    (a) $v_1 = (1, 1, 1)$, $v_2 = (0, 2, 3)$, $v_3 = (1, 0, 2)$
    (b) $v_1 = (1, 0, 1)$, $v_2 = (2, 4, 8)$
    (c) $v_1 = (3, 0, 1)$, $v_2 = (1, 1, 1)$, $v_3 = (4, 1, 2)$
    (d) $v_1 = (0, 0, 1)$, $v_2 = (0, 1, 1)$, $v_3 = (-1, 2, 2)$

*3. Find a basis for the row space of the given matrix.

(a) $\begin{bmatrix} 2 & 3 \\ 4 & 6 \end{bmatrix}$ (b) $\begin{bmatrix} 1 & 2 \\ 2 & 4 \\ 3 & 1 \\ 4 & 3 \end{bmatrix}$ (c) $\begin{bmatrix} 1 & 0 & 2 \\ 2 & 0 & 4 \\ 3 & 1 & 0 \end{bmatrix}$

4. Suppose that the vectors v_1, v_2, v_3 are a basis for a vector space V. Show that the vectors $u_1 = v_1 + v_2$, $u_2 = v_2 + v_3$, and $u_3 = 2v_3$ form a basis for V.

5. Let V be an n-dimensional vector space. Show that V has subspaces of dimension k for $k = 0, 1, \ldots, n$. (*Hint:* Let $v_1, v_2, \ldots, v_n$ be a basis for V and consider the subspaces spanned by $v_1, v_2, \ldots, v_k$, $k = 1, \ldots, n$.)

6. Suppose that $v_1, v_2, \ldots, v_k$ is an independent set of vectors in a vector space V. Show that if these vectors do not form a basis for V, then there must exist a vector v_{k+1} in V such that the set $v_1, v_2, \ldots, v_k, v_{k+1}$ is independent. (*Hint:* Problem 8 in Section 2.3.)

7. (a) Show that if $v_1, v_2, \ldots, v_k$ is an independent set of vectors in an n-dimensional vector space V and $k < n$, then there exist $n - k$ vectors $v_{k+1}, \ldots, v_n$ such that the vectors $v_1, \ldots, v_k, v_{k+1}, \ldots, v_n$ form a basis for V. (*Hint:* Use Problem 6.)

   (b) Show that if the vectors $v_1, v_2, \ldots, v_r$ span an n-dimensional vector space V and $r > n$, then we can find a basis for V by discarding $r - n$ vectors from the spanning set. (*Hint:* It is always possible to discard some vector from a dependent set of vectors without changing the space spanned by the vectors.)

8. Let V be a nontrivial subspace of R^n and let $v_1 \neq 0$ be a vector in V. Use Problem 6 to show that V has a basis which includes v_1.

2.5   THE RANK OF A MATRIX

In Section 2.4 we defined the dimension of a vector space to be the number of vectors in a basis for the space. Thus to find the dimension of a vector space V we must find a basis for V. In this section we discuss some computational procedures for finding a basis for the row space, the column space, and the null space.

Recall that we have two general methods for describing a vector space V. First, we can describe V by a set of vectors that span V. If we let A be the matrix whose rows (columns) are these spanning vectors, then V is the row (column) space of A. Second, we can describe a vector space V by prescribing a homogeneous system of equations (constraints) which the vectors in V must satisfy. In this case V is the null space of a matrix. Thus a vector space can always be viewed as the row space, the column space, or the null space of an appropriate matrix. Therefore, our main task in this section is to develop procedures for constructing bases for the row space, the column space, and the null space of a matrix A. We will see that these procedures all have the same first step, the reduction of A to an echelon matrix.

Let us begin with the row space of a matrix A. Suppose that we have used row operations to reduce A to an echelon matrix E. In Section 2.4 we showed that the nonzero rows of E form a basis for the row space of A.

EXAMPLE 1   Row operations reduce the matrix

$$A = \begin{bmatrix} 1 & 2 & 0 & 2 \\ 1 & 2 & 1 & 3 \\ 3 & 6 & 2 & 8 \end{bmatrix} \quad \text{to} \quad E = \begin{bmatrix} 1 & 2 & 0 & 2 \\ 0 & 0 & 1 & 1 \\ 0 & 0 & 0 & 0 \end{bmatrix}.$$

Thus the vectors $(1, 2, 0, 2)$ and $(0, 0, 1, 1)$ form a basis for the row space of A. $R(A)$ is a two-dimensional subspace of R^4.

EXERCISE 1   Find a basis for the row space of the given matrix.

$$\text{(a)} \begin{bmatrix} 1 & 1 & -1 \\ 2 & 1 & 0 \\ 2 & 2 & 3 \end{bmatrix} \quad \text{(b)} \begin{bmatrix} 2 & -3 & 1 \\ 4 & -6 & 2 \\ 6 & -9 & 3 \end{bmatrix}$$

There are two ways to find a basis for the column space of a matrix A: Since the columns of A are rows of A^T, we can find a basis for $C(A)$ by using the procedure described above to find a basis for $R(A^T)$. In other words, we can find a basis for the column space of A by reducing the transpose of A to echelon form. We illustrate this procedure in the following example.

EXAMPLE 2   Let

$$A = \begin{bmatrix} 1 & 2 & 2 \\ 2 & 4 & 3 \\ -3 & -6 & 1 \end{bmatrix}.$$

Row operations reduce A^T to the echelon matrix

$$E = \begin{bmatrix} 1 & 2 & -3 \\ 0 & -1 & 7 \\ 0 & 0 & 0 \end{bmatrix}.$$

Therefore, the two vectors $(1, 2, -3)$ and $(0, -1, 7)$ form a basis for $C(A)$.

The basis for $C(A)$ that was constructed in Example 2 includes the vector $(0, -1, 7)$, which is not a column in the matrix A. We should be able to choose a basis for $C(A)$ from the columns of A. This, of course, requires a different procedure.

Let us consider an example that illustrates this procedure. Row operations reduce the matrix

$$A = \begin{bmatrix} 1 & 3 & -2 & 1 \\ 2 & 6 & -2 & 8 \\ -1 & -3 & 8 & 17 \end{bmatrix} \quad \text{to} \quad E = \begin{bmatrix} 1 & 3 & 0 & 7 \\ 0 & 0 & 1 & 3 \\ 0 & 0 & 0 & 0 \end{bmatrix}.$$

Notice that A and E do *not* have the same column spaces. Every vector in $C(E)$ has its third component equal to zero. This is certainly not true for all vectors in the column space of A.

Denote the columns of A by v_1, v_2, v_3, v_4. We claim that v_1 and v_3, the columns of A that correspond to the pivot columns of E, form a basis for the column space of A. First, we show that these columns are independent. Consider the equation

$$x_1 v_1 + x_3 v_3 = 0. \tag{1}$$

Letting $x = (x_1, 0, x_3, 0)$, this equation becomes $Ax = 0$. Since $Ax = 0$ and $Ex = 0$ have the same solutions, equation (1) is equivalent to $Ex = 0$. The only solution of $Ex = 0$ having $x_2 = x_4 = 0$ is the trivial solution $x = 0$. Thus the only solution of $Ax = 0$ with $x_2 = x_4 = 0$ is $x = 0$. It follows that the only solution of $x_1 v_1 + x_3 v_3 = 0$ is $x_1 = x_3 = 0$. Hence the vectors v_1 and v_3 are independent.

We must now show that the other columns of A are linear combinations of v_1 and v_3. First we show that v_2 is a linear combination of v_1 and v_3. Since the equation $Ex = 0$ has a solution with $x_2 = 1$ and $x_4 = 0$, the equation $Ax = 0$ also has a solution with $x_2 = 1$ and $x_4 = 0$. Thus there are numbers x_1 and x_3 such that

$$x_1 v_1 + v_2 + x_3 v_3 = 0.$$

Solving this equation for v_2 shows that v_2 is a linear combination of v_1 and v_3. Similarly, since $Ax = 0$ has a solution with $x_2 = 0$ and $x_4 = 1$, there are numbers x_1 and x_3 such that

$$x_1 v_1 + x_3 v_3 + v_4 = 0.$$

This shows that v_4 is also a linear combination of v_1 and v_3. Thus v_1 and v_3 span $C(A)$ and since they are independent, they form a basis for $C(A)$.

We now use this reasoning to prove that if A is an $m \times n$ matrix and E is its echelon form, then the columns of A that correspond to the pivot columns of E form basis for $C(A)$. Let us call these columns of A the pivot columns of A. First we show that the pivot columns are independent. Expressing the zero vector as a linear combination of the pivot columns is equivalent to finding a solution to $Ax = 0$ having all free variables equal to zero. But $Ax = 0$ and $Ex = 0$ have the same solutions, and the only solution of $Ex = 0$ having all free variables equal to zero is the zero solution. Thus it is impossible to express the zero vector as a nontrivial linear combination of the pivot columns of A.

To prove that the pivot columns span $C(A)$, it is sufficient to show that any other column of A is a linear combination of the pivot columns of A. Every other column of A corresponds to a free variable, say x_i. There is a solution of $Ax = 0$ with $x_i = 1$ and all other free variables equal to zero. This solution allows us to express the ith column of A as a linear combination of the pivot columns of A.

We can summarize this argument as follows. The pivot columns of A are independent because the only solution of $Ax = 0$ having all free variables equal to zero is the zero solution. These pivot columns span $C(A)$, because for each free variable there is a solution of $Ax = 0$ having this free variable equal to 1 and all other free variables equal to zero.

EXAMPLE 3   Gaussian elimination reduces the matrix

$$A = \begin{bmatrix} 2 & 0 & -1 & 1 \\ 4 & 1 & 0 & 5 \\ 0 & 3 & 6 & 8 \end{bmatrix} \quad \text{to} \quad E = \begin{bmatrix} 2 & 0 & -1 & 1 \\ 0 & 1 & 2 & 3 \\ 0 & 0 & 0 & -1 \end{bmatrix}.$$

The pivot columns of E are columns 1, 2, and 4. So columns 1, 2, and 4 of A form a basis for the column space of A. Hence $C(A)$ is a three-dimensional subspace of R^3 and it follows that $C(A) = R^3$.

EXERCISE 2   Find a basis for the column space of the given matrix.

$$\text{(a)} \begin{bmatrix} 1 & 0 & -1 \\ 2 & 1 & 3 \\ 0 & 5 & -2 \end{bmatrix} \quad \text{(b)} \begin{bmatrix} 0 & 3 \\ 1 & -1 \\ 1 & 2 \end{bmatrix} \quad \text{(c)} \begin{bmatrix} 1 & 0 & -1 & 0 & 1 \\ 1 & -1 & -2 & 2 & -2 \\ 1 & 1 & 0 & 1 & 1 \end{bmatrix}$$

Finally, in order to find a basis for the null space of an $m \times n$ matrix A, we again reduce A to an echelon matrix E. Then $N(A) = N(E)$ because the systems $Ax = 0$ and $Ex = 0$ have the same solutions. Let r be the number of pivot columns of E. Then $Ex = 0$ has $n - r$ free variables. If $r = n$, then there are no free variables and hence $Ax = 0$ has no nontrivial solution, that is, $N(A) = \{0\}$. If $r < n$, then to each free variable x_i there corresponds a solution vector obtained by setting x_i equal to 1 and all other free variables equal to 0. By Result 4 in Section 2.3, the $n - r$ vectors obtained in this way are linearly independent and span the null space of A. Therefore, they form a basis for $N(A)$.

EXAMPLE 4 To find a basis for the null space of the matrix

$$A = \begin{bmatrix} 1 & 2 & 1 & -1 & 3 \\ 2 & 4 & 3 & 0 & 2 \\ 3 & 6 & 4 & -1 & 5 \end{bmatrix},$$

we reduce A to the echelon matrix

$$E = \begin{bmatrix} 1 & 2 & 0 & -3 & 7 \\ 0 & 0 & 1 & 2 & -4 \\ 0 & 0 & 0 & 0 & 0 \end{bmatrix}.$$

The general solution of $Ax = 0$ is $x_1 = -2x_2 + 3x_4 - 7x_5$, $x_3 = -2x_4 + 4x_5$, where x_2, x_4, and x_5 are free variables. For each of these free variables, there is a solution with this free variable equal to 1 and all other free variables equal to 0. In this way we obtain three independent solutions of $Ax = 0$.

$$
\begin{array}{ll}
u = (-2, 1, 0, 0, 0) & x_2 = 1, x_4 = x_5 = 0 \\
v = (3, 0, -2, 1, 0) & x_4 = 1, x_2 = x_5 = 0 \\
w = (-7, 0, 4, 0, 1) & x_5 = 1, x_2 = x_4 = 0
\end{array}
$$

In vector form, the general solution is $x = x_2 u + x_4 v + x_5 w$. The vectors u, v, and w form a basis for the null space of the matrix A. Thus $N(A)$ is a three-dimensional subspace of R^5.

EXAMPLE 5 To find a basis for the subspace V of R^5 consisting of all vectors that satisfy the constraint equations

$$
\begin{array}{l}
x_1 + 2x_2 + x_3 - x_4 + 3x_5 = 0 \\
2x_1 + 4x_2 + 3x_3 \phantom{{}-x_4} + 2x_5 = 0 \\
3x_1 + 6x_2 + 4x_3 - x_4 + 5x_5 = 0,
\end{array}
$$

we form the system $Ax = 0$, where A is the matrix in Example 4. The subspace V is then the null space of A and a basis for the subspace is the basis for $N(A)$ computed in Example 4.

EXERCISE 3   Find a basis for the null space of the given matrix.

$$\text{(a)} \begin{bmatrix} 1 & 2 & 3 & 4 \\ 2 & 6 & 0 & 6 \end{bmatrix} \qquad \text{(b)} \begin{bmatrix} 1 & 1 & 2 \\ 2 & 2 & 1 \\ 3 & 3 & 2 \end{bmatrix}$$

Let us now summarize the procedure for finding bases for the row space, column space, and null space of a matrix.

Theorem 2.5   *Let A be an m × n matrix and suppose that A has been reduced to an echelon matrix E.*

(a) *The nonzero rows of E form a basis for the row space of A.*
(b) *The columns of A corresponding to the pivot columns of E form a basis for the column space of A.*
(c) *If the system Ex = 0 has no free variables, then N(A) = {0}. If there are free variables, then to each free variable there corresponds a solution vector of Ax = 0 obtained by setting this free variable equal to 1 and all other free variables equal to 0. These vectors form a basis for the null space of A.*

The preceding theorem contains some remarkable relationships between the dimensions of the three subspaces associated with an $m \times n$ matrix A. Since the number of nonzero rows of an echelon matrix E is equal to the number of pivot columns of E, the row space and column space of A have the same dimension. This dimension is equal to the number of nonzero rows and also the number of pivot columns of the echelon matrix E. We call this number the **rank** of A. Evidently, the rank of A cannot exceed the number of rows of A or the number of columns of A; that is, rank $A \leq m$ and rank $A \leq n$. The dimension of the null space of A is called the **nullity** of A and is equal to the number of free variables in the system $Ex = 0$. Since the number of free variables plus the number of pivot columns of E is equal to the number of columns of A (see Theorem 1.1), it follows that the dimension of the null space of A is equal to the number of columns of A minus the rank of A,

$$\text{nullity } A = n - \text{rank } A.$$

Finally, the matrix A and its transpose have the same rank because $R(A^T) = C(A)$ and hence

$$\text{rank } A^T = \dim R(A^T) = \dim C(A) = \text{rank } A.$$

Let us summarize this discussion in the following theorem.

Theorem 2.6   *Suppose that A is an m × n matrix. Then:*

(a) *rank A = dim $R(A)$ = dim $C(A)$.*
(b) *nullity A = dim $N(A)$ = n − rank A.*
(c) *rank A = rank A^T.*

Theorem 2.6(b) can be written in the form

$$\text{rank } A + \text{nullity } A = n,$$

where n is the number of columns of A. For this reason it is often referred to as the **rank-plus-nullity theorem.**

Since the rank r of a matrix A is the dimension of the column space of A, no set of more than r columns of A can be independent (Theorem 2.3) and every set of precisely r independent columns of A must be a basis for the column space of A (Theorem 2.4). Similarly, since $r = \dim R(A)$, no set of more than r rows of A can be independent and every set of precisely r independent rows of A must be a basis for the row space of A. In other words, the rank of A is equal to the maximum number of independent columns of A and also equal to the maximum number of independent rows of A. This proves the remarkable fact that the number of independent rows of a matrix is equal to the number of independent columns of the matrix. In particular, if A is an $m \times n$ matrix, then the rows of A are independent if and only if rank $A = m$ and the columns of A are independent if and only if rank $A = n$.

EXAMPLE 6   The rows of the matrix

$$A = \begin{bmatrix} 1 & -1 & 2 \\ -2 & 2 & -4 \\ 1 & 1 & 0 \\ 2 & 0 & 1 \end{bmatrix}$$

must be linearly dependent since the rank of A is at most 3. Forward elimination reduces A to the echelon matrix

$$E = \begin{bmatrix} 1 & -1 & 2 \\ 0 & 2 & -2 \\ 0 & 0 & -1 \\ 0 & 0 & 0 \end{bmatrix}.$$

Thus the rank of A is 3 and the row space is a three-dimensional subspace of R^3. Consequently, $R(A) = R^3$. The vectors $(1, -1, 2)$, $(0, 2, -2)$, and $(0, 0, -1)$ form a basis for the row space of A. This basis contains only one of the rows of A. Any three independent rows of A also form a basis for $R(A)$. For instance, the first, third, and fourth rows of A are independent and hence form a basis for $R(A)$. However, the first three rows of A are not independent. (Why?) The column space of A is a three-dimensional subspace of R^4 and the columns of A form a basis for $C(A)$.

Suppose that V is a vector space which is spanned by a given set of vectors $v_1, v_2, \ldots, v_n$. Theorem 2.5 provides us with two methods for finding a basis and the dimension of V.

1. Let A be the matrix with rows $v_1, v_2, \ldots, v_n$. Hence V is the row space of A. Reduce A to an echelon matrix E. Then the nonzero rows of E form a basis for V and $\dim V = \text{rank } A$.

2. Let A be the matrix with columns $v_1, v_2, \ldots, v_n$. Hence V is the column space of A. Reduce A to an echelon matrix E. Then the columns of A that correspond to the pivot columns of E form a basis for V and $\dim V = \operatorname{rank} A$.

Notice that the basis vectors for V obtained by method 1 are usually not among the original spanning vectors $v_1, v_2, \ldots, v_n$ because the rows of a matrix change during the reduction to echelon form. However, the basis for V obtained by method 2 is a subset of the original spanning set $v_1, v_2, \ldots, v_n$ because the basis is chosen from the columns of the matrix A itself. Thus method 2 contracts a spanning set of V to a basis for V.

EXAMPLE 7   Let V be the subspace of R^4 spanned by the vectors $v_1 = (1, 1, 2, 2)$, $v_2 = (2, 3, 7, 6)$, $v_3 = (1, 2, 5, 4)$, $v_4 = (1, 0, 1, 2)$, and $v_5 = (2, 1, 0, 1)$. We reduce the matrix

$$A = \begin{bmatrix} 1 & 1 & 2 & 2 \\ 2 & 3 & 7 & 6 \\ 1 & 2 & 5 & 4 \\ 1 & 0 & 1 & 2 \\ 2 & 1 & 0 & 1 \end{bmatrix} \quad \text{to} \quad E = \begin{bmatrix} 1 & 1 & 2 & 2 \\ 0 & 1 & 3 & 2 \\ 0 & 0 & 2 & 2 \\ 0 & 0 & 0 & 0 \\ 0 & 0 & 0 & 0 \end{bmatrix}.$$

Since $V = R(A)$, the nonzero rows of the echelon matrix E, that is, the vectors $(1, 1, 2, 2)$, $(0, 1, 3, 2)$, and $(0, 0, 2, 2)$, form a basis for V. Thus V is a three-dimensional subspace of R^4. Note that only the first of the three basis vectors is one of the original five spanning vectors. If we want a basis for V consisting of three of the original spanning vectors, we let A be the matrix with columns v_1, v_2, v_3, v_4, v_5 and reduce

$$A = \begin{bmatrix} 1 & 2 & 1 & 1 & 2 \\ 1 & 3 & 2 & 0 & 1 \\ 2 & 7 & 5 & 1 & 0 \\ 2 & 6 & 4 & 2 & 1 \end{bmatrix} \quad \text{to} \quad E = \begin{bmatrix} 1 & 2 & 1 & 1 & 2 \\ 0 & 1 & 1 & -1 & -1 \\ 0 & 0 & 0 & 2 & -1 \\ 0 & 0 & 0 & 0 & 0 \end{bmatrix}.$$

Then $V = C(A)$ and the columns of A corresponding to the pivot columns of the echelon matrix E form a basis for V. Thus v_1, v_2, and v_4 are a basis for V.

We remarked in Section 2.4 that an independent set of vectors in a vector space can always be expanded to a basis for the space. In Problem 8 of this section we outline a practical procedure for this expansion process.

We conclude by applying the techniques of this section to some of the examples given in Chapter 1.

EXAMPLE 8   In Section 1.2 we asked whether there always exists a meaningful set of prices for the goods in a closed economy. The example we considered consisted of three producers and we saw that in order to have a state of equilibrium, the price vector $p = (p_1, p_2, p_3)$ must be a solution of the system

$$\begin{bmatrix} \frac{2}{5} & \frac{1}{4} & \frac{1}{2} \\ \frac{1}{5} & \frac{1}{2} & \frac{1}{4} \\ \frac{2}{5} & \frac{1}{4} & \frac{1}{4} \end{bmatrix} \begin{bmatrix} p_1 \\ p_2 \\ p_3 \end{bmatrix} = \begin{bmatrix} p_1 \\ p_2 \\ p_3 \end{bmatrix}.$$

As explained in Section 1.2, the entries in the ith column of the matrix describe how the goods produced by astronaut i are divided among the three astronauts. Therefore, the sum of the entries in any column is 1. If we call the coefficient matrix A, we can rewrite the equation as $Ap = Ip$ or $(A - I)p = 0$.

$$\begin{bmatrix} -\frac{3}{5} & \frac{1}{4} & \frac{1}{2} \\ \frac{1}{5} & -\frac{1}{2} & \frac{1}{4} \\ \frac{2}{5} & \frac{1}{4} & -\frac{3}{4} \end{bmatrix} \begin{bmatrix} p_1 \\ p_2 \\ p_3 \end{bmatrix} = \begin{bmatrix} 0 \\ 0 \\ 0 \end{bmatrix}.$$

The entries in each column of $A - I$ sum to zero. Therefore, the rows of $A - I$ are linearly dependent and hence the rank of $A - I$ is 2 or less. It follows that the columns of $A - I$ are linearly dependent. This means that there is a nontrivial solution p. The general solution of $(A - I)p = 0$ is $p = p_3(\frac{5}{4}, 1, 1)$ and the null space of $A - I$ is one-dimensional. We conclude that in this economy all sets of prices are essentially the same. All solutions except the trivial one assign the same relative values to the goods. Doubling p_3 is pure inflation. The only thing that changes is the "value" of the monetary unit.

EXAMPLE 9   In Section 1.2 we saw that an electric circuit gives rise to a system of linear equations. The particular circuit that was discussed had four junctions and six branches. We deduced from Kirchhoff's law that the current vector $I = (I_1, I_2, I_3, I_4, I_5, I_6)$ has to be the solution of the homogeneous linear system $GI = 0$, where

$$G = \begin{bmatrix} 1 & -1 & -1 & 0 & 0 & 0 \\ 0 & 1 & 0 & -1 & -1 & 0 \\ 0 & 0 & 1 & 1 & 0 & -1 \\ -1 & 0 & 0 & 0 & 1 & 1 \end{bmatrix}.$$

This 4×6 matrix (called the *incidence matrix* for the circuit) contains information about the design of the circuit. Each row of the matrix corresponds to a junction. For instance, the third row tells us that currents I_3 and I_4 flow into the third junction J_3 and that I_6 flows out of the junction. Each column of the matrix corresponds to a current. For instance, the fifth column indicates that I_5 flows from junction J_2 to junction J_4. In particular, each column contains besides zero entries exactly one $+1$ entry and one -1 entry. Therefore, the sum of the rows is a zero row and hence the rows of the matrix G are linearly dependent. Since the first three rows of G are linearly independent, the rank of G is 3. Thus the null space of G has dimension $6 - 3 = 3$. In fact, solving the system $GI = 0$ leads to the general solution

$$I = I_4(0, 1, -1, 1, 0, 0) + I_5(1, 1, 0, 0, 1, 0) + I_6(1, 0, 1, 0, 0, 1).$$

We now see that in this particular circuit we have to measure at least three currents before we can calculate all currents.

PROBLEMS 2.5

*1. In each part the matrix is in echelon form. Find its rank, a basis for the column space, and a basis for the row space.

(a) $\begin{bmatrix} 3 & -1 & 2 \\ 0 & 0 & 1 \\ 0 & 0 & 0 \end{bmatrix}$
(b) $\begin{bmatrix} 1 & 2 & 0 & 9 & 3 \\ 0 & 3 & 1 & -5 & 8 \\ 0 & 0 & 0 & 0 & 2 \end{bmatrix}$

(c) $\begin{bmatrix} 6 & 3 & 1 & 3 & 1 & 0 \\ 0 & 0 & 0 & 2 & 0 & 2 \\ 0 & 0 & 0 & 0 & 0 & 0 \\ 0 & 0 & 0 & 0 & 0 & 0 \end{bmatrix}$

2. In each part find the rank of the given matrix, a basis for its row space, and a basis for its column space.

(a) $\begin{bmatrix} -1 & 3 \\ 2 & 5 \end{bmatrix}$
(b) $\begin{bmatrix} 1 & -1 & 3 \\ 3 & -2 & 6 \\ -2 & 3 & 9 \end{bmatrix}$
(c) $\begin{bmatrix} 6 & 5 \\ 2 & 1 \\ 6 & 4 \\ 0 & -3 \end{bmatrix}$

(d) $\begin{bmatrix} 3 & -1 \\ -6 & 2 \\ 2 & 5 \end{bmatrix}$
(e) $\begin{bmatrix} 1 & 2 & -1 \\ 2 & 4 & -2 \\ 1 & 2 & 2 \end{bmatrix}$
(f) $\begin{bmatrix} 1 & -1 & 0 & 2 \\ -2 & 2 & 0 & -4 \\ 3 & -3 & 0 & 6 \end{bmatrix}$

(g) $\begin{bmatrix} 1 & -2 & 1 & 0 & 5 \\ -1 & 0 & 1 & -2 & 2 \\ 1 & -6 & 3 & -2 & 12 \\ 2 & -3 & 0 & 2 & 3 \end{bmatrix}$

*3. Find the rank, a basis for the column space, and a basis for the row space of the following matrices.

(a) $\begin{bmatrix} 2 & -1 & 5 \\ 4 & 5 & 8 \end{bmatrix}$
(b) $\begin{bmatrix} 1 & 6 & 4 \\ 0 & 2 & 2 \\ 3 & 5 & -1 \\ -1 & 1 & 3 \end{bmatrix}$

(c) $\begin{bmatrix} 1 & 2 & -1 & 1 & 2 \\ 1 & 2 & -1 & 1 & 2 \\ 2 & 4 & -2 & 2 & 5 \\ 2 & 4 & 0 & 1 & -1 \end{bmatrix}$

*4. In each part find the nullity of the given matrix. If the nullity is not zero, find a basis for the null space.

(a) $\begin{bmatrix} 1 & -1 & 1 \\ 2 & 0 & 3 \\ 2 & 1 & 4 \end{bmatrix}$

(b) $\begin{bmatrix} 1 & 0 & 1 & 0 \\ -2 & 4 & 1 & -2 \\ -3 & 8 & 3 & -4 \end{bmatrix}$

(c) $\begin{bmatrix} 3 & 4 & -7 \end{bmatrix}$

(d) $\begin{bmatrix} 0 \\ 9 \\ 3 \end{bmatrix}$

(e) $\begin{bmatrix} 1 & 6 & 0 & 7 \\ 0 & -2 & 1 & -3 \\ 0 & 3 & 4 & -1 \\ 0 & 2 & 5 & -3 \end{bmatrix}$

5. Find a basis for the subspace V of R^4 consisting of all vectors x that satisfy the given constraint equations.

(a) $x_1 + x_2 + x_3 + x_4 = 0$
$\quad x_1 + x_2 - x_3 - x_4 = 0$
(b) $x_1 + 2x_2 - x_3 = 0$
(c) $x_1 + x_2 = 3x_3 - x_4$
$\quad 2x_1 + 3x_2 = x_4$

*6. In each part let V be the vector space spanned by the given vectors. Find the dimension of V and a basis for V by deleting some of the vectors if necessary.

(a) $(1, -1, 1)$, $(3, 5, 1)$, $(5, 3, 3)$
(b) $(0, 3, 1)$, $(1, 1, 1)$, $(-3, 0, -2)$, $(1, 0, 5)$
(c) $(1, -1, 2, 0)$, $(2, -2, 4, 0)$, $(0, 1, 1, 1)$, $(4, -3, 9, 1)$

7. In each part let V be the vector space spanned by the given vectors. Find a basis for V and the dimension of V.

(a) $(-2, 3, 1)$, $(0, 1, 2)$, $(-1, 0, 3)$
(b) $(1, 1, 1, 1)$, $(1, 2, 3, 4)$, $(2, 3, 4, 5)$, $(0, 1, 1, 0)$
(c) $(1, -2, 1, 1)$, $(2, 0, 3, 1)$, $(4, -4, 5, 3)$, $(3, 1, 0, 1)$
(d) $(-1, 0, 1, 1, 0)$, $(2, 1, 3, 0, 1)$, $(0, 1, 5, 2, 1)$, $(2, 2, 8, 2, 2)$

8. In this problem we develop a procedure for expanding an independent set of vectors to a basis. Suppose that V is subspace of R^m spanned by the vectors $v_1, v_2, \ldots, v_s$ and $u_1, u_2, \ldots, u_k$ are k linearly independent vectors in V. Let A be the $m \times (k + s)$ matrix with columns $u_1, u_2, \ldots, u_k, v_1, v_2, \ldots, v_s$ (in that order).

(a) Show that V is the column space of A.
(b) Show that if A is reduced to an echelon matrix E, then the first k columns of E are pivot columns.
(c) Deduce that the columns of A corresponding to the pivot columns of E form a basis for V which includes the vectors $u_1, u_2, \ldots, u_k$.

*9. Use Problem 8 to find a basis for R^3 which includes the vector $(1, 1, 2)$. (*Hint:* The standard basis vectors e_1, e_2, and e_3 span R^3.)

10. Use Problem 8 to find a basis for R^4 which includes the independent vectors $u_1 = (1, 2, 2, 4)$ and $u_2 = (1, 1, 1, 2)$. (*Hint:* The standard basis vectors e_1, e_2, e_3, and e_4 span R^4.)

*11. Find a basis for R^4 which includes the vectors $(1, 1, 0, 0)$ and $(1, 2, 3, 1)$.

12. Find all possible price vectors for the closed economy described in Problem 4 of Section 1.2. Explain why all price vectors for this economy are essentially the same.

*13. The price vectors $p = (p_1, p_2, p_3, p_4)$ in a certain closed economy satisfy the equation $Ap = p$, where

$$A = \begin{bmatrix} \frac{1}{2} & 0 & \frac{1}{4} & 0 \\ 0 & \frac{1}{3} & 0 & \frac{2}{3} \\ \frac{1}{2} & 0 & \frac{3}{4} & 0 \\ 0 & \frac{2}{3} & 0 & \frac{1}{3} \end{bmatrix}.$$

Find all possible price vectors for this economy. Are they all essentially the same? Can you interpret your answer in terms of noninteracting subeconomies?

14. For each of the circuits in Problem 2 of Section 1.2, find its incidence matrix and determine how many currents must be measured before all currents can be calculated.

15. Show that an $m \times n$ matrix A has rank 1 if and only if there is a nonzero $m \times 1$ matrix B and a nonzero $1 \times n$ matrix C such that $A = BC$.

16. Let A be an $m \times n$ matrix and B be an $n \times r$ matrix.
    (a) Show that the null space of B is a subspace of the null space of AB.
    (c) Show that the column space of AB is a subspace of the column space of A.
    (c) Use parts (a) and (b) to show that rank(AB) $\leq$ rank B and that rank(AB) $\leq$ rank A.

2.6   LINEAR SYSTEMS REVISITED

The set of all solutions of a homogeneous linear system $Ax = 0$ is a vector space, namely the null space of the matrix. In Section 2.5 we saw how to find a basis and the dimension of the null space of A. We pointed out in Section 2.1 that the set of all solutions of a nonhomogeneous system $Ax = b$, $b \neq 0$, is never a vector space. In fact, this set may be empty because the system $Ax = b$ may be inconsistent. However, if $Ax = b$ is consistent and v_0 is one solution of the system, then all solutions can be expressed in terms of v_0 and the vectors in the null space of A.

EXAMPLE 1   Consider the linear system $Ax = b$, where $A = \begin{bmatrix} 1 & 2 \\ 2 & 4 \end{bmatrix}$ and $b = (3, 6)$. This system is consistent and its general solution is $x = (x_1, x_2) = (3, 0) + x_2(-2, 1)$. The vector $v_0 = (3, 0)$ is a solution vector (for the choice $x_2 = 0$). The general solution of the corresponding homogeneous system $Ax = 0$ is $x = (x_1, x_2) = x_2(-2, 1)$. Thus a vector v is a solution of $Ax = b$ if and only if v can be expressed in the form $v = v_0 + u$, where u is a vector in the null space of A. The geometric picture is as follows. The null space of

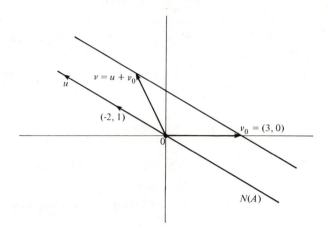

Figure 2.9

A is a line through the origin. The solutions of $Ax = b$ also form a straight line in R^2 which is parallel to the null space but displaced from it by the particular solution v_0 (see Figure 2.9). The solution vector v_0 is one of infinitely many vectors on the line of all solutions. It is the one corresponding to the choice $x_2 = 0$ of the free variable. Any other solution vector can be used to displace the null space, generating the line of solutions of $Ax = b$.

Example 1 is typical of the general situation. Let v_0 be a solution of a consistent linear system $Ax = b$. If u is any vector in the null space of the matrix A, then $v_0 + u$ is also a solution of $Ax = b$ because

$$A(v_0 + u) = Av_0 + Au = b + 0 = b.$$

In this way we obtain all solutions of $Ax = b$. To see this, suppose that $Av = b$. Then the vector $u = v - v_0$ is in the null space of A because

$$Au = A(v - v_0) = Av - Av_0 = b - b = 0.$$

Since $v = v_0 + u$, we have expressed v as a sum of v_0 and a vector in the null space of A. Thus all solutions of $Ax = b$ can be found by displacing the null space of A by the solution vector v_0.

Result 1 Suppose that the linear system $Ax = b$ is consistent and v_0 is a solution of the system. Then the collection of all solutions consists of all vectors v that can be expressed in the form $v = v_0 + u$, where u belongs to the null space of A.

EXAMPLE 2 Consider the system $Ax = b$, where $b = (-4, 8)$ and $A = \begin{bmatrix} 1 & -2 & -3 \\ -2 & 4 & 6 \end{bmatrix}$. The null space of A is the plane in R^3 spanned by the vectors $u_1 = (2, 1, 0)$ and $u_2 = (3, 0, 1)$. The vector $v_0 = (1, 1, 1)$ is a solution

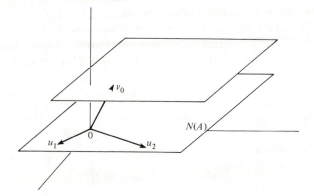

Figure 2.10

of $Ax = b$. Thus the solutions of $Ax = b$ are precisely the vectors $v = v_0 + \alpha u_1 + \beta u_2$, where α and β are arbitrary scalars. These vectors form a plane that is parallel to the null space of A (see Figure 2.10).

EXERCISE 1   Describe the solution sets of the following linear systems geometrically.

(a) $\begin{bmatrix} 3 & 0 & 1 \\ 5 & -1 & 2 \end{bmatrix} \begin{bmatrix} x_1 \\ x_2 \\ x_3 \end{bmatrix} = \begin{bmatrix} 3 \\ 4 \end{bmatrix}$ (b) $\begin{bmatrix} 1 & 1 \\ 8 & 4 \end{bmatrix} \begin{bmatrix} x_1 \\ x_2 \end{bmatrix} = \begin{bmatrix} 1 \\ -1 \end{bmatrix}$

Let A be an $m \times n$ matrix. If b is a vector in R^m, then we know that the system $Ax = b$ is consistent if and only if b lies in the column space of A. Suppose that b is a vector in the column space of A. Then by Result 1 all solutions of $Ax = b$ can be obtained by displacing the null space of A by a single solution v_0 of $Ax = b$. The null space of A is a vector space of dimension $n - r$, where r is the rank of the matrix A. Hence we might even be tempted to speak of the set of all solutions of $Ax = b$ as having dimension $n - r$. Of course, strictly speaking, we have defined the concept of dimension only for vector spaces, and the solutions of $Ax = b$ do not form a vector space unless $b = 0$. The important point is that the "size" of the solution set of $Ax = b$ depends only on the matrix A and not on the particular vector b in the column space of A. If b and c are vectors in the column space of A, then the solution sets of $Ax = b$ and $Ax = c$ differ only in their displacement from the null space of A and not in their "size." In particular, we see that the system $Ax = b$ has a unique solution for every vector b in the column space of A if and only if the null space of A is trivial (i.e., $r = n$).

In the preceding discussion we considered only consistent systems. Whether or not a specific vector b in R^m lies in the column space of A usually depends on both the matrix A and the vector b. However, if $C(A) = R^m$, then $Ax = b$ is consistent for every vector b in R^m. Conversely, if $Ax = b$ is consistent for every vector b in R^m, then $C(A) = R^m$. Since the column space of A is an r-dimensional subspace of R^m, we see that $Ax = b$ has at least one solution for every vector b in R^m if and only if the rank of A is equal to m. We now have proved the following theorem.

Theorem 2.7 *Suppose that A is an m × n matrix.*

(a) (*Uniqueness of solutions*) *The system Ax = b has a unique solution for every vector b in the column space of A if and only if the rank of A is equal to n.*

(b) (*Existence of solutions*) *The system Ax = b has at least one solution for every vector b in R^m [C(A) = R^m] if and only if the rank of A is equal to m.*

EXERCISE 2 We saw in Chapter 1 that a linear system may have no solutions, a unique solution, or infinitely many solutions. Let A be an $m \times n$ matrix of rank r and let b be a vector in R^m. Use Theorem 2.7 to verify the following statements.

(a) If $r = n$, then $Ax = b$ has at most one solution.
(b) If $r < n$, then $Ax = b$ has no solutions or infinitely many solutions.
(c) If $m < n$, then $Ax = b$ has no solutions or infinitely many solutions.

EXERCISE 3 Recall that if we reduce a matrix A to an echelon matrix E, then the rank of A is equal to the number of pivots in E. Convince yourself that the conclusion of Theorem 2.7 and Exercise 2 follow from this interpretation of the rank of A.

We can combine the uniqueness and existence parts of Theorem 2.7 and answer the question: Which $m \times n$ matrices have the property that the system $Ax = b$ has a unique solution for every vector b in R^m? By the theorem these matrices must have rank equal to m and rank equal to n; that is, these matrices are the square matrices with rank as large as possible. Our next goal is to show that these matrices are precisely the nonsingular matrices defined in Chapter 1.

Recall that an $n \times n$ matrix A is called nonsingular or invertible if there exists an $n \times n$ matrix B such that $AB = I$ and $BA = I$, where I is the $n \times n$ identity matrix. Such a matrix B is called an inverse of A. If A does not have an inverse, then A is called singular. The terms "nonsingular" and "singular" apply only to square matrices. We can quickly settle the question of how many inverses a nonsingular matrix A has. If B and C are inverses of A, then

$$B = BI = B(AC) = (BA)C = IC = C.$$

It follows that a nonsingular matrix A has exactly one inverse, which we denote by A^{-1}. Then the equation $AA^{-1} = A^{-1}A = I$ shows that A^{-1} is also nonsingular and that the inverse of A^{-1} is A, that is, $(A^{-1})^{-1} = A$.

To show that an $n \times n$ matrix A is nonsingular if and only if its rank is equal to n, we need a preliminary result.

Result 2 Let A be a $m \times n$ matrix and let I be the $m \times m$ identity matrix. Then the rank of A is equal to m if and only if there is an $n \times m$ matrix B such that $AB = I$.

Proof The columns of I are the standard basis $e_1, e_2, \ldots, e_m$ of R^m. Suppose that rank $A = m$. Then for every vector v in R^m the system $Ax = v$ has at least one solution (Theorem 2.7). In particular, we find vectors $b_1, b_2, \ldots, b_m$ in R^n such that $Ab_i = e_i$, $i = 1, \ldots, m$. It follows that $AB = I$, where B is the $n \times m$ matrix whose columns are $b_1, b_2, \ldots, b_m$.

Conversely, suppose that B is an $n \times m$ matrix such that $AB = I$. If $b_1, b_2, \ldots, b_m$ are the columns of B, then

$$Ab_i = e_i \qquad \text{for } i = 1, \ldots, m$$

and hence each e_i belongs to the column space of A. Thus $C(A) = R^m$ and it follows that rank $A = \dim C(A) = m$.

Theorem 2.8 *If A is an $n \times n$ matrix, then the following three conditions are equivalent. That is, if one of these conditions is true, then all three conditions are true.*

(a) *A is nonsingular.*
(b) *The rank of A is n.*
(c) *$Ax = b$ has a unique solution for every vector b in R^n.*

In particular, if A is nonsingular, then the rows (and also the columns) form a basis for R^n and

$$C(A) = R(A) = R^n, \qquad N(A) = \{0\}.$$

Proof We already know that conditions (b) and (c) are equivalent. Thus it is sufficient to show that conditions (a) and (b) are equivalent. Suppose that A is nonsingular and that b is any vector in R^n. Then $x = A^{-1}b$ is a solution of $Ax = b$ because $A(A^{-1}b) = (AA^{-1})b = Ib = b$. It is also the only solution of $Ax = b$ because $Ax = b$ implies that $x = A^{-1}Ax = A^{-1}b$. Since for every vector b in R^n, the system $Ax = b$ has a unique solution, namely $x = A^{-1}b$, it follows that rank $A = n$. Conversely, let us assume that the rank of A is n and show that A must be nonsingular. Since $m = n$, we can conclude from Result 2 that there is an $n \times n$ matrix B such that $AB = I$, the $n \times n$ identity matrix. The transpose of A is also an $n \times n$ matrix and its rank is equal to n because A and its transpose have the same rank.

Therefore, by Result 2, there is an $n \times n$ matrix C such that $A^T C = I$. Now

$$C^T A = C^T (A^T)^T = (A^T C)^T = I^T = I.$$

Hence $B = IB = (C^T A)B = C^T(AB) = C^T I = C^T$ and we conclude that $BA = C^T A = I = AB$. It follows that B is the inverse of A. Hence A is nonsingular.

EXERCISE 4   Determine which of the following matrices are nonsingular.

(a) $\begin{bmatrix} -3 & 8 \\ 5 & 4 \end{bmatrix}$ (b) $\begin{bmatrix} 1 & -1 & 0 \\ 3 & 0 & 5 \\ -2 & 4 & 1 \end{bmatrix}$ (c) $\begin{bmatrix} 1 & 0 & 1 & 1 \\ -2 & 0 & -2 & 0 \\ 1 & 2 & 3 & 4 \\ 1 & -1 & 0 & 0 \end{bmatrix}$

In Section 1.7 we presented a method for finding the inverse of an $n \times n$ matrix A (provided that the inverse exists). Recall that this method produces an $n \times n$ matrix B such that $AB = I$. At the time we claimed that B was, in fact, the inverse of A, but we did not show that $BA = I$. Now we can show this. Since $AB = I$, it follows from Result 2 that the rank of A is equal to n. Hence by Theorem 2.8, A is nonsingular and therefore has an inverse A^{-1}. Since $B = IB = A^{-1}AB = A^{-1}I = A^{-1}$, we see that B is indeed the inverse of A.

Since a matrix and its transpose have the same rank, Theorem 2.8 implies that the transpose of a nonsingular matrix is nonsingular. This can also be proved directly and the proof gives a formula for the inverse of A^T in terms of the inverse of A. Suppose that A is nonsingular. Then

$$(A^{-1})^T A^T = (AA^{-1})^T = I^T = I \quad \text{and} \quad A^T(A^{-1})^T = (A^{-1}A)^T = I^T = I.$$

Hence A^T is nonsingular and $(A^{-1})^T$ must be the inverse of A^T, that is,

$$(A^T)^{-1} = (A^{-1})^T.$$

Now we consider products of nonsingular matrices. Let A and B be two nonsingular $n \times n$ matrices. We claim that their product AB is nonsingular and that $(AB)^{-1} = B^{-1}A^{-1}$. Indeed, $B^{-1}A^{-1}AB = B^{-1}IB = B^{-1}B = I$. This shows that $B^{-1}A^{-1}$ is the inverse of the matrix AB and that AB is nonsingular. Note that in general $A^{-1}B^{-1}$ is not the inverse of AB because we cannot change the order of the factors in the product $A^{-1}B^{-1}AB$.

We can give a simple formula for the inverse of a nonsingular 2×2 matrix. By Theorem 2.8 the matrix

$$A = \begin{bmatrix} a & b \\ c & d \end{bmatrix}$$

is nonsingular if and only if A has rank 2. Suppose that $a \neq 0$. Subtracting

c/a times the first row from the second reduces A to the echelon matrix

$$\begin{bmatrix} a & b \\ 0 & d - \dfrac{bc}{a} \end{bmatrix}.$$

Thus A has rank 2 if and only if

$$0 \neq d - \frac{bc}{a} = \frac{ad - bc}{a},$$

that is, if and only if $ad - bc \neq 0$. This is also true when $a = 0$. In that case, by merely interchanging the rows of A, we reduce A to the echelon matrix

$$\begin{bmatrix} c & d \\ 0 & b \end{bmatrix}.$$

Hence A has rank 2 if and only if both $c \neq 0$ and $b \neq 0$, that is, $0 \neq -bc = ad - bc$. Thus the 2×2 matrix A is nonsingular if and only if $ad - bc \neq 0$. The number $ad - bc$ is called the determinant of A. (Determinants are discussed in Chapter 7.) If A is nonsingular, an easy calculation shows that $AB = BA = I$, where

$$B = \frac{1}{ad - bc} \begin{bmatrix} d & -b \\ -c & a \end{bmatrix}.$$

Thus B is the inverse of A. For example, the matrix $\begin{bmatrix} 2 & 3 \\ 4 & 6 \end{bmatrix}$ is singular since $2 \cdot 6 - 3 \cdot 4 = 0$. The matrix $\begin{bmatrix} 3 & 2 \\ 5 & 1 \end{bmatrix}$ is nonsingular since $3 \cdot 1 - 2 \cdot 5 = -7 \neq 0$. Its inverse is

$$\begin{bmatrix} 3 & 2 \\ 5 & 1 \end{bmatrix}^{-1} = \frac{1}{-7} \begin{bmatrix} 1 & -2 \\ -5 & 3 \end{bmatrix} = \begin{bmatrix} -\frac{1}{7} & \frac{2}{7} \\ \frac{5}{7} & -\frac{3}{7} \end{bmatrix}.$$

EXERCISE 5   Determine whether the matrix $\begin{bmatrix} -2 & 1 \\ 17 & -6 \end{bmatrix}$ is nonsingular. Find the inverse of the matrix if possible. Do the same for the matrix $\begin{bmatrix} 4 & 9 \\ 8 & 18 \end{bmatrix}$.

We conclude this section with a discussion of an application that involves finding the inverse of a certain nonsingular matrix. In Section 1.2 we introduced a model of a closed economy. In that model we assumed that each producing unit had an unlimited supply of raw materials. In a normal economy the output of each producing unit depends upon the availability of certain goods which are produced by others. If we wish to analyze the interrelationships between various producing units, we must use a model of the economy that focuses on production and consumption rather than on pricing. Such a model (called the input–output model) has been developed by W. W. Leontief (b. 1906), winner of the Nobel Prize in Economic Science in 1973.

To describe the Leontief input–output model, we first divide the economy into certain numbered sectors where all production takes place, together with an "open" sector where the consumption unrelated to production takes place. This open sector accounts for all consumption by individuals in the economy as contrasted with the consumption necessary for production. Suppose that the production sectors are (1) agriculture, (2) manufacturing, and (3) service. We let d_i denote the quantity of sector i production that is demanded by the open sector and x_i denote the total production of sector i. The model describes the production and consumption of all goods by means of a system of equations. Each one of these equations expresses the rather obvious fact that the total production of any sector is the sum of that part of the sector's output used in production and that part of the sector's output consumed (demanded) by the individuals.

$$\begin{aligned} x_1 &= a_{11}x_1 + a_{12}x_2 + a_{13}x_3 + d_1 \\ x_2 &= a_{21}x_1 + a_{22}x_2 + a_{23}x_3 + d_2 \\ x_3 &= a_{31}x_1 + a_{32}x_2 + a_{33}x_3 + d_3 \end{aligned} \tag{1}$$

The coefficients a_{ij}, the input coefficients, measure the production that is used as input for the production of goods. Specifically, a_{ij} is the number of units of sector i production consumed in producing a unit of sector j production. All of the input coefficients are nonnegative. Furthermore, the x_i's and the d_i's are all nonnegative. If $x = (x_1, x_2, x_3)$ is the production vector, $d = (d_1, d_2, d_3)$ is the demand vector, and $A = [a_{ij}]$ is the matrix of input coefficients, then the Leontief input–output model states that

$$x = Ax + d$$

or, equivalently,

$$(I - A)x = d. \tag{2}$$

The model is used to compute the production x necessary to satisfy a given demand d. We hope that the economy as described by our model will have a production vector x for every possible demand vector d. This says that $I - A$ is a nonsingular matrix and that equation (2) has a unique solution given by

$$x = (I - A)^{-1}d. \tag{3}$$

Since for each demand vector d the production vector x obtained from (3) must have nonnegative components, the entries of $(I - A)^{-1}$ are all nonnegative.

We suspect that if the economy described by the equations (1) can actually have all its sectors producing (all x_i's positive) and have something left over for the open sector (at least one d_i positive), then it could somehow manage to satisfy any demand from the open sector. This leads us to make the following conjecture.

Conjecture Given a matrix A with nonnegative entries. If there exists one nonzero vector d (with nonnegative entries) such that the equation $(I - A)x = d$ has a solution x with all entries positive, then the matrix $I - A$ has an inverse and every entry in the inverse is nonnegative.

This conjecture is indeed true, but its proof requires advanced techniques.

PROBLEMS 2.6

*1. Give a geometric description of the solution sets of the following linear systems.

(a) $\begin{bmatrix} 2 & 1 \\ 4 & 2 \end{bmatrix} \begin{bmatrix} x_1 \\ x_2 \end{bmatrix} = \begin{bmatrix} 3 \\ 6 \end{bmatrix}$

(b) $\begin{bmatrix} 2 & 3 & 1 \\ 0 & 2 & -2 \\ -1 & 1 & -3 \end{bmatrix} \begin{bmatrix} x_1 \\ x_2 \\ x_3 \end{bmatrix} = \begin{bmatrix} 4 \\ 0 \\ -2 \end{bmatrix}$

(c) $\begin{bmatrix} 1 & -2 & 4 \end{bmatrix} \begin{bmatrix} x_1 \\ x_2 \\ x_3 \end{bmatrix} = [9]$

*2. Determine which of the following matrices are nonsingular.

(a) $\begin{bmatrix} -3 & 5 \\ 0 & 2 \end{bmatrix}$

(b) $\begin{bmatrix} 6 & -18 \\ -5 & 15 \end{bmatrix}$

(c) $\begin{bmatrix} 1 & -1 & 0 \\ 0 & 0 & 2 \\ 3 & 2 & 1 \end{bmatrix}$

(d) $\begin{bmatrix} -1 & 0 & 5 \\ 2 & 0 & 1 \\ 3 & 0 & -4 \end{bmatrix}$

(e) $\begin{bmatrix} 0 & 0 & 2 \\ 0 & 1 & 2 \\ 3 & 4 & 5 \end{bmatrix}$

(f) $\begin{bmatrix} 1 & 2 & 0 & 1 \\ 1 & 3 & 0 & 0 \\ 1 & 4 & 1 & 0 \\ 1 & 5 & 0 & 1 \end{bmatrix}$

(g) $\begin{bmatrix} 1 & 0 & 1 & 0 & 0 \\ -1 & 0 & -1 & 1 & 0 \\ 2 & 1 & 1 & 1 & 1 \\ 0 & 1 & -1 & -1 & 1 \\ 0 & 2 & 1 & -2 & 2 \end{bmatrix}$

*3. If possible, find the inverses of the following matrices.

(a) $\begin{bmatrix} 6 & 1 \\ 3 & -2 \end{bmatrix}$

(b) $\begin{bmatrix} 5 & 10 \\ 3 & 6 \end{bmatrix}$

(c) $\begin{bmatrix} 0 & -8 \\ 9 & 2 \end{bmatrix}$

4. Find the inverses of A, B, A^T, and AB, where

$$A = \begin{bmatrix} 1 & -1 & 1 \\ 0 & 1 & 2 \\ 1 & 0 & 2 \end{bmatrix} \quad \text{and} \quad B = \begin{bmatrix} 0 & 2 & 0 \\ 1 & 1 & 2 \\ 2 & 0 & 1 \end{bmatrix}.$$

5. Suppose that A is a nonsingular $m \times m$ matrix and B is an $m \times n$ matrix.
(a) Show that $N(AB) = N(B)$.
(b) Use part (a) to show that rank $(AB) =$ rank B.

6. Suppose that A is an $m \times n$ matrix and B is a nonsingular $n \times n$ matrix. Show that $C(AB) = C(A)$ and conclude that rank $(AB) =$ rank A.

7. Let A be an $m \times n$ matrix and let I be the $n \times n$ identity matrix. Show that the rank of A is equal to n if and only if there is an $n \times m$ matrix B such that $BA = I$.

8. Let $A = \begin{bmatrix} a & b \\ c & d \end{bmatrix}$, where a, b, c, and d are all nonnegative, be the matrix of input coefficients for a two-production-sector economy.
   (a) Show that the matrix $I - A$ will have an inverse with nonnegative entries if and only if the following two conditions hold:
      (1) $a < 1, d < 1$.
      (2) $(1 - a)(1 - d) > bc$.
   (b) Discuss the economic meaning of each condition above.

Chapter 3

Inconsistent Systems, Inner Products, and Projections

In Chapter 1 we saw that Gaussian elimination not only solves any consistent system of equations, but that it also identifies any inconsistent system by producing a false equation. In Chapter 2 we saw how to characterize consistent and inconsistent systems of equations geometrically. If the vector of constant terms lies in the column space of the coefficient matrix, the system is consistent. Otherwise, it is inconsistent. The algebraic question of whether a system of equations is consistent is equivalent to the geometric question of whether a certain vector lies in a subspace. In this chapter we use this geometric characterization of consistent systems together with the concepts of length and angle to investigate the question of what can be done with inconsistent systems.

3.1 INCONSISTENT SYSTEMS OF EQUATIONS

We begin our discussion of inconsistent systems with several examples of situations where these systems arise naturally.

EXAMPLE 1 In Section 1.2 we studied the problem of finding the equation of a line $y = a + bx$ passing through the m points $(x_1, y_1), \ldots, (x_m, y_m)$, with $x_1 < x_2 < \cdots < x_m$. The coefficients a and b must satisfy the system of m

equations

$$a + bx_1 = y_1$$
$$\vdots \qquad \vdots \qquad \vdots \qquad \text{or}$$
$$a + bx_m = y_m$$

$$\begin{bmatrix} 1 & x_1 \\ \vdots & \vdots \\ 1 & x_m \end{bmatrix} \begin{bmatrix} a \\ b \end{bmatrix} = \begin{bmatrix} y_1 \\ \vdots \\ y_m \end{bmatrix}.$$

When $m = 2$ these equations always have a unique solution. Geometrically, the reason for this is that any two distinct points determine a straight line. However, when m is greater than 2, the system may not have a solution. Three or more distinct points do not always lie on a line. In fact, the more points we have, the more likely it is that they do not lie on a straight line.

EXAMPLE 2   Suppose that an object is traveling along a straight line and that we have reason to believe that it is moving with a constant velocity. To determine its velocity v, we measure the position of the object at times $t_1, \ldots, t_m$. If y_i denotes the position at time t_i and y_0 denotes the initial position of the object (which of course is unknown), then the following system of equations must be satisfied:

$$y_0 + vt_1 = y_1$$
$$\vdots \qquad \vdots \qquad \vdots \qquad \text{or}$$
$$y_0 + vt_m = y_m$$

$$\begin{bmatrix} 1 & t_1 \\ \vdots & \vdots \\ 1 & t_m \end{bmatrix} \begin{bmatrix} y_0 \\ v \end{bmatrix} = \begin{bmatrix} y_1 \\ \vdots \\ y_m \end{bmatrix}.$$

Since measurements of physical quantities are *never* exact, we *must expect* this system of equations to be inconsistent. We could, of course, use only two of the measurements and thereby be assured that the equations have a unique solution. But this is usually not a good idea. The line L_1 in Figure 3.1 is determined by two of the data points, but it certainly does not "fit the data" very well. On the other hand, even though the line L_2 does not pass through *any* of the data points, it does "fit the data" well. When we use only two of the data points, the errors in the individual measurements will usually affect our results significantly. By using more data points than are needed, we reduce the effects on our results of the errors in individual measurements. But when we do this we must be prepared to deal with an inconsistent system of equations.

Figure 3.1

The problems that we considered in Examples 1 and 2 are mathematically identical. Both examples led to an $m \times 2$ system of equations with $m > 2$. In the general case we consider a system of equations $Ax = b$, where A is an $m \times n$ matrix of rank $r < m$ and b is an m-vector. We have seen in Chapter 2 that this system will have a solution if and only if the components of the vector b satisfy $m - r$ constraint equations (one for each zero row of the echelon matrix). Obviously, when $m - r$ is large, it will be difficult for the components of an m-vector b to satisfy all the constraints. In other words, the larger the value of $m - r$, the more likely it is that the system $Ax = b$ is inconsistent. From an algebraic point of view, then, the question of the consistency of the system $Ax = b$ is equivalent to the question of whether the vector b satisfies the constraint equations.

Now consider the question of the consistency of the system $Ax = b$ (A an $m \times n$ matrix of rank $r < m$) from a geometric point of view. In Chapter 2 we saw that the equation $Ax = b$ is consistent if and only if the vector b lies in the column space of A, an r-dimensional subspace of R^m. If $m - r$ is large, this r-dimensional subspace is a rather "thin" subspace of R^m and most m-vectors will not lie in it. Moreover, the larger $m - r$ is, the "thinner" the subspace is. (A line is a "thin" subspace of R^2 and an even "thinner" subspace of R^3.) Thus the larger $m - r$ is, the more likely it is that the vector b will not be in the column space of A and hence the more likely it is that the system is inconsistent.

We now turn to the question of what can be done with inconsistent systems. As above, let A be an $m \times n$ matrix of rank $r < m$ and let b be an m-vector. Suppose that the system $Ax = b$ is inconsistent. In this case the m-vector b does not lie in the column space of A. We cannot solve $Ax = b$ as it stands. Perhaps by changing the system slightly we can obtain a system that can be solved. One way to do this is to replace b by some vector p that *is* in the column space of A. The new system $Ax = p$ will be consistent and any solution $\bar{x}$ to $Ax = p$ can be considered in some sense an "approximate solution" to the original inconsistent system $Ax = b$. But for this process to be reasonable, we must choose the "best possible" replacement for b. Since our objective is to obtain something that is close to being a solution of $Ax = b$, it is reasonable to choose p to be as close to b as possible (while still being in the column space of A). This p is the "best possible" replacement for b, the consistent system $Ax = p$ is the "best possible" replacement for $Ax = b$, and the "best" we can obtain as an approximate solution to $Ax = b$ is a vector $\bar{x}$ such that $A\bar{x} = p$.

It is clear from the discussion above that we must develop a procedure for finding the vector p in the column space of A that is closest to b. Let us first consider this problem in the familiar setting of three-dimensional space. Suppose that A is a 3×2 matrix and that the rank of A is either 1 or 2. Hence the column space of A is a line or plane in R^3. From Figure 3.2 it is clear that the vector p that is closest to b is found by dropping a perpendicular from b to the line or plane.

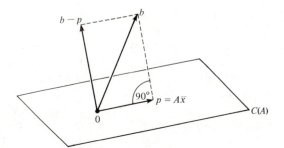

Figure 3.2

The geometry of three-dimensional space has suggested a method that can be used to attack the problem of "solving" an inconsistent system of equations. This method depends on the concepts of length and perpendicularity. Hence to extend this method to the general case of an $m \times n$ system, we must first extend the notions of length and perpendicularity from R^3 to R^m. This is the topic of the next section.

PROBLEMS 3.1

*1. Suppose we know that an object is traveling with constant velocity. The position of the object is measured at various times and the following data are obtained.

t	1	2	3	4	5
d	3	5	9	11	12

What equations must the velocity and the initial position satisfy? Show that the equations are inconsistent.

*2. Suppose that we know from certain theoretical considerations that the variables t and y satisfy a relationship of the form

$$y = a + bt + ct^2.$$

An experiment is conducted that gives the data in the following table. What equations must the vector $x = (a, b, c)$ satisfy? Show that the equations are inconsistent.

t	-1	0	1	2
y	-2	-1	0	3

3. Repeat Problem 2 when the relationship is of the form:
   (a) $y = at + bt^2 + ct^3$ (b) $y = a + bt^2$

4. Let $(x_1, y_1), \ldots, (x_m, y_m)$ be m points in R^2 with $x_1 < x_2 < \cdots < x_m$, and let $p(x) = a_0 + a_1 x + \cdots + a_n x^n$ be a polynomial of degree n.
   (a) Show that the matrix form of the system of equations which the coefficients $a_0, a_1, \ldots, a_n$ of the polynomial p must satisfy in order that

$$p(x_i) = y_i, \qquad i = 1, \ldots, m,$$

is $Ax = b$, where

$$A = \begin{bmatrix} 1 & x_1 & \cdots & x_1^n \\ \vdots & & & \\ 1 & x_m & \cdots & x_m^n \end{bmatrix}, \qquad x = \begin{bmatrix} a_0 \\ \vdots \\ a_n \end{bmatrix}, \qquad b = \begin{bmatrix} y_1 \\ \vdots \\ y_m \end{bmatrix}.$$

(b) Show that if the rank of the matrix A in part (a) is m, then there is a solution to the equation $Ax = b$. Hence there is a polynomial of degree n which passes through the given points (x_i, y_i), $i = 1, \ldots, m$.

(c) Show that there is one and only one polynomial of degree $m - 1$ which passes through the given set of m points. (*Hint:* Show that if $n = m - 1$ then the rank of A is m. Do this for $m = 2, 3,$ and 4.)

(d) Show that if the rank of A is $n + 1$, then $m \geq n + 1$, and the equation has a solution if and only if b lies in the column space of A. Therefore, there may not be a polynomial of degree n which passes through the given points. As we have suggested, the larger m gets, the less likely it is that the equations have a solution.

3.2 LENGTH AND ORTHOGONALITY

The Pythagorean theorem states that the square of the length of the hypotenuse of a right triangle is equal to the sum of the squares of the lengths of the other two sides. If $x = (x_1, x_2)$ is a vector in R^2, we find that [see Figure 3.3(a)]

$$\text{length of } x = \sqrt{x_1^2 + x_2^2}.$$

Similarly, if $x = (x_1, x_2, x_3)$ is a vector in R^3, then two applications of the Pythagorean theorem give [Figure 3.3(b)]

$$\text{length of } x = \sqrt{\left(\sqrt{x_1^2 + x_2^2}\right)^2 + x_3^2}$$

$$= \sqrt{x_1^2 + x_2^2 + x_3^2}.$$

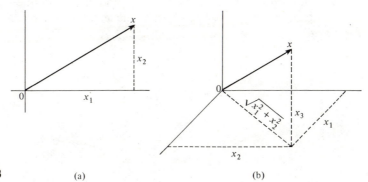

Figure 3.3   (a)   (b)

It should be clear at this point how we should define the length of a vector in R^m when $m > 3$.

Definition If x is a vector in R^m, then the **length** (or **norm**) of x, denoted by $\|x\|$, is defined by the equation

$$\|x\| = \sqrt{x_1^2 + x_2^2 + \cdots + x_m^2}.$$

EXAMPLE 1

$$\|(0, 1, 2)\| = \sqrt{0^2 + 1^2 + 2^2} = \sqrt{5}.$$

$$\|(1, 2, 3, -2)\| = \sqrt{1^2 + 2^2 + 3^2 + (-2)^2} = \sqrt{18} = 3\sqrt{2}.$$

If $e_1, \ldots, e_m$ are the standard basis vectors for R^m, then $\|e_j\| = 1$, $j = 1, \ldots, m$. Vectors of length 1 are called **unit vectors**. Given any nonzero vector x, the vector $y = \dfrac{x}{\|x\|}$ is a unit vector in the direction of x.

EXERCISE 1 Find the length of each of the following vectors.

(a) $(1, -1)$ (b) $(1, -1, 1, 2)$

Since the length of a vector in R^m is the square root of a sum of squares, it is nonnegative. In other words, $\|x\| \geq 0$ for all x. Since $\|0\|^2 = 0^2 + \cdots + 0^2$, the zero vector has zero length. It is the only vector with zero length. To prove this, suppose that $x = (x_1, \ldots, x_m)$ is a nonzero vector. At least one of its components is not zero, say x_i. Then

$$0 < x_i^2 \leq x_1^2 + \cdots + x_i^2 + \cdots + x_m^2 = \|x\|^2,$$

so that $\|x\| > 0$. Thus we have shown that for all x,

$$\|x\| \geq 0, \text{ and } \|x\| = 0 \text{ if and only if } x = 0. \tag{1}$$

EXERCISE 2 If x is a vector in R^m and α is a scalar, show that $\|\alpha x\| = |\alpha| \cdot \|x\|$.

Again motivated by the geometry of R^2 and R^3, we define the distance between two vectors x and y in R^m to be the length of the vector $x - y$ (see Figure 3.4).

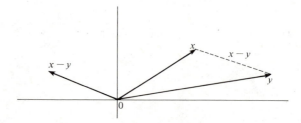

Figure 3.4

Definition   If $x = (x_1, \ldots, x_m)$ and $y = (y_1, \ldots, y_m)$ are two vectors in R^m, then the **distance** between x and y is defined to be

$$\| x - y \| = \sqrt{(x_1 - y_1)^2 + (x_2 - y_2)^2 + \cdots + (x_m - y_m)^2}.$$

Note that the length of a vector is equal to its distance from the origin (the zero vector).

EXAMPLE 2   The distance between $(0, 2, -2, 1)$ and $(-2, 0, -2, 2)$ is 3 since $[0 - (-2)]^2 + (2 - 0)^2 + [-2 - (-2)]^2 + (1 - 2)^2 = 9$.

EXERCISE 3   Find the distance between each of the following pairs of vectors.

(a) $(1, 1, 1)$ and $(0, 0, 1)$ (b) $(1, \frac{1}{2}, \frac{1}{3}, \frac{1}{4})$ and $(1, 0, \frac{1}{3}, \frac{1}{4})$

Suppose that $x = (x_1, x_2, x_3)$ and $y = (y_1, y_2, y_3)$ are two nonzero perpendicular vectors in R^3. Then the triangle in Figure 3.5 is a right triangle and we conclude from the Pythagorean theorem that

$$\| x - y \|^2 = \| x \|^2 + \| y \|^2.$$

Since

$$\| x - y \|^2 = (x_1 - y_1)^2 + (x_2 - y_2)^2 + (x_3 - y_3)^2$$
$$= \| x \|^2 + \| y \|^2 - 2(x_1 y_1 + x_2 y_2 + x_3 y_3),$$

the equality $\| x - y \|^2 = \| x \|^2 + \| y \|^2$ holds if and only if

$$x_1 y_1 + x_2 y_2 + x_3 y_3 = 0. \qquad (2)$$

Thus x and y are perpendicular if and only if equation (2) is true. The reader should use a similar argument to check that the vectors $x = (x_1, x_2)$ and $y = (y_1, y_2)$ are perpendicular if and only if $x_1 y_1 + x_2 y_2 = 0$. This last equation and equation (2) suggest a method for generalizing the notion of perpendicularity to R^m.

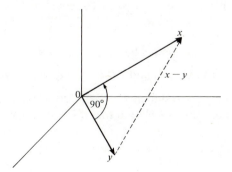

Figure 3.5

Definition   Two m-vectors $x = (x_1, \ldots, x_m)$ and $y = (y_1, \ldots, y_m)$ are called **orthogonal** (or **perpendicular**) if

$$x_1 y_1 + \cdots + x_m y_m = 0. \tag{3}$$

EXAMPLE 3   The 4-vectors $(1, 1, 0, 1)$ and $(1, -1, 3, 0)$ are orthogonal since $1 \cdot 1 + 1 \cdot (-1) + 0 \cdot 3 + 1 \cdot 0 = 0$. Similarly, $(1, 0, 0)$ and $(0, 5, 6)$ are orthogonal in R^3.

EXERCISE 4   Show that each of the following pairs of vectors is orthogonal.

   (a) $(1, 3)$ and $(3, -1)$ (b) $(1, 1, 1, 1)$ and $(1, -6, 2, 3)$

     The left-hand side of equation (3) is a sum of products of corresponding components of the vectors x and y. We have seen sums of this type before. They occur when we multiply a matrix times a vector and when we multiply a matrix times a matrix. This sum is so important that we will give it a special name and a special notation.

Definition   If $x = (x_1, \ldots, x_m)$ and $y = (y_1, \ldots, y_m)$ are m-vectors, then their **inner product** (or **dot product**), denoted by $\langle x, y \rangle$, is the number defined by the equation

$$\langle x, y \rangle = x_1 y_1 + \cdots + x_m y_m.$$

EXERCISE 5   Let A be an $m \times n$ matrix with rows $a_1, \ldots, a_m$, let B be an $n \times p$ matrix with columns $b_1, \ldots, b_p$, and let x be an n-vector. Show that

$$Ax = \begin{bmatrix} \langle a_1, x \rangle \\ \vdots \\ \langle a_m, x \rangle \end{bmatrix} \quad \text{and} \quad AB = \begin{bmatrix} \langle a_1, b_1 \rangle & \cdots & \langle a_1, b_p \rangle \\ \vdots & & \vdots \\ \langle a_m, b_1 \rangle & \cdots & \langle a_m, b_p \rangle \end{bmatrix}.$$

     In this notation we express the fact that x and y are orthogonal by the equation $\langle x, y \rangle = 0$. Also, the length of a vector and the inner product of the vector with itself are related by the equation

$$\| x \| = \sqrt{\langle x, x \rangle}.$$

Finally, in this notation (1) becomes: for all x,

$$\langle x, x \rangle \geq 0, \text{ and } \langle x, x \rangle = 0 \text{ if and only if } x = 0. \tag{4}$$

If we interpret this result in terms of orthogonality, it says that the zero vector is orthogonal to itself and that it is the only vector with this property.

EXERCISE 6   Show that the zero m-vector is orthogonal to every m-vector.

In Theorem 3.1 we list some important properties of the inner product. The first three are simple consequences of the definition. The fourth is simply a restatement of (4).

Theorem 3.1   *Let x, y, and z be m-vectors and α a scalar.*

(a) $\langle x, y \rangle = \langle y, x \rangle$.
(b) $\langle x, y + z \rangle = \langle x, y \rangle + \langle x, z \rangle$.
(c) $\langle \alpha x, y \rangle = \alpha \langle x, y \rangle$.
(d) $\langle x, x \rangle \geq 0$, and $\langle x, x \rangle = 0$ if and only if $x = 0$.

EXERCISE 7   Verify properties (a), (b), and (c) in Theorem 3.1. Why does it follow from (d) that the only vector that is orthogonal to itself is the zero vector?

EXAMPLE 4   In this example we will show that the Pythagorean theorem holds in R^m. This should come as no surprise for two reasons. First, we used the Pythagorean theorem to motivate our definition of length in R^m, $m > 3$. Second, the Pythagorean theorem is concerned only with the relationship between the lengths of the sides of a right triangle, and hence it should be independent of the dimension of the space in which the triangle is situated.

Let us now prove the Pythagorean theorem. Suppose that x and y are m-vectors. Using (a) and (b) of Theorem 3.1 we obtain

$$\|x + y\|^2 = \langle x + y, x + y \rangle$$
$$= \langle x, x \rangle + \langle x, y \rangle + \langle y, x \rangle + \langle y, y \rangle$$
$$= \|x\|^2 + 2\langle x, y \rangle + \|y\|^2.$$

Thus $\|x + y\|^2 = \|x\|^2 + \|y\|^2$ if and only if $\langle x, y \rangle = 0$, that is, if and only if x and y are orthogonal. Thus the square of the length of the hypotenuse ($\|x + y\|^2$) is equal to the sum of the squares of the lengths of the other two sides ($\|x\|^2 + \|y\|^2$) if and only if the triangle in Figure 3.6 is a right triangle.

Definition   A set $v_1, \ldots, v_k$ of m-vectors is called **orthogonal** if $\langle v_i, v_j \rangle = 0$ whenever $i \neq j$.

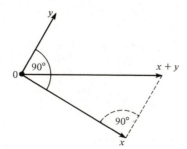

Figure 3.6

That is, a set of vectors is orthogonal if any two distinct vectors in the set are orthogonal to each other. We also say that the vectors $v_1, \ldots, v_k$ are mutually orthogonal. If, in addition, all the vectors in the set have unit length, then the set of vectors is called **orthonormal**.

EXERCISE 8   Show that if $v_1, \ldots, v_k$ is an orthogonal set of nonzero vectors, and if $w_1 = v_1/\|v_1\|, \ldots, w_k = v_k/\|v_k\|$, then $w_1, \ldots, w_k$ is an orthonormal set of vectors. Thus it is always possible to *normalize* an orthogonal set of nonzero vectors to obtain an orthonormal set by simply dividing each vector in the set by its length.

The standard basis vectors $e_1, \ldots, e_m$ in R^m are the most familiar example of an orthonormal set of vectors. They provide us with a picture of R^m which suggests that R^m cannot contain more than m mutually orthogonal nonzero vectors, one orthogonal vector for each dimension. The geometry of R^m also suggests the more general result that nonzero orthogonal vectors must be independent.

Result 1   If $v_1, \ldots, v_k$ are nonzero orthogonal vectors in R^m, then they are linearly independent.

Proof   Suppose that

$$\alpha_1 v_1 + \cdots + \alpha_k v_k = 0. \tag{5}$$

Taking the inner product of v_1 with both sides of this equation, we obtain

$$\alpha_1 \langle v_1, v_1 \rangle + \cdots + \alpha_k \langle v_1, v_k \rangle = \langle v_1, 0 \rangle.$$

By Exercise 6, $\langle v_1, 0 \rangle = 0$. Since the v's are orthogonal, $\langle v_1, v_j \rangle = 0$ for $j = 2, \ldots, k$ and the equation reduces to $\alpha_1 \langle v_1, v_1 \rangle = 0$. Finally, since $v_1 \neq 0$, $\langle v_1, v_1 \rangle = \|v_1\|^2 \neq 0$ [Theorem 3.1(d)], and hence $\alpha_1 = 0$. Thus equation (5) reduces to $\alpha_2 v_2 + \cdots + \alpha_k v_k = 0$. Now form the inner product of v_2 with both sides of this equation and conclude in the same way that $\alpha_2 = 0$. Repeating this argument we find that each of the α's is equal to zero. Hence $v_1, \ldots, v_k$ are linearly independent.

EXERCISE 9   Why do we need to assume that the vectors are nonzero in the previous result? Also, why does it follow from this result that R^m cannot contain more than m mutually orthogonal nonzero vectors?

EXERCISE 10   Show that each of the following collections of vectors is independent.

(a) $(1, 1, 1)$, $(1, -1, 0)$, and $(1, 1, -2)$
(b) $(1, 1, 1, -1)$, $(3, -1, -1, 1)$, and $(0, 1, 1, 2)$

PROBLEMS 3.2

1. Find the length of each of the following vectors.
   *(a) $(1, -1, 2)$ *(b) $(\frac{1}{3}, -\frac{1}{3}, \frac{1}{6}, -\frac{1}{12})$
   (c) $(\sqrt{2}, 5, 0, -2)$ (d) $(\frac{1}{2}, \frac{1}{3}, \frac{1}{4}, \frac{2}{3})$
   *(e) $(-1, 0, \pi, 0)$ *(f) $(1/a, b)$
   (g) $(5, 0, 1, 0, 1, 3)$ (h) $(\frac{1}{2}, \frac{5}{8})$

2. Find the distance between each of the following pairs of vectors.
   *(a) $(1, 0, 1)$ and $(0, -1, 0)$
   (b) $(1, 1, \sqrt{3}, 1)$ and $(1, 0, 0, 1)$
   *(c) $(1, \frac{1}{2}, \frac{1}{3}, \frac{2}{3})$ and $(2, \frac{1}{3}, \frac{1}{6}, -\frac{1}{3})$
   (d) $(-1, 2, 0, 1, 1)$ and $(-1, -2, 0, \frac{1}{2}, 0)$

*3. Which of the following pairs of vectors are orthogonal?
   (a) $(1, 2)$ and $(2, 1)$
   (b) $(1, 2, 3)$ and $(3, 0, -1)$
   (c) $(\sqrt{2}, 5, 0, -1)$ and $(\sqrt{2}, -1, 7, -3)$
   (d) $(1, 0, 0, 0, 1)$ and $(0, 3, 7, -2, 0)$

4. Expand each of the following using Theorem 3.1.
   (a) $\langle a + \beta b, c \rangle$
   *(b) $\langle a + \beta b, a + \beta b \rangle$
   (c) $\langle \alpha a + \beta b, \alpha a + \beta b \rangle$

5. Use Theorem 3.1 to prove that $x \neq 0$ if and only if $\langle x, x \rangle > 0$.

*6. Determine which of the following sets of vectors are independent.
   (a) $(1, 1)$ and $(-1, 1)$ (b) $(-1, 2, 3)$ and $(6, 0, 2)$
   (c) $(0, 0, 1, 0, 0)$, $(0, 0, 0, 1, 0)$, $(0, 0, 0, 0, 1)$, and $(0, 0, 0, 0, 0)$

7. Find a nonzero vector that is orthogonal to:
   *(a) $(1, 0, 1)$ and $(1, 1, -1)$ (b) $(1, -1, 2)$ and $(2, 1, -1)$

8. Find two nonzero 3-vectors that are orthogonal to each other and to the vector:
   *(a) $(1, 1, 1)$ (b) $(1, 1, 0)$ (c) $(1, 2, 1)$

9. Find two nonzero 4-vectors that are orthogonal to each other and to the vector:
   *(a) $(1, 2, 1, -3)$ (b) $(1, 1, 0, 0)$ (c) $(1, 0, 2, 1)$

10. Let $a = (1, 3, 5)$ and $b = (1, 1, 1)$. Find vectors x and y such that $a = x + y$, x is collinear with b, and y is orthogonal to b. Interpret this procedure geometrically.

11. *(a) Show that $\| x + y \| = \| x - y \|$ if and only if x is orthogonal to y.
    (b) Show that the vector $x - y$ is orthogonal to the vector $x + y$ if and only if $\| x \| = \| y \|$.

12. (a) Show that if x and y are orthogonal unit vectors in R^m, then $\| x - y \| = \sqrt{2}$.
    (b) Show that if x and y are orthogonal vectors in R^m, then $\| x - y \|^2 = \| x \|^2 + \| y \|^2$.

13. (a) Show that if x and y are m-vectors, then $\langle x, y \rangle = x^T y$.
    (b) Show that if A is an $m \times n$ matrix, then for any n-vector x and m-vector y,

$$\langle Ax, y \rangle = \langle x, A^T y \rangle.$$

(*Hint:* Use part (a) and the fact that $(AB)^T = B^T A^T$.)

*14. If x is a unit vector in R^m, show that the matrix $A = xx^T$ is symmetric and has the property that $A^2 = A$.

15. Let $a = (a_1, a_2, a_3)$ and $b = (b_1, b_2, b_3)$ be two noncollinear vectors.
    (a) Show that there is a nonzero vector $c = (c_1, c_2, c_3)$ that is orthogonal to both a and b. One way of showing this is to solve the system of equations

$$\begin{matrix} \langle a, c \rangle = 0 \\ \langle b, c \rangle = 0 \end{matrix} \qquad \text{or} \qquad \begin{bmatrix} -a- \\ -b- \end{bmatrix} \begin{bmatrix} | \\ c \\ | \end{bmatrix} = 0.$$

    (b) Show that $(a_2 b_3 - a_3 b_2, a_3 b_1 - a_1 b_3, a_1 b_2 - a_2 b_1)$ is a solution to these equations. This particular solution is called the **cross product** of a and b, and is denoted by $a \times b$.
    (c) Show that any vector orthogonal to a and b is a scalar multiple of $a \times b$.

*16. Show that

$$\| x + y \|^2 + \| x - y \|^2 = 2\| x \|^2 + 2\| y \|^2$$

for any two m-vectors x and y. Give a geometric interpretation of this result in terms of parallelograms when x and y are vectors in R^3. It is because of this interpretation that this equation is called the *parallelogram law*.

17. Let A be a $n \times n$ matrix. Then A is called an **orthogonal matrix** if its columns form an orthonormal set of vectors in R^n.
    *(a) Show that A is orthogonal if and only if $A^{-1} = A^T$. Therefore A is orthogonal if and only if $AA^T = I$.
    *(b) Show that A is orthogonal if and only if its rows form an orthonormal set of vectors in R^n. This proves the remarkable fact that the rows of a matrix form an orthonormal set if and only if the columns form an orthonormal set.
    (c) Show that A is orthogonal if and only if A^{-1} is orthogonal.
    (d) Show that if A and B are orthogonal, then AB and BA are orthogonal.
    *(e) Show that if A is an orthogonal matrix, then

$$\begin{matrix} \| Ax \| = \| x \| & \text{for all } x, & \text{and} \\ \langle Ax, Ay \rangle = \langle x, y \rangle & \text{for all } x \text{ and } y. \end{matrix}$$

    Thus orthogonal matrices preserve length and orthogonality.
    *(f) Show that if x is a unit vector, then $I - 2xx^T$ is an orthogonal matrix.
    (g) Show that if A is orthogonal, then A is symmetric if and only if $A^2 = I$.
    *(h) Find all orthogonal matrices that are triangular.
    (i) Find all 3×3 orthogonal matrices whose first two columns are:

$$(1)\ \begin{bmatrix} 1 & 0 \\ 0 & 1 \\ 0 & 0 \end{bmatrix} \qquad *(2)\ \begin{bmatrix} 1/\sqrt{2} & 1/\sqrt{3} \\ 0 & 1/\sqrt{3} \\ 1/\sqrt{2} & -1/\sqrt{3} \end{bmatrix} \qquad (3)\ \begin{bmatrix} 1/\sqrt{2} & 0 \\ 1/\sqrt{3} & 1/\sqrt{3} \\ -1/\sqrt{6} & 2/\sqrt{6} \end{bmatrix}$$

(j) Find an orthogonal matrix whose first column is:
(1) $(1, 0, 0)$ (2) $(0, 1, 0)$
(3) $(1/\sqrt{2}, 0, 1/\sqrt{2})$ (4) $(1/\sqrt{3}, -1/\sqrt{3}, 1/\sqrt{3})$
(k) How many 2×2 orthogonal matrices are there all of whose entries are zeros and ones? List them.
(l) How many 3×3 orthogonal matrices are there all of whose entries are zeros and ones? List them.
(m) Generalize parts (k) and (l).

3.3   THE CAUCHY–SCHWARZ INEQUALITY AND ANGLES

In this section we discuss some deeper properties of the inner product. We begin with the Cauchy–Schwarz inequality, which is undoubtedly one of the most important inequalities in higher mathematics. However, its proof, although difficult, requires only those properties of the inner product given in Theorem 3.1.

Result 1   (Cauchy–Schwarz Inequality) If x and y are vectors in R^m, then

$$|\langle x, y \rangle| \leq \|x\| \cdot \|y\|.$$

Moreover, equality holds if and only if one of the vectors is a scalar multiple of the other.

Proof   If either x or y are zero, the result is obvious. Hence let us suppose that both x and y are nonzero. Then for any scalar α,

$$
\begin{aligned}
\|x - \alpha y\|^2 &= \langle x - \alpha y, x - \alpha y \rangle \\
&= \langle x, x \rangle - 2\alpha \langle x, y \rangle + \alpha^2 \langle y, y \rangle \\
&= \|x\|^2 - 2\alpha \langle x, y \rangle + \alpha^2 \|y\|^2.
\end{aligned}
$$

Since the length of a vector is nonnegative,

$$0 \leq \|x\|^2 - 2\alpha \langle x, y \rangle + \alpha^2 \|y\|^2.$$

If we choose $\alpha = \dfrac{\langle x, y \rangle}{\|y\|^2}$, we obtain

$$
\begin{aligned}
0 &\leq \|x\|^2 - 2\frac{\langle x, y \rangle^2}{\|y\|^2} + \frac{\langle x, y \rangle^2}{\|y\|^4} \|y\|^2 \\
&= \|x\|^2 - \frac{\langle x, y \rangle^2}{\|y\|^2} \\
&= \frac{\|x\|^2 \|y\|^2 - \langle x, y \rangle^2}{\|y\|^2}.
\end{aligned}
$$

Thus $\|x\|^2 \|y\|^2 - \langle x, y \rangle^2 \geq 0$ and therefore $|\langle x, y \rangle| \leq \|x\| \cdot \|y\|$. The reader should check that the equality sign holds for nonzero vectors x and y

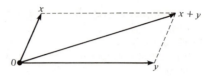

Figure 3.7

if and only if

$$\| x - \alpha y \| = 0 \qquad \text{where} \qquad \alpha = \langle x, y \rangle / \| y \|^2.$$

Without vector notation, this inequality does not look quite so simple. It becomes

$$| x_1 y_1 + \cdots + x_m y_m | \leq \sqrt{\left(x_1^2 + \cdots + x_m^2 \right)} \sqrt{\left(y_1^2 + \cdots + y_m^2 \right)}.$$

Using the Cauchy–Schwarz inequality we can prove another important inequality, the *triangle inequality*. It states that the length of any one side of a triangle cannot exceed the sum of the lengths of the other two (see Figure 3.7).

Result 2 (Triangle Inequality) For any two vectors x and y in R^m, $\| x + y \| \leq \| x \| + \| y \|.$

Proof To prove this inequality we expand the quantity $\| x + y \|^2$ and use the Cauchy–Schwarz inequality to obtain

$$\begin{aligned}
\| x + y \|^2 &= \langle (x + y), (x + y) \rangle \\
&= \| x \|^2 + 2\langle x, y \rangle + \| y \|^2 \\
&\leq \| x \|^2 + 2\| x \| \, \| y \| + \| y \|^2 \\
&= (\| x \| + \| y \|)^2.
\end{aligned}$$

Taking the square root of both sides of this inequality completes the proof.

The Cauchy–Schwarz inequality also enables us to extend the concept of angle to m-dimensional space. Every calculus student has encountered the formula

$$\langle x, y \rangle = \| x \| \cdot \| y \| \cos \theta, \tag{1}$$

where x and y are vectors in R^2 and θ is the angle between these vectors. It is an immediate consequence of the law of cosines (see Problem 7). It also follows from the addition formula for the cosine. From Figure 3.8 we obtain

$$\begin{aligned}
\cos \theta &= \cos(\alpha - \beta) \\
&= \cos \alpha \cos \beta + \sin \alpha \sin \beta \\
&= \frac{x_1 y_1 + x_2 y_2}{\| x \| \cdot \| y \|} \\
&= \frac{\langle x, y \rangle}{\| x \| \cdot \| y \|}.
\end{aligned}$$

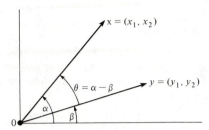

Figure 3.8

Equation (1) suggests that we define the cosine of the angle θ between two nonzero vectors x and y by the equation

$$\cos \theta = \frac{\langle x, y \rangle}{\|x\| \cdot \|y\|}.$$

Of course, this definition makes sense if and only if

$$-1 \le \frac{\langle x, y \rangle}{\|x\| \cdot \|y\|} \le 1.$$

But this is exactly the statement of the Cauchy–Schwarz inequality. Therefore, given two vectors x and y in R^m, it follows from the Cauchy–Schwarz inequality that there is a unique number θ, $0 \le \theta \le \pi$, such that

$$\langle x, y \rangle = \|x\| \cdot \|y\| \cos \theta.$$

θ is called the **angle** between x and y.

This definition of angle in R^m is consistent with our definition of orthogonality in R^m. For if x and y are nonzero vectors, then it follows directly from the equation $\langle x, y \rangle = \|x\| \cdot \|y\| \cos \theta$ that $\langle x, y \rangle = 0$ if and only if $\cos \theta = 0$, and this holds if and only if $\theta = \pi/2$.

EXERCISE 1 Find the cosine of the angle θ between each of the following pairs of vectors.

(a) $(1, 1, 1)$, $(1, 1, -1)$ (b) $(1, 0, 1, 1, 1)$, $(2, 3, 1, 1, 1)$

PROBLEMS 3.3

1. Find the angle between each of the following pairs of vectors.
   *(a) $(2, 1)$ and $(1, 1)$ (b) $(1, 1, 0)$ and $(0, 1, 0)$
   *(c) $(1, 1, 0, 1)$ and $(1, 1, 1, 1)$ (d) $(-1, 0, 2, -2, 0)$ and $(0, 1, 1, 0, 5)$
   (e) $(1, 1, 1)$ and $(1, 1, 0)$

2. Find all unit vectors that make an angle of $\pi/3$ with:
   *(a) $(1, 0)$ (b) $(1, 1)$ (c) $(1, 2)$

3. Find all unit vectors that make an angle of $\pi/4$ with:
   *(a) $(1, 0)$ (b) $(1, 1)$ (c) $(1, 2)$

4. Find all unit vectors that make an angle of $\pi/3$ with:
   *(a) $(1,0,0)$ and $(0,1,0)$ (b) $(1,0,1)$ and $(0,1,0)$

5. Let $v = (1,4,2)$. Let α, β, and γ be the angles that this vector makes with each of the coordinate vectors e_1, e_2, and e_3. The numbers $\cos \alpha$, $\cos \beta$, and $\cos \gamma$ are called the *direction cosines* of v. Find the direction cosines of the vector v and show that the vector

$$(\cos \alpha, \cos \beta, \cos \gamma)$$

   is a unit vector in the direction of v.

*6. Show that if x and y are both unit vectors in R^m with $\langle x, y \rangle = 1$, then $x = y$. What can you conclude when $\langle x, y \rangle = -1$?

7. Verify formula (1) in this section using the law of cosines. In case you have forgotten (as one of the authors did), the law of cosines states that $c^2 = a^2 + b^2 - 2ab \cos \theta$.

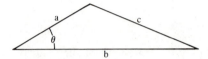

8. Show that $|\ \|x\| - \|y\|\ | \le \|x + y\|$. (*Hint:* Apply the triangle inequality to $x = x + y - y$ and $y = y + x - x$.)

3.4   ORTHOGONAL COMPLEMENTS

In our previous discussion of orthogonality we have restricted ourselves to talking about what it means for two vectors to be orthogonal. We discuss next what it means for a vector to be orthogonal to an entire subspace.

The following definition is clearly motivated by the geometry of R^2 and R^3 (see Figure 3.9).

Definition   A vector u in R^m is **orthogonal to a subspace** V of R^m, written $u \perp V$, if u is orthogonal to every vector in V, that is,

$$\langle u, v \rangle = 0 \qquad \text{for all } v \text{ in } V. \tag{1}$$

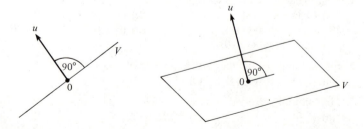

Figure 3.9

It follows immediately from the definition of the inner product that the zero vector is orthogonal to all of R^m. Are there any other vectors in R^m that are orthogonal to all of R^m? Clearly, such a vector must be orthogonal to itself and hence must be the zero vector [Theorem 3.1(d)]. Thus the only vector that is orthogonal to all of R^m is the zero vector.

Result 1 Let u be an m-vector. Then $\langle u, x \rangle = 0$ for all x in R^m if and only if $u = 0$.

Given a vector u and a subspace V, it is difficult to verify condition (1) directly since it requires that we check an infinite number of equations. However, the situation is actually not as bad as it seems; we need only check that the vector u is orthogonal to every vector in a spanning set for V.

Result 2 Let u be a vector in R^m and let V be a subspace of R^m. Then u is orthogonal to V if and only if u is orthogonal to a spanning set for V. That is, if $v_1, \ldots, v_n$ span V, then

$$u \perp V \quad \text{if and only if} \quad u \perp v_i, \quad i = 1, \ldots, n.$$

Proof First suppose that u is orthogonal to V. Then u is orthogonal to every vector in V and hence to each of the vectors $v_1, \ldots, v_n$.

Conversely, suppose that u is orthogonal to each of the vectors $v_1, \ldots, v_n$. If v is any vector in V, then $v = \alpha_1 v_1 + \cdots + \alpha_n v_n$. Since $\langle u, v_1 \rangle = \cdots = \langle u, v_n \rangle = 0$,

$$\langle u, v \rangle = \alpha_1 \langle u, v_1 \rangle + \cdots + \alpha_n \langle u, v_n \rangle$$
$$= 0.$$

Thus u is orthogonal to v. Since v is an arbitrary vector in V, it follows that u is orthogonal to V.

Since a basis is a spanning set, it follows immediately that a vector is orthogonal to a subspace if and only if it is orthogonal to every vector in a basis for the subspace.

EXERCISE 1 Verify that the vector $(2, 1, 1, -1)$ is orthogonal to the column space of the matrix

$$A = \begin{bmatrix} 1 & -1 & -1 & 1 \\ 0 & 0 & 2 & 1 \\ 0 & 2 & 0 & 1 \\ 2 & 0 & 0 & 4 \end{bmatrix}.$$

EXERCISE 2 Verify that the vector $(-1, 1, 2)$ is orthogonal to the subspace of R^3 consisting of all vectors (x, y, z) such that $x - y - 2z = 0$.

We can use this last result to find all vectors that are orthogonal to a given subspace. Let V be an r-dimensional subspace of R^m and suppose that $v_1, \ldots, v_n$ span V. By the previous result, an m-vector x is orthogonal to V if and only if the system of equations

$$\langle v_i, x \rangle = 0, \qquad i = 1, \ldots, n,$$

is satisfied. If A is the matrix whose columns are the vectors $v_1, \ldots, v_n$, then A has rank r and this system of equations becomes $A^T x = 0$. Thus an m-vector x is orthogonal to V if and only if x is in the null space of the matrix A^T. Since the null space of A^T is a subspace of R^m of dimension $m - r$, the set of all vectors orthogonal to V is a subspace of R^m of dimension $m - r$. We call this subspace the **orthogonal complement** of V and denote it by $V^\perp$ (read V perp). Since $V^\perp$ contains *all* vectors in R^m that are orthogonal to V, any vector contained in both V and $V^\perp$ would have to be orthogonal to itself. By Theorem 3.1(d), the only vector that is orthogonal to itself is the zero vector. Hence the only vector contained in both V and $V^\perp$ is the zero vector. We summarize our findings in parts (a), (b), and (c) of the following theorem.

Theorem 3.2   *Let V be a subspace of R^m.*

   (*a*) *The orthogonal complement $V^\perp$ of V is a subspace of R^m.*
   (*b*) *If the dimension of V is r, then the dimension of $V^\perp$ is $m - r$.*
   (*c*) *The only vector in R^m that lies in both V and $V^\perp$ is the zero vector.*
   (*d*) *If $v_1, \ldots, v_r$ form a basis for V and $w_1, \ldots, w_{m-r}$ form a basis for $V^\perp$, then $v_1, \ldots, v_r, w_1, \ldots, w_{m-r}$ form a basis for R^m.*

Proof of (d)   Since there are m vectors in this collection, it suffices to show that these vectors are independent. Suppose that

$$0 = \alpha_1 v_1 + \cdots + \alpha_r v_r + \beta_1 w_1 + \cdots + \beta_{m-r} w_{m-r}.$$

Then

$$\alpha_1 v_1 + \cdots + \alpha_r v_r = -(\beta_1 w_1 + \cdots + \beta_{m-r} w_{m-r}).$$

The vector on the left-hand side of this equation is in V, and the vector on the right-hand side is in $V^\perp$. Thus both vectors (being equal) must be in both V and $V^\perp$. Thus by part (c)

$$\alpha_1 v_1 + \cdots + \alpha_r v_r = 0 \qquad \text{and} \qquad \beta_1 w_1 + \cdots + \beta_{m-r} w_{m-r} = 0.$$

Since $v_1, \ldots, v_r$ are independent, all of the α_i's are 0. Similarly, all of the β_j's are 0. This completes the proof.

Note that the proof of parts (a) and (b) of Theorem 3.2 provides us with a computational procedure for actually determining the orthogonal complement of a subspace. First we express the subspace V as the column space of a matrix A (the columns of A consist of vectors that span V). Then the orthogonal complement of V is the null space of A^T. Since any subspace can be expressed as the column space of a matrix in this way, we can summarize our procedure in terms of matrices.

Result 3   The orthogonal complement of the column space $C(A)$ of a matrix A is the null space of its transpose A^T. In symbols,

$$C(A)^\perp = N(A^T).$$

EXAMPLE 1   To find the orthogonal complement of the subspace V spanned by $(1, 2, 2, 1)$ and $(1, 0, 2, 0)$, we form the matrix

$$A = \begin{bmatrix} 1 & 1 \\ 2 & 0 \\ 2 & 2 \\ 1 & 0 \end{bmatrix}$$

and compute the null space of A^T. The reduced echelon form of A^T is $E = \begin{bmatrix} 1 & 0 & 2 & 0 \\ 0 & 1 & 0 & \frac{1}{2} \end{bmatrix}$ and hence a basis for the null space is $(-2, 0, 1, 0)$ and $(0, -1, 0, 2)$. Thus the orthogonal complement $V^\perp$ is the two-dimensional subspace of R^4 spanned by these two vectors.

EXERCISE 3   Find a basis for the orthogonal complement of the subspace of R^4 spanned by $(1, 1, 2, -1)$.

Let us find the orthogonal complements of the various subspaces of R^3. The orthogonal complement of the trivial subspace $\{0\}$ is all of R^3, since any vector is orthogonal to the zero vector. The orthogonal complement of a straight line through the origin is the plane through the origin that is perpendicular to the given line [see Figure 3.10(a)]. The orthogonal complement of a plane through the origin is the line through the origin perpendicular to the plane [see Figure 3.10(b)]. Finally, the orthogonal complement of R^3 is the zero vector, since the only vector orthogonal to every vector is the zero vector (see Result 1).

EXERCISE 4   Verify the statements made in the preceding paragraph by using Theorem 3.2(b) to determine the dimension of each orthogonal complement.

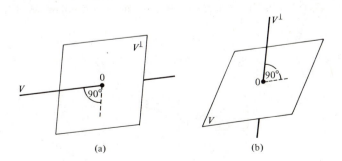

Figure 3.10   (a)   (b)

PROBLEMS 3.4

1. Find a basis for the orthogonal complement of the subspace spanned by:
   (a) $(1, 1)$ *(b) $(1, 0, 1)$
   (c) $(1, 1, 3)$ and $(1, 1, 2)$ *(d) $(1, 1, 0, 0)$
   *(e) $(1, 1, 0, 0)$ and $(0, 2, 4, 5)$ *(f) $(2, 2, 1, 0)$, $(2, 4, 0, 1)$, and $(4, -2, 1, -1)$
   *(g) $(1, -1, 1, -1, 1)$ *(h) $(1, 1, 1, 1, 1)$ and $(1, 0, 1, 0, 1)$
   *(i) $(1, -1, 1, 1, 1)$, $(2, 0, 3, -2, 0)$, and $(3, -3, 4, 5, 2)$

2. For each subspace in Problem 1, verify parts (b) and (d) of Theorem 3.2.

3. Verify that the vector $(4, 2, -1)$ is orthogonal to the column space of the matrix

$$A = \begin{bmatrix} 1 & 0 & -1 \\ -1 & 1 & 2 \\ 2 & 2 & 0 \end{bmatrix}.$$

   What is the orthogonal complement of $C(A)$?

4. Determine the orthogonal complement of each of the subspaces of R^2. Also, when the subspace is a straight line, determine the equation of the orthogonal complement in terms of the equation of the line.

5. Given a straight line (or a plane) in R^3 through the origin, determine the equation of the orthogonal complement in terms of the equation of the line (or plane).

6. Let S be a subset of R^m.
   (a) Show that the set $S^\perp$ of all vectors in R^m which are orthogonal to S is a subspace of R^m.
   (b) Let V be the subspace of R^m which is spanned by the vectors in S. Show that $V^\perp = S^\perp$.
   (c) Let $S^{\perp\perp}$ be the set of all vectors that are orthogonal to $S^\perp$. Show that S is a subset of $S^{\perp\perp}$.

7. Let V be a subspace of R^m.
   *(a) Show that $V = V^{\perp\perp}$. (*Hint:* What is the dimension of $V^{\perp\perp}$?)
   (b) Show that V is the orthogonal complement of $V^\perp$.

8. Let A be an $m \times n$ matrix.
   (a) Show that the null space of A is the orthogonal complement of the row space of A.
   (b) What is the orthogonal complement of the null space of A^T? [*Hint:* Use Exercise 6(b) and Result 3.]
   (c) What is the orthogonal complement of the null space of A?

*9. Find the orthogonal complement of the column space and the row space of each of the following matrices.

   (a) $\begin{bmatrix} 1 & -3 \\ -2 & 6 \end{bmatrix}$ (b) $\begin{bmatrix} 1 & -1 \\ 0 & 2 \\ 2 & 1 \end{bmatrix}$ (c) $\begin{bmatrix} 1 & 1 & 1 \\ 0 & 2 & 1 \\ 0 & 0 & 3 \end{bmatrix}$

(d) $\begin{bmatrix} 1 \\ 2 \\ 3 \\ 4 \end{bmatrix}$ (e) $\begin{bmatrix} 1 & 1 \\ -1 & -2 \\ 1 & -1 \\ -1 & 1 \end{bmatrix}$

10. Two subspaces V and W of R^m are called *orthogonal subspaces* if every vector in V is orthogonal to every vector in W. Thus V and W are orthogonal subspaces if and only if $\langle v, w \rangle = 0$ for every v in V and w in W.

   (a) Show that the subspace V spanned by $(-5, 0, 0, 0, 1)$ and $(0, -2, 0, 1, 0)$ is orthogonal to the subspace W spanned by $(1, 2, 3, 4, 5)$ and $(1, -2, 3, -4, 5)$.

   *(b) Is the subspace W in part (a) the orthogonal complement of the subspace V?

   (c) Show that a subspace V and its orthogonal complement $V^{\perp}$ are orthogonal subspaces.

   *(d) Show that if V and W are orthogonal subspaces of R^m with the property that every vector x in R^m can be written as a sum of a vector in V and a vector in W, then $W = V^{\perp}$.

   (e) Show that if V and W are orthogonal subspaces of R^m with the property that $\dim V + \dim W = m$, then $W = V^{\perp}$.

3.5   PROJECTIONS

In Section 3.1 we discussed the question of what can be done with an inconsistent system of equations. We saw that one way to approach such a system was to replace it by a consistent system which could then be solved. The consistent system that is the "best possible" replacement for the inconsistent system $Ax = b$ is the system $Ax = p$, where p is the vector in the column space of A which is closest to b. In this section we develop a method for computing p.

Let V be a straight line or plane through the origin in R^3. If a vector b does not belong to the subspace V, then from geometry we know that the vector p in the subspace that is closest to b is found by dropping a perpendicular from b to the subspace (see Figure 3.11). This means that p is the vector in the subspace V with the property that the vector $b - p$ is orthogonal to V. Therefore, $b - p$ is in $V^{\perp}$ because $V^{\perp}$ contains all vectors

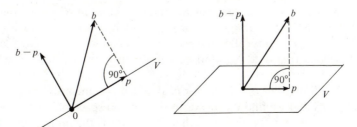

Figure 3.11

orthogonal to V. Since

$$b = p + (b - p),$$

we see that b can be expressed algebraically as a sum of a vector in V and a vector in $V^\perp$. Thus the subspace V and its orthogonal complement $V^\perp$ yield an interesting (and, as we shall see, very useful) algebraic description of R^3. Specifically, we are able to decompose any vector in R^3 into a sum of two vectors, one in V and the other in $V^\perp$.

In our next theorem, we generalize this discussion to higher-dimensional spaces.

Theorem 3.3 *Let V be a subspace of R^m and let $V^\perp$ be its orthogonal complement. Given any vector b in R^m, there is a unique vector p in V such that $b - p$ is in $V^\perp$. Since*

$$b = p + (b - p),$$

the vectors p and $b - p$ give a unique decomposition of b into a sum of two vectors, one in V and the other in $V^\perp$.

Proof By Theorem 3.2(d), if $v_1, \ldots, v_n$ is a basis for V and $w_1, \ldots, w_{m-n}$ is a basis for $V^\perp$, then $v_1, \ldots, v_n, w_1, \ldots, w_{m-n}$ is a basis for R^m. If b is an m-vector and

$$b = \alpha_1 v_1 + \cdots + \alpha_n v_n + \beta_1 w_1 + \cdots + \beta_{m-n} w_{m-n}$$

is the expansion of b in terms of this basis, then

$$
\begin{aligned}
p &= \alpha_1 v_1 + \cdots + \alpha_n v_n && \text{is in } V, && \text{and} \\
q &= \beta_1 w_1 + \cdots + \beta_{m-n} w_{m-n} && \text{is in } V^\perp.
\end{aligned}
$$

Thus there is a vector p in V such that $b - p$ is in $V^\perp$.

It remains to show that the vector p is unique. Suppose that there is another vector p_1 in V such that $b - p_1$ is also in $V^\perp$. Since $V^\perp$ is a subspace, the vector

$$(b - p_1) - (b - p) = p - p_1$$

is in $V^\perp$. But it is also in V since p and p_1 are both in V. By Theorem 3.2(c), $p = p_1$. Thus p is unique.

Definition Let V be a subspace of R^m and b a vector in R^m. Then the **projection of b onto V** is the unique vector p in V such that $b - p$ is orthogonal to V. The vector $b - p$ is called the **projection of b onto $V^\perp$**. Thus a vector p is the projection of b onto V if and only if p is in V and $b - p$ is in $V^\perp$.

Recall that our procedure for dealing with an inconsistent system $Ax = b$ requires that we find the vector in the column space of A which is closest to b. The following result shows that this vector is the projection of b onto the column space of A. This result also shows that in R^3 the projection of a

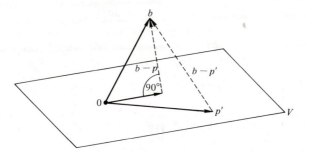

Figure 3.12

vector onto a subspace corresponds to the geometrical construction of dropping a perpendicular from a vector to a subspace.

Result 1   Let V be a subspace of R^m and b a vector in R^m. The projection of b onto V is the vector in V that is closest to b.

Proof   Let p be the projection of b onto V. If p' is any other vector in V, $p' - p$ is in V and hence is orthogonal to $b - p$ (see Figure 3.12). By the Pythagorean theorem

$$\|b - p'\|^2 = \|b - p\|^2 + \|p - p'\|^2.$$

Since $\|p - p'\| \geq 0$, it follows that $\|b - p\| \leq \|b - p'\|$. This completes the proof.

The proof of Theorem 3.3 provides us with a method for computing the projection of a vector onto a subspace. Given an n-dimensional subspace V of R^m, first determine a basis for V, say $v_1, \ldots, v_n$. As we have seen in Section 3.4, we can use this basis for V to find a basis for $V^\perp$. We form the matrix A having the vectors $v_1, \ldots, v_n$ as its columns. Then $V^\perp$ is the null space of A^T, and hence a basis $w_1, \ldots, w_{m-n}$ for $V^\perp$ can be read off from the echelon form of A^T. The vectors $v_1, \ldots, v_n, w_1, \ldots, w_{m-n}$ form a basis for R^m. Now given any vector b in R^m, express b in terms of this basis,

$$b = \alpha_1 v_1 + \cdots + \alpha_n v_n + \beta_1 w_1 + \cdots + \beta_{m-n} w_{m-n}.$$

The vector

$$p = \alpha_1 v_1 + \cdots + \alpha_n v_n$$

is the projection of b onto V and the vector

$$b - p = \beta_1 w_1 + \cdots + \beta_{m-n} w_{m-n}$$

is the projection of b onto $V^\perp$.

EXAMPLE 1   Let V be the subspace of R^3 spanned by the vectors $(2, 0, -1)$ and $(1, 1, 0)$. Since these vectors are independent, they form a basis for V. To find a basis for $V^{\perp}$, we let A be the matrix

$$A = \begin{bmatrix} 2 & 1 \\ 0 & 1 \\ -1 & 0 \end{bmatrix}$$

and find the null space of A^T. The reduced echelon form of A^T is

$$E = \begin{bmatrix} 1 & 0 & -\frac{1}{2} \\ 0 & 1 & \frac{1}{2} \end{bmatrix}.$$

Hence a basis for the null space of A^T is $(1, -1, 2)$, so this vector is a basis for $V^{\perp}$. To express the 3-vector $b = (7, -1, 5)$ in terms of the basis $(2, 0, -1)$, $(1, 1, 0)$, and $(1, -1, 2)$ of R^3, we use Gaussian elimination to reduce the augmented matrix

$$\begin{bmatrix} 2 & 1 & 1 & : & 7 \\ 0 & 1 & -1 & : & -1 \\ -1 & 0 & 2 & : & 5 \end{bmatrix}$$

to the echelon form

$$\begin{bmatrix} 1 & 0 & 0 & : & 1 \\ 0 & 1 & 0 & : & 2 \\ 0 & 0 & 1 & : & 3 \end{bmatrix}.$$

Thus the projection of $(7, -1, 5)$ onto V is

$$p = 1(2, 0, -1) + 2(1, 1, 0)$$

and the projection of $(7, -1, 5)$ onto $V^{\perp}$ is

$$(7, -1, 5) - p = 3(1, -1, 2).$$

EXERCISE 1   Let V be the subspace spanned by $(1, 1, 1)$. Find the projections of $(1, 1, 2)$ onto V and onto $V^{\perp}$.

This procedure is very long. However, if we look closely at it we see that some improvements can be made. We began with a subspace V of R^m and a basis $v_1, \ldots, v_n$ for V. We then formed the matrix A whose columns are $v_1, \ldots, v_n$. The subspaces V and $V^{\perp}$ and the matrix A are related as follows:

1. V is the column space of A.
2. $V^{\perp}$ is the null space of A^T.

Now p is the projection of b onto V if and only if p is in V and $b - p$ is in $V^{\perp}$. Using this and 1 and 2 above we see that p is the projection of b onto V if and only if the following two conditions hold.

1. $p = A\bar{x}$ for some $\bar{x}$ in R^n.
2. $A^T(b - p) = A^T(b - A\bar{x}) = 0$.

Finally, observe that our argument depends only on the fact that the columns of A span V and does not require that the columns of A be independent. Thus we have proved the following result.

Result 2 Let V be a subspace of R^m and let A be a matrix whose columns span V. A vector p is the projection of an m-vector b onto V if and only if the following two conditions hold.

1. $p = A\bar{x}$ for some $\bar{x}$ in R^n.
2. $A^T A\bar{x} = A^T b$.

The equations $A^T A\bar{x} = A^T b$ in condition 2 of Result 2 are called the **normal equations**. A solution to this system gives us a vector $\bar{x}$ such that $p = A\bar{x}$ is the projection of b onto V. In the next section we will have much more to say about the solution $\bar{x}$ to this system. See also Example 5 below.

The normal equations are obtained by simply multiplying both sides of the equation $Ax = b$ by A^T. In view of this remark it should be impossible to forget them.

PROJECTION ONTO A SUBSPACE

Given

      A subspace V of R^m and a vector b in R^m.

Goal

      To find the projection of b onto V.

Procedure

    1. Let A be a matrix whose columns span V.
    2. Use Gaussian elimination to find the solution $\bar{x}$ to the normal equations $A^T A\bar{x} = A^T b$.
    3. The projection of b onto V is

$$p = A\bar{x}.$$

EXAMPLE 2 Consider the problem of finding the projection of $(4, -1, 1)$ onto the subspace V of R^3 spanned by $(1, 0, 1)$ and $(1, 1, 0)$. Letting A be the matrix whose columns are these vectors, we have

$$A = \begin{bmatrix} 1 & 1 \\ 0 & 1 \\ 1 & 0 \end{bmatrix}, \quad A^T A = \begin{bmatrix} 2 & 1 \\ 1 & 2 \end{bmatrix}, \quad A^T b = \begin{bmatrix} 5 \\ 3 \end{bmatrix}.$$

Gaussian elimination applied to the augmented matrix $[A^T A \ : \ A^T b]$ yields the echelon matrix

$$\begin{bmatrix} 1 & 0 & : & \frac{7}{3} \\ 0 & 1 & : & \frac{1}{3} \end{bmatrix}.$$

Hence $\bar{x} = (\frac{7}{3}, \frac{1}{3})$ and the projection is given by

$$p = A\bar{x} = \begin{bmatrix} 1 & 1 \\ 0 & 1 \\ 1 & 0 \end{bmatrix} \begin{bmatrix} \frac{7}{3} \\ \frac{1}{3} \end{bmatrix} = \begin{bmatrix} \frac{8}{3} \\ \frac{1}{3} \\ \frac{7}{3} \end{bmatrix} = \frac{1}{3} \begin{bmatrix} 8 \\ 1 \\ 7 \end{bmatrix}.$$

The projection onto $V^\perp$ is

$$b - p = (4, -1, 1) - \tfrac{1}{3}(8, 1, 7) = \left(\tfrac{4}{3}, -\tfrac{4}{3}, -\tfrac{4}{3}\right).$$

EXERCISE 2   Find the projection of $(1, -1, 4)$ onto the subspace V of R^3 spanned by $(1, 0, -1)$ and $(2, 1, 0)$. Also find the projection onto $V^\perp$.

EXAMPLE 3   Suppose that we wish to find the projection of one m-vector b onto another m-vector v. This problem is equivalent to finding the projection of b onto the one-dimensional subspace V of R^m that is spanned by v. Let A be the matrix whose single column is v. Then

$$A^T A = v^T v = \langle v, v \rangle,$$
$$A^T b = v^T b = \langle v, b \rangle,$$

and hence the normal equations $A^T A \bar{x} = A^T b$ become

$$\langle v, v \rangle \bar{x} = \langle v, b \rangle.$$

Thus

$$\bar{x} = \frac{\langle v, b \rangle}{\langle v, v \rangle}$$

and

$$p = \frac{\langle v, b \rangle}{\langle v, v \rangle} v.$$

EXAMPLE 4   If $v = (1, 1, 1, 1)$ and $b = (2, 1, 6, 3)$, then $\langle v, b \rangle = 12$ and $\langle v, v \rangle = 4$. Thus the projection of b onto v is

$$p = \frac{\langle v, b \rangle}{\langle v, v \rangle} v = \frac{12}{4} (1, 1, 1, 1) = (3, 3, 3, 3).$$

EXERCISE 3   Show that if p is the projection of b onto v and θ is the angle between v and b, then

$$\| p \| = \| b \| \; |\cos \theta| \qquad \text{and} \qquad p = \| b \| \cos \theta \frac{v}{\| v \|}.$$

Think about this result geometrically (see Figure 3.13).

EXERCISE 4   Find the projection of b onto v if:
  (a) $v = (1, 2), \qquad b = (1, 1)$
  (b) $v = (1, 0, 1), \quad b = (0, 1, 1)$

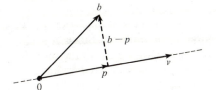

Figure 3.13

Up to this point in our discussion of projections we have been careful to point out that the columns of the matrix A are only required to span the subspace V; they are not required to be independent. If they are independent (i.e., when the spanning set for V is in fact a basis), then (as we will prove) the matrix A^TA is nonsingular. Thus the normal equations $A^TA\bar{x} = A^Tb$ have the unique solution

$$\bar{x} = (A^TA)^{-1}A^Tb$$

and the projection of b onto V is given by

$$p = A\bar{x} = A(A^TA)^{-1}A^Tb.$$

Result 3   Let A be an $m \times n$ matrix of rank n. Then A^TA is an $n \times n$ nonsingular matrix.

Proof   The fact that A^TA is an $n \times n$ matrix follows immediately from the definition of matrix multiplication. To prove that it is nonsingular, it is sufficient to show that the nullity of A^TA is zero. Let x be in the null space of A^TA. Then $A^TAx = 0$ and hence Ax, a vector in the column space $C(A)$ of A, is also in the null space $N(A^T)$ of A^T. By Result 3 of Section 3.4, $C(A)^{\perp} = N(A^T)$ and since the only vector common to a subspace and its orthogonal complement is the zero vector, $Ax = 0$. Finally, since A is an $m \times n$ matrix of rank n, its null space is trivial. Thus $x = 0$.

The important point here is that when the columns of A are independent, the normal equations $A^TA\bar{x} = A^Tb$ have a unique solution. The important point is not that the solution is given by the formula

$$\bar{x} = (A^TA)^{-1}A^Tb.$$

To find $\bar{x}$ by means of this formula involves the computation of the inverse of a matrix, and as we pointed out in Chapter 1, this is a time-consuming and hence expensive operation. It is better to solve the normal equations by Gaussian elimination.

Warning: When A is not a square matrix, $(A^TA)^{-1}$ is not equal to $A^{-1}(A^T)^{-1}$. Why?

EXAMPLE 5   Let V be a subspace of R^m, let $v_1, \ldots, v_n$ be a basis for V, and let A be the $m \times n$ matrix having $v_1, \ldots, v_n$ as its columns. Since the columns of

A are independent, A^TA is nonsingular (Result 3) and hence the normal equations $A^TA\bar{x} = A^Tb$ have a unique solution for all m-vectors b. Now let p be the projection of b onto V. Since $p = A\bar{x}$, it follows that $\bar{x}$ is the coordinate vector of p with respect to the basis we have chosen. If b happens to be in the subspace V, then clearly $p = b$ (i.e., the projection of b onto V is b itself). Hence in this case $\bar{x}$ is the coordinate vector of b itself with respect to the basis $v_1, \ldots, v_n$.

EXAMPLE 6   Let A be an $m \times n$ matrix of rank n. By Result 3 A^TA is nonsingular. Consider the matrix

$$P = A(A^TA)^{-1}A^T.$$

If b is an m-vector, and if we let

$$\bar{x} = (A^TA)^{-1}A^Tb,$$

then the projection of b onto $C(A)$ is

$$Pb = A(A^TA)^{-1}A^Tb = A\bar{x} = p.$$

Hence applying P to a vector b gives the projection of b onto the column space of A. For this reason, the matrix P is called the **projection matrix** for the subspace $C(A)$. See Problem 15 for some properties of this matrix.

PROBLEMS 3.5

1. Find the projection of b onto v for each of the following.
   *(a) $v = (0, 1)$, $b = (2, 1)$
   (b) $v = (1, 1, 1)$, $b = (5, 1, 1)$
   (c) $v = (1, 2, -1)$, $b = (0, 1, 2)$
   *(d) $v = (1, -1, 0, 1)$, $b = (3, -2, 8, 5)$
   (e) $v = (1, 0, 2, 0, 3)$, $b = (4, 5, 5, 5, 0)$

2. Find the projection of $b = (1, 1, 1)$ onto the subspace V of R^3 spanned by $(1, 0, 1)$ and $(0, 1, 1)$. Also find the projection of b onto $V^\perp$.

*3. (a) Find the projection of $u = (1, 2, 3, 1, 1)$ onto the subspace V spanned by $(1, 0, 1, 0, 1)$ and $(0, 1, 0, 1, 0)$. Also find the projection of u onto $V^\perp$.
   (b) What is the distance from the vector u to the subspace V?
   (c) What vector in the subspace V is nearest to u?

4. Find the projection of the vector $u = (2, 1, 0, 1)$ onto the subspace V of R^4 consisting of all vectors (x_1, x_2, x_3, x_4) such that $x_1 + x_2 + x_3 + x_4 = 0$. Also answer parts (b) and (c) of Problem 3.

*5. Find the projection of $u = (1, 1, 1)$ onto the subspace V of R^3 spanned by $(0, 2, 3)$, $(1, 0, 1)$, $(-1, 2, 2)$. Also answer parts (b) and (c) of Problem 3.

6. Find the projection of the vector $(1, -1, 4)$ onto the plane $x - 2y + z = 0$. Find the perpendicular distance from the vector to the plane.

7. Let u and a be m-vectors. The equation of the line in R^m through a in the direction of u is $x = a + tu$ where t is a real number.

   (a) If $m = 2$, show that the equation of the line through $a = (x_0, y_0)$ in the direction of $u = (-b, a)$ can be written in the form $a(x - x_0) + b(y - y_0) = 0$.

   (b) If b is an m-vector which is not on the line $x = a + tu$, the distance from b to the line is defined to be the distance from $b - a$ to the subspace spanned by u. Illustrate this definition with a picture when $m = 2$ and $m = 3$.

   (c) If p is the projection of $b - a$ onto the subspace spanned by u, show that the point on the line nearest the point b is $a + p$.

8. Use the previous problem to find the distance from the point b to the line $x = a + tu$ and the point on the line which is nearest to b if:

   (a) $b = (1, 1)$, $a = (-1, 1)$, and $u = (-3, 2)$
   (b) $b = (0, 1)$, $a = (0, -1)$, and $u = (-1, 1)$
   (c) $b = (2, 4)$, $a = (4, 1)$, and $u = (1, 1)$
   (d) $b = (3, 1, 1)$, $a = (1, 1, 1)$, and $u = (1, 0, 1)$
   *(e) $b = (1, 1, 1)$, $a = (0, 1, 2)$, and $u = (1, 2, 3)$
   *(f) $b = (1, 2, 3)$, $a = (0, 2, 1)$, and $u = (2, 1, 2)$
   *(g) $b = (1, 1, 0, 1)$, $a = (1, 0, 1, 0)$, and $u = (0, 1, 1, 1)$
   *(h) $b = (1, 2, -1, 0, 1)$, $a = (0, 2, 1, 1, 1)$, and $u = (2, 1, 1, -1, 1)$

9. Let n and a be m-vectors. The equation of the plane in R^m through a and orthogonal to n is
$$\langle x - a, n \rangle = 0.$$

   (a) If $m = 3$, show that if $a = (x_0, y_0, z_0)$ and $n = (a, b, c)$ then the equation of the plane can be written in the form
$$a(x - x_0) + b(y - y_0) + c(z - z_0) = 0.$$

   (b) If b is an m-vector which is not on the plane, the distance from b to the plane is defined to be the distance from $b - a$ to the orthogonal complement of the subspace spanned by n. Illustrate this definition with a picture when $m = 3$.

   (c) Show that the distance from b to the plane $\langle x - a, n \rangle = 0$ is
$$\frac{|\langle b - a, n \rangle|}{\|n\|}.$$

   (d) If p is the projection of $b - a$ onto the orthogonal complement of the subspace spanned by n, show that the point on the plane nearest to b is
$$b - \frac{\langle b - a, n \rangle}{\langle n, n \rangle} n.$$

10. Use the previous problem to find the distance from the point b to the plane $\langle x - a, u \rangle = 0$ and the point on the plane which is nearest to b if:

    *(a) $b = (2, 1, -1)$, $a = (0, 2, 2)$, and $n = (5, -2, 3)$
    (b) $b = (1, 1, 1)$, $a = (1, 0, 1)$, and $n = (1, -1, 1)$
    *(c) $b = (1, -4, -3)$, $a = (1, -1, -1)$, and $n = (2, -3, 6)$
    (d) $b = (0, 1, 6)$, $a = (0, 1, 0)$, and $n = (1, 2, -1)$
    *(e) $b = (2, 3, -1)$, $a = (1, 0, 0)$, and $n = (1, 1, 1)$
    (f) $b = (1, 2, 3, 4)$, $a = (1, 2, 3, 0)$, and $n = (1, 0, 1, 0)$
    *(g) $b = (-1, 1, 1, -1)$, $a = (2, 2, 1, 2)$, and $n = (3, 0, 1, 3)$

(h) $b = (-3, -1, 0, 1, 2, 1)$, $a = (0, 1, 0, 1, 0, 1)$, and $n = (1, 1, 1, 0, 1, 1)$

*(i) $b = (4, 3, 2, 1, 0, -1)$, $a = (3, 2, 1, 0, -1, -2)$, and $n = (2, -1, 1, 0, 0, 1)$

11. Let V be the subspace spanned by $(1, 0, 1, 0)$, $(1, 1, -1, 0)$, and $(1, -2, -1, 1)$.
   *(a) Find the projection matrix P for the projection onto the subspace V.
   *(b) Find the projection of $(1, 0, 0, 0)$ and $(1, 1, 0, 0)$ onto the subspace V.
   (c) What is the null space of P?

12. Let V be a subspace of R^m and b an m-vector.
   (a) Show that b is in V if and only if the projection of b onto V is b itself.
   (b) Show that b is orthogonal to V if and only if the projection of b onto V is the zero vector.

13. Let v be any nonzero m-vector.
   (a) Show that if b is an m-vector and p is the projection of b onto v, then $p = \alpha v$ for some scalar α.
   (b) Use the fact that $(b - p) \perp v$ to show that $\alpha = \dfrac{\langle b, v \rangle}{\langle v, v \rangle}$. Hence $p = \dfrac{\langle b, v \rangle}{\langle v, v \rangle} v$.
       Compare this proof with the one given in the text.

14. Let A be an $m \times n$ matrix of rank r.
   (a) Give a proof, analogous to that of Result 3, that $A^T A$ is a symmetric $n \times n$ matrix of rank r.
   (b) Prove this same fact by showing that the null space of $A^T A$ is equal to the null space of A. (*Hint:* If x is in the null space of $A^T A$, then $\langle x, A^T A x \rangle = \langle Ax, Ax \rangle = 0$.)
   (c) Show that if $A^T A$ is nonsingular ($r = n$), then A has rank n. This proves the converse of Result 3.

15. Let $P = A(A^T A)^{-1} A^T$ be the projection matrix associated with an $m \times n$ matrix A of rank n. Prove each of the following.
   *(a) $P^2 = P$
   (b) $P = P^T$
   *(c) The matrix $I - P$ has properties (a) and (b).
   (d) For any vector y in R^m, $(I - P)y$ is the projection of y onto the orthogonal complement of the column space of A.

16. Let P be a symmetric matrix with the property that $P^2 = P$.
   (a) Show that P is the projection matrix for the projection onto the column space of P. (*Hint:* Show that for any x, $x - Px$ is orthogonal to the column space of P.)
   *(b) What is the null space of P?

17. Show that a vector u is orthogonal to a subspace V if and only if the distance from u to V is $\|u\|$.

18. Let u be a nonzero m-vector. Show that any m-vector x can be written in one and only one way as $x = a + b$, where a and u are collinear and b is orthogonal to u. Compare with Problem 6 in Section 3.2.

19. The Cauchy–Schwarz inequality states that for any two m-vectors u and v, $|\langle u, v \rangle| \le \|u\| \cdot \|v\|$. Prove this inequality by computing the length of $u - p$, where p is the projection of u onto v. Can you now give a reason for the choice of α in the proof of the Cauchy–Schwarz inequality?

20. Show that if A is an $n \times n$ matrix of rank n, then
$$\left(A^T A\right)^{-1} = A^{-1} \left(A^T\right)^{-1}.$$

21. Show that for every $m \times n$ matrix A and every m-vector b, the normal equations $A^T A \bar{x} = A^T b$ always have a solution.

22. Let V be a subspace of R^m and let b_1, b_2 be m-vectors. If p_1 and p_2 are the projections of b_1 and b_2 onto V, what are the projections of $b_1 + b_2$ and αb_1 onto V?

23. In this problem we outline a different proof of Result 2. Let A be an $m \times n$ matrix and b an m-vector.
   (a) Show that p is the projection of b onto $C(A)$ if and only if $p = A\bar{x}$ and $(b - A\bar{x}) \perp C(A)$.
   (b) Use part (a) and Problem 13 of Section 3.2 to show that $(b - A\bar{x}) \perp C(A)$ if and only if $\langle x, A^T b - A^T A\bar{x}\rangle = 0$ for all x in R^n.
   (c) Now use Result 1 of Section 3.4 to show that $\langle x, A^T b - A^T A\bar{x}\rangle = 0$ for all x in R^n if and only if $A^T A\bar{x} = A^T b$.
   (d) Use parts (a), (b), and (c) to prove Result 2.

24. Let A be an $m \times n$ matrix. Show that if b is any vector in the column space of A, then there is a unique vector x in the row space of A such that $Ax = b$. [*Hint:* Use Problem 8(a) of Section 3.4 and Theorem 3.3.]

3.6   THE LEAST-SQUARES SOLUTION OF INCONSISTENT SYSTEMS

In this section we return to and complete our discussion of the problem of finding the best possible solution of an inconsistent system of equations. Suppose that $Ax = b$ is an inconsistent system. Then b does not lie in the column space of A and (as we have seen in Section 3.5) the best we can do is to replace b by its projection p onto the column space of A. The system $A\bar{x} = p$ is consistent and a solution $\bar{x}$ to this system is called (for reasons to be discussed below) a least-squares solution to the inconsistent system $Ax = b$. We have seen in Section 3.5 the best way to find p is to solve the normal equations

$$A^T A\bar{x} = A^T b.$$

But since it is $\bar{x}$ and not the projection $p = A\bar{x}$ that we are interested in now, we do not need to find p!

Definition   A **least-squares solution** to an inconsistent system $Ax = b$ is a solution to the normal equations $A^T A\bar{x} = A^T b$.

EXAMPLE 1   The system of equations

$$\begin{array}{rcr} x_1 - x_2 &=& -3 \\ 2x_1 + x_2 &=& 0 \\ 3x_1 + 2x_2 &=& 6 \end{array} \quad \text{or} \quad \begin{bmatrix} 1 & -1 \\ 2 & 1 \\ 3 & 2 \end{bmatrix}\begin{bmatrix} x_1 \\ x_2 \end{bmatrix} = \begin{bmatrix} -3 \\ 0 \\ 6 \end{bmatrix}$$

is inconsistent. Since

$$A^TA = \begin{bmatrix} 1 & 2 & 3 \\ -1 & 1 & 2 \end{bmatrix} \begin{bmatrix} 1 & -1 \\ 2 & 1 \\ 3 & 2 \end{bmatrix} = \begin{bmatrix} 14 & 7 \\ 7 & 6 \end{bmatrix}$$

and

$$A^Tb = \begin{bmatrix} 1 & 2 & 3 \\ -1 & 1 & 2 \end{bmatrix} \begin{bmatrix} -3 \\ 0 \\ 6 \end{bmatrix} = \begin{bmatrix} 15 \\ 15 \end{bmatrix},$$

the normal equations are

$$\begin{bmatrix} 14 & 7 \\ 7 & 6 \end{bmatrix} \begin{bmatrix} x_1 \\ x_2 \end{bmatrix} = \begin{bmatrix} 15 \\ 15 \end{bmatrix}.$$

The solution of this system is $\bar{x} = (-\frac{3}{7}, 3)$ which is the least-squares solution of the inconsistent system.

EXAMPLE 2   Consider the inconsistent system

$$\begin{bmatrix} 1 & 0 & 2 \\ 0 & 1 & -2 \\ 1 & 1 & 0 \\ 1 & 2 & -2 \end{bmatrix} \begin{bmatrix} x_1 \\ x_2 \\ x_3 \end{bmatrix} = \begin{bmatrix} -1 \\ -1 \\ 1 \\ 1 \end{bmatrix}.$$

Notice that in this example the third column is a combination of the first two. Since

$$A^TA = \begin{bmatrix} 3 & 3 & 0 \\ 3 & 6 & -6 \\ 0 & -6 & 12 \end{bmatrix} \quad \text{and} \quad A^Tb = \begin{bmatrix} 1 \\ 2 \\ -2 \end{bmatrix}$$

the normal equations are

$$\begin{bmatrix} 3 & 3 & 0 \\ 3 & 6 & -6 \\ 0 & -6 & 12 \end{bmatrix} \begin{bmatrix} x_1 \\ x_2 \\ x_3 \end{bmatrix} = \begin{bmatrix} 1 \\ 2 \\ -2 \end{bmatrix}.$$

The reduced echelon form of this system is

$$\begin{bmatrix} 1 & 0 & 2 & \vdots & 0 \\ 0 & 1 & -2 & \vdots & \frac{1}{3} \\ 0 & 0 & 0 & \vdots & 0 \end{bmatrix}.$$

The solution is $(x_1, x_2, x_3) = (0, \frac{1}{3}, 0) + x_3(-2, 2, 1)$. There are infinitely many solutions to the normal equations in this case. Any of these solutions is a least-squares solution to the inconsistent system.

EXERCISE 1   Find the least-squares solution to the following inconsistent system.

$$\begin{aligned} x_1 + 2x_2 + x_3 &= 4 \\ x_1 + 5x_2 \phantom{{} + x_3} &= 19 \\ x_1 - x_2 + 2x_3 &= 1 \end{aligned}$$

Let A be an $m \times n$ matrix $(m \geq n)$. If A has rank n, the normal equations have a unique solution. If A has rank less than n, then there are infinitely many solutions (see Example 2). In most applications this situation will not arise. For a discussion of how to choose the optimal solution when it does arise, we refer the reader to Problem 10.

We now discuss the reason for calling a solution $\bar{x}$ to the normal equations a least-squares solution. Since the distance between two vectors in R^m is given by the length of their difference, the projection p has the property that

$$\|b - p\| \leq \|b - y\|$$

for all y in the column space of A. Equivalently, since $p = A\bar{x}$ for some $\bar{x}$ in R^n, the vector $\bar{x}$ has the property that

$$\|b - A\bar{x}\| \leq \|b - Ax\| \qquad \text{for all } x \text{ in } R^n.$$

If we let $r(x) = b - Ax$, then we see that the minimum of the function $\|r(x)\|$ occurs at $\bar{x}$; that is, $\|r(\bar{x})\| \leq \|r(x)\|$ for all x in R^n. Let $(Ax)_i$ denote the ith component of the vector Ax and $r_i(x)$ denote the ith component of the vector $r(x)$. Then we have $r_i(x) = b_i - (Ax)_i$. Since

$$\|r(x)\|^2 = r_1(x)^2 + \cdots + r_m(x)^2,$$

and since $r_i(x)$ measures the error in the ith component, the vector $\bar{x}$ has the property that the sum of the squares of the errors $r_i(\bar{x})$ in each of the components is as small as possible. For this reason $\bar{x}$ is called a least-squares solution.

We have two equivalent ways of viewing a least-squares solution to an inconsistent system. From the geometric point of view, a least-squares solution to an inconsistent system $Ax = b$ is a vector $\bar{x}$ with the property that $p = A\bar{x}$ is the projection of b onto the column space of A. From the analytic point of view, a least-squares solution to an inconsistent system $Ax = b$ is a vector $\bar{x}$ with the property that the sum of the squares of the errors in each of the components of $b - A\bar{x}$ is as small as possible.

EXAMPLE 3   In the examples in Section 3.1 we considered the problem of passing a straight line $y = a + bx$ through a collection of points (x_1, y_1), $\ldots, (x_m, y_m)$ with $x_1 < x_2 < \cdots < x_m$. In the unlikely event that all these points lie on a straight line, the answer to our problem is obvious; the straight line that gives the best possible fit is this straight line. When the data points do not lie on a straight line, then we are faced with the inconsistent system of equations

$$Ax = \begin{bmatrix} 1 & x_1 \\ 1 & x_2 \\ \vdots & \vdots \\ 1 & x_m \end{bmatrix} \begin{bmatrix} a \\ b \end{bmatrix} = \begin{bmatrix} y_1 \\ y_2 \\ \vdots \\ y_m \end{bmatrix}.$$

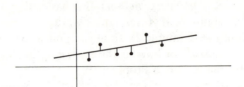

Figure 3.14

Let $\bar{x} = (\bar{a}, \bar{b})$ be the least-squares solution to this problem. Then $A\bar{x}$ is the vector in the column space of A which is nearest to b and the line $y = \bar{a} + \bar{b}x$ gives the best possible fit to the data in the sense of least squares. It is called the *regression line* in economics and statistics. From our discussion above,

$$r(a, b) = (y_1 - (a + bx_1), \ldots, y_m - (a + bx_m))$$

and the least-squares solution has the property that the sum of the squares of the errors

$$r_i(\bar{x}) = y_i - (\bar{a} + \bar{b}x_i)$$

in each of the components is as small as possible. The errors $r_i(\bar{x})$ are the differences between the observed values y_i and the corresponding values of y that lie on the line. In Figure 3.14 these errors are represented by the vertical-line segments joining the points to the line. The regression line has the property that the sum of the squares of the length of each of these segments is as small as possible.

EXAMPLE 4   In Example 2 of Section 3.1 we supposed that we had good reason to believe that the object was traveling with a constant velocity and hence that the variables satisfied a linear relationship $y = y_0 + vt$. Finding the coefficients of the best straight-line fit to the data will give us an estimate for the velocity of the object. It will also give us a way of predicting the position of the object at future times.

Suppose now that we do not know what relationship (if any) exists between the variables with which we are concerned. For example, we may be looking for a relationship between the price of a product and the amount sold, or high school grade-point average and college grade-point average, or the amount of fertilizer applied and the yield of a crop, or smoking and cancer. The usual procedure in this situation is to collect data and plot them. We hope that the graph of the data will suggest a relationship between the variables. For example, the graph in Figure 3.15(a) suggests that the variables are in a linear relationship $y = a + bx$ and the graph in Figure 3.15(b) suggests a quadratic relationship $y = a + bx + cx^2$. Figure 3.15(c), on the other hand, suggests that there is no relationship between the

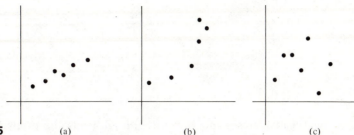

Figure 3.15   (a)   (b)   (c)

variables. In the first two cases, we are led to the inconsistent systems

$$
\begin{bmatrix} 1 & x_1 \\ 1 & x_2 \\ \vdots & \vdots \\ 1 & x_m \end{bmatrix} \begin{bmatrix} a \\ b \end{bmatrix} = \begin{bmatrix} y_1 \\ y_2 \\ \vdots \\ y_m \end{bmatrix} \quad \text{and} \quad \begin{bmatrix} 1 & x_1 & x_1^2 \\ 1 & x_2 & x_2^2 \\ \vdots & \vdots & \vdots \\ 1 & x_m & x_m^2 \end{bmatrix} \begin{bmatrix} a \\ b \\ c \end{bmatrix} = \begin{bmatrix} y_1 \\ y_2 \\ \vdots \\ y_m \end{bmatrix},
$$

respectively. The least-squares solutions to these inconsistent systems will give us the coefficients of the best possible linear or quadratic fit to the data. In the third case, we might decide that even though the variables appear to be unrelated, we want to find the *best* possible linear (or quadratic,...) fit to the data. Again, we will generate an inconsistent system of equations.

EXAMPLE 5   *Multiple regression* involves finding the linear equation
$$y = a_0 + a_1 x_1 + \cdots + a_n x_n$$
which gives the best fit (in the sense of least squares) to the data
$$(x_{11}, x_{12}, \ldots, x_{1n}, y_1), \ldots, (x_{m1}, x_{m2}, \ldots, x_{mn}, y_m).$$
This problem is equivalent to finding the least-squares solution $\bar{x} = (\bar{a}_0, \bar{a}_1, \ldots, \bar{a}_n)$ to the (usually) inconsistent system $Ax = b$, where

$$
A = \begin{bmatrix} 1 & x_{11} & \cdots & x_{1n} \\ 1 & x_{21} & \cdots & x_{2n} \\ \vdots & \vdots & & \vdots \\ 1 & x_{m1} & \cdots & x_{mn} \end{bmatrix}, \quad x = \begin{bmatrix} a_0 \\ a_1 \\ \vdots \\ a_n \end{bmatrix}, \quad \text{and} \quad b = \begin{bmatrix} y_1 \\ y_2 \\ \vdots \\ y_m \end{bmatrix}.
$$

In the case where y depends only on two parameters, x_1 and x_2, that is,
$$y = a_0 + a_1 x_1 + a_2 x_2,$$
the problem becomes one of finding the plane in R^3 that best fits the data $(x_{11}, x_{12}, y_1), \ldots, (x_{m1}, x_{m2}, y_m)$. In this case,
$$r(x) = (y_1 - (a_0 + a_1 x_{11} + a_2 x_{12}), \ldots, y_m - (a_0 + a_1 x_{m1} + a_2 x_{m2}))$$

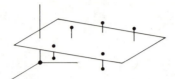

Figure 3.16

and the least-squares solution $\bar{x} = (\bar{a}_0, \bar{a}_1, \bar{a}_2)$ has the property that the sum of the squares of the errors

$$r_i(\bar{x}) = y_i - (\bar{a}_0 + \bar{a}_1 x_{i1} + \bar{a}_2 x_{i2})$$

in each of the components is a minimum. In Figure 3.16 these errors are represented by the vertical line segments joining the points to the plane.

PROBLEMS 3.6

1. Find a least-squares solution to each of the following inconsistent systems of equations.

   *(a) $x_1 + x_2 = 1$
          $x_1 - x_2 = 0$
          $2x_1 \qquad = 2$

   *(b) $x_1 + 2x_2 \qquad\quad = 0$
          $2x_1 + 3x_2 + 2x_3 = 1$
          $x_1 + 3x_2 - 2x_3 = 1$
          $\qquad x_2 - 2x_3 = 0$

   (c) $-x_1 + 2x_2 - x_3 = 1$
         $2x_1 - 4x_2 + 3x_3 = 1$
         $\qquad x_2 + x_3 = 2$
         $2x_1 - 4x_2 + x_3 = 1$

   (d) $x_1 \qquad\quad + 2x_3 = -1$
         $\qquad x_2 - 2x_3 = -2$
         $x_1 + x_2 + x_3 = 1$
         $x_1 + 2x_2 - 2x_3 = 1$

2. Find the best fit by a line of the form $y = a$ to the data

t	1	2	3	4	5
y	7	4	5	3	8

*3. Find the best fit to the data

t	-1	0	1	2
y	1	1	4	10

   by a quadratic function $y = a + bt + ct^2$.

*4. Find the least-squares solutions to Problems 1, 2, and 3 in Section 3.1.

*5. Find the best fit to the data

x_1	0	1	1	0	-1
x_2	0	1	0	1	1
y	2	3	-4	1	1

by a linear equation of the form:

(a) $y = a_0 + a_1 x_1 + a_2 x_2$

(b) $y = a_1 x_1 + a_2 x_2$

6. By solving the normal equations, show that the coefficients $(\bar{a}, \bar{b})$ of the least-squares line $y = \bar{a} + \bar{b}x$ for the data $(x_1, y_1), \ldots, (x_m, y_m)$ are given by

$$\bar{b} = \frac{(x_1 y_1 + \cdots + x_m y_m) - m\bar{x}\bar{y}}{(x_1^2 + \cdots + x_m^2) - m\bar{x}^2}$$

$$\bar{a} = \bar{y} - \bar{b}\bar{x},$$

where $\bar{x} = (1/m)(x_1 + \cdots + x_m)$ and $\bar{y} = (1/m)(y_1 + \cdots + y_m)$ are the averages of the x_i's and y_i's, respectively. Therefore, the least-squares line is

$$y = \bar{y} + \bar{b}(x - \bar{x}).$$

7. Using calculus, show that the minimum of the expression

$$r(a, b) = (y_1 - a - bx_1)^2 + \cdots + (y_m - a - bx_m)^2$$

is $(\bar{a}, \bar{b})$, where $\bar{a}$ and $\bar{b}$ are given in Problem 6.

*8. Suppose that a plot of the data $(x_1, y_1), \ldots, (x_m, y_m)$ indicates that the data can be best represented by an exponential curve of the form

$$y = \alpha \beta^x.$$

Taking logarithms (to the base e) of both sides, we obtain

$$\log y = \log \alpha + (\log \beta)x.$$

This is a linear equation of the form

$$z = a + bx,$$

where $z = \log y$, $a = \log \alpha$, and $b = \log \beta$. If we find the coefficients $\bar{a}$ and $\bar{b}$ of the regression line for the data $(x_1, z_1), \ldots, (x_m, z_m)$, where $z_i = \log y_i$, then we can determine α and β from the equations

$$\alpha = e^{\bar{a}} \quad \text{and} \quad \beta = e^{\bar{b}}.$$

Carry out this procedure for the following data.

x	0	1	2
y	e^2	e^5	e^{13}

9. The Malthusian law of population growth predicts that under ideal conditions the population N at time t is

$$N = N_0 e^{\lambda t}$$

where N_0 is the population at time $t = 0$ and λ is a constant (called the population growth rate). Suppose that a population begins with 6 members and that observations are made to collect the following data:

t	3	5	6
N	15	25	40

(a) What is the population growth rate?

(b) What is the predicted population when $t = 8$?

10. If A is an $m \times n$ matrix with rank less than n, then the normal equations $A^T A \bar{x} = A^T b$ do not have a unique solution. But all of the solutions $\bar{x}$ have the property that $p = A\bar{x}$ is the projection of b onto the column space of A.

(a) Use Theorem 3.3 and Problem 8 of Section 3.4 to show that any solution $\bar{x}$ can be decomposed into

$$\bar{x} = \bar{x}_1 + \bar{x}_2$$

where $\bar{x}_1$ is in the null space of A and $\bar{x}_2$ is in the row space of A.

(b) Use Problem 24 of Section 3.5 to show that for any m-vector b there is a unique vector $\bar{x}_2$ in the row space of A such that $A\bar{x}_2$ is the projection of b onto the column space of A.

(c) Show that if $\bar{y}$ is any other solution to the normal equations $A^T A \bar{x} = A^T b$, then $\| \bar{x}_2 \| < \| \bar{y} \|$. Thus among all solutions to the normal equations there is one with smallest length. This solution is called the optimal solution.

3.7 ORTHOGONAL BASES AND PROJECTIONS

In Section 3.5 we saw that if the columns of the matrix A form a basis for a subspace V (and not simply a spanning set for V), then the normal equations $A^T A \bar{x} = A^T b$ have a unique solution. In this section we investigate what follows from the assumption that the columns of A not only form a basis for V but form an orthonormal basis for V. This naturally leads us to investigate the possibility of constructing an orthonormal basis for a vector space.

Suppose that V is a subspace of R^m and that $w_1, \ldots, w_n$ is an orthonormal basis for V. As always, we let A be the matrix whose columns are these basis vectors. Since the columns of A are independent, given any m-vector b the normal equations

$$A^T A \bar{x} = A^T b \tag{1}$$

have a unique solution. This much we already know. But the columns of A are not only independent vectors, they are also orthonormal vectors. Hence

$$A^T A = \begin{bmatrix} -w_1- \\ \vdots \\ -w_n- \end{bmatrix} \begin{bmatrix} | & & | \\ w_1 & \cdots & w_n \\ | & & | \end{bmatrix} = I,$$

and the normal equations reduce to

$$\bar{x} = A^T b.$$

Thus the normal equations not only have a unique solution, they have a solution that is trivial to find. It is

$$\bar{x} = A^T b = \begin{bmatrix} \langle w_1, b \rangle \\ \vdots \\ \langle w_n, b \rangle \end{bmatrix}.$$

Therefore, the projection of an m-vector b onto V is given by

$$p = A\bar{x} = \langle w_1, b \rangle w_1 + \cdots + \langle w_n, b \rangle w_n. \tag{2}$$

We recognize the terms $\langle w_i, b \rangle w_i$ in the sum on the right-hand side of this equation as the projections of b onto the unit vectors w_i. Thus p is the sum of the projections of b onto the orthonormal basis $w_1, \ldots, w_n$. In particular, if b happens to be a vector in the subspace V, then $p = b$ and (2) gives the expansion of b in terms of this orthonormal basis for V. We summarize our findings in the following theorem.

Theorem 3.4   *Let $w_1, \ldots, w_n$ be an orthonormal basis for a subspace V of R^m, let b be a vector in R^m, and let A be the matrix whose columns are $w_1, \ldots, w_n$.*

(a) *The projection of b onto V is*

$$p = A\bar{x} = \langle w_1, b \rangle w_1 + \cdots + \langle w_n, b \rangle w_n.$$

   Thus p is the sum of the projections of b onto the orthonormal vectors $w_1, \ldots, w_n$.

(b) *If b is a vector in the subspace V, then the projection of b onto V is b itself. Hence the expansion of b in terms of the orthonormal basis $w_1, \ldots, w_n$ is given by*

$$b = \langle w_1, b \rangle w_1 + \cdots + \langle w_n, b \rangle w_n.$$

   That is, b is the sum of its projections onto each vector in the orthonormal basis.

Note that the formulas in (a) and (b) are exactly the same. If b is not in the subspace V, then the formula gives the expansion of the projection of b onto V in terms of the orthonormal basis. If b is in the subspace, then the formula gives the expansion of b itself in terms of the orthonormal basis.

EXAMPLE 1   Suppose that V is a two-dimensional vector space and that w_1 and w_2 form an orthonormal basis for V. If b is any vector in V, it is geometrically obvious [see Figure 3.17(a)] that the expansion of b in terms of this basis is given by the formula

$$b = (\text{proj of } b \text{ onto } w_1) + (\text{proj of } b \text{ onto } w_2)$$
$$= \langle w_1, b \rangle w_1 + \langle w_2, b \rangle w_2.$$

Similarly, if b is a vector not in V, then the projection of b onto the subspace

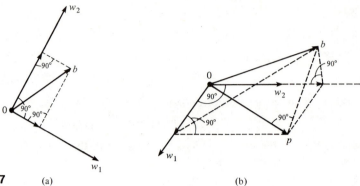

Figure 3.17   (a)   (b)

V is obviously given by [Figure 3.17(b)]

$$p = \text{proj of } b \text{ onto } V$$
$$= (\text{proj of } b \text{ onto } w_1) + (\text{proj of } b \text{ onto } w_2)$$
$$= \langle w_1, b \rangle w_1 + \langle w_2, b \rangle w_2.$$

The only difference between an orthonormal basis and an orthogonal basis is that in the former all vectors have unit length. The vectors in an orthogonal basis can always be normalized to form an orthonormal basis by dividing each of the vectors by its length. However, it is often easier to perform hand calculations with an orthogonal basis than an orthonormal basis, especially if the vectors have integer components. In this case the components of orthonormal vectors usually involve square roots which result from the division of the vector by its length. For example, the vectors

$$(1, 1, 0), \quad (1, -1, 1), \quad \text{and} \quad (-1, 1, 2)$$

form an orthogonal basis for R^3. If we normalize these vectors to obtain an orthonormal basis, we obtain

$$\left(1/\sqrt{2}, 1/\sqrt{2}, 0\right), \left(1/\sqrt{3}, -1/\sqrt{3}, 1/\sqrt{3}\right), \text{ and } \left(-1/\sqrt{6}, 1/\sqrt{6}, 2/\sqrt{6}\right).$$

What happens to the results in Theorem 3.4 if we replace the orthonormal basis $w_1, \ldots, w_n$ by an orthogonal basis $v_1, \ldots, v_n$? Very little. The columns of A become $v_1, \ldots, v_n$. Since they are orthogonal,

$$A^TA = \begin{bmatrix} -v_1- \\ \vdots \\ -v_n- \end{bmatrix} \begin{bmatrix} | & & | \\ v_1 & \cdots & v_n \\ | & & | \end{bmatrix} = \begin{bmatrix} \langle v_1, v_1 \rangle & & \\ & \ddots & \\ & & \langle v_n, v_n \rangle \end{bmatrix}.$$

Thus, although A^TA is no longer the identity matrix, it is a diagonal matrix. Since

$$A^Tb = \begin{bmatrix} -v_1- \\ \vdots \\ -v_n- \end{bmatrix} b = \begin{bmatrix} \langle v_1, b \rangle \\ \vdots \\ \langle v_n, b \rangle \end{bmatrix},$$

the normal equations $A^T A \bar{x} = A^T b$ become

$$\begin{bmatrix} \langle v_1, v_1 \rangle & & \\ & \ddots & \\ & & \langle v_n, v_n \rangle \end{bmatrix} \bar{x} = \begin{bmatrix} \langle v_1, b \rangle \\ \vdots \\ \langle v_n, b \rangle \end{bmatrix}.$$

Hence

$$\bar{x} = \begin{bmatrix} \langle v_1, b \rangle / \langle v_1, v_1 \rangle \\ \vdots \\ \langle v_n, b \rangle / \langle v_n, v_n \rangle \end{bmatrix}.$$

Therefore,

$$p = A\bar{x} = \begin{bmatrix} | & & | \\ v_1 & \cdots & v_n \\ | & & | \end{bmatrix} \begin{bmatrix} \langle v_1, b \rangle / \langle v_1, v_1 \rangle \\ \vdots \\ \langle v_n, b \rangle / \langle v_n, v_n \rangle \end{bmatrix}$$

$$= \frac{\langle v_1, b \rangle}{\langle v_1, v_1 \rangle} v_1 + \cdots + \frac{\langle v_n, b \rangle}{\langle v_n, v_n \rangle} v_n.$$

Thus when we have an orthogonal basis for V, parts (a) and (b) of Theorem 3.4 become:

(a') *The projection of b onto V is*

$$p = A\bar{x} = \frac{\langle v_1, b \rangle}{\langle v_1, v_1 \rangle} v_1 + \cdots + \frac{\langle v_n, b \rangle}{\langle v_n, v_n \rangle} v_n.$$

Thus p is the sum of the projections of b onto the orthogonal vectors $v_1, \ldots, v_n$.

(b') *If b is a vector in the subspace V, then the expansion of b in terms of the orthogonal basis $v_1, \ldots, v_n$ is given by*

$$b = A\bar{x} = \frac{\langle v_1, b \rangle}{\langle v_1, v_1 \rangle} v_1 + \cdots + \frac{\langle v_n, b \rangle}{\langle v_n, v_n \rangle} v_n.$$

Thus b is the sum of its projections onto each vector in the orthogonal basis.

If we now normalize the v's to obtain an orthonormal basis

$$w_1 = \frac{v_1}{\|v_1\|}, \ldots, w_n = \frac{v_n}{\|v_n\|},$$

then (a') and (b') reduce to (a) and (b) of Theorem 3.4. Thus given an orthogonal basis, it makes no difference whether we first normalize it and then compute the projection or whether we first compute the projection and then normalize. The result is exactly the same.

EXAMPLE 2   The three vectors $(1, 1, 0)$, $(1, -1, 1)$, and $(-1, 1, 2)$ are orthogonal vectors in R^3. Hence they form a basis for R^3, and any 3-vector $x = (x_1, x_2, x_3)$ is given (uniquely) by the sum of its projections onto these vectors.

$$x = (x_1, x_2, x_3) = \frac{x_1 + x_2}{2}(1, 1, 0) + \frac{x_1 - x_2 + x_3}{3}(1, -1, 1)$$
$$+ \frac{(-x_1 + x_2 + 2x_3)}{6}(-1, 1, 2).$$

If we had first normalized the vectors to obtain an orthonormal basis, our result would have been

$$x = (x_1, x_2, x_3) = \frac{x_1 + x_2}{\sqrt{2}}\left(1/\sqrt{2}, 1/\sqrt{2}, 0\right)$$
$$+ \frac{x_1 - x_2 + x_3}{\sqrt{3}}\left(1/\sqrt{3}, -1/\sqrt{3}, 1/\sqrt{3}\right)$$
$$+ \frac{-x_1 + x_2 + 2x_3}{\sqrt{6}}\left(-1/\sqrt{6}, 1/\sqrt{6}, 2/\sqrt{6}\right).$$

Of course, these two expressions for x are the same.

EXERCISE 1   Express $(1, 2, 3)$ as a linear combination of the vectors $(1, 1, 1)$, $(2, -3, 1)$, and $(4, 1, -5)$.

EXAMPLE 3   Let V be the subspace of R^3 spanned by the two vectors $(1, 1, 0)$ and $(1, -1, 1)$. Since these vectors are orthogonal, the projection of $(1, 1, 1)$ onto V is

$$p = \frac{2}{2}(1, 1, 0) + \frac{1}{3}(1, -1, 1) = \left(\tfrac{4}{3}, \tfrac{2}{3}, \tfrac{1}{3}\right).$$

EXERCISE 2   Show that the vector $(1, 1, 0, 0)$ is not in the subspace V spanned by $(1, 1, 1, 1)$, $(1, -1, 1, -1)$, and $(1, 2, -1, -2)$, and then find its projection onto V.

EXAMPLE 4   Consider the following inconsistent system of equations

$$\begin{bmatrix} 1 & 4 \\ -1 & 2 \\ 1 & 0 \\ -1 & 2 \end{bmatrix} \begin{bmatrix} x_1 \\ x_2 \end{bmatrix} = \begin{bmatrix} -3 \\ -3 \\ 0 \\ -3 \end{bmatrix}.$$

Since the columns of A are orthogonal, the projection of $(-3, -3, 0, -3)$ onto the column space of A is given by

$$p = \frac{3}{4}(1, -1, 1, -1) + \frac{-24}{24}(4, 2, 0, 2).$$

Hence the least-squares solution to the inconsistent system is $\bar{x} = \left(\tfrac{3}{4}, -1\right)$.

EXERCISE 3 Find a least-squares solution to the inconsistent system

$$\begin{bmatrix} 1 & 1 \\ 1 & -2 \\ 0 & 6 \\ 1 & 1 \end{bmatrix} \begin{bmatrix} x_1 \\ x_2 \end{bmatrix} = \begin{bmatrix} 1 \\ 0 \\ 2 \\ 0 \end{bmatrix}.$$

PROBLEMS 3.7

*1. Find the least-squares solution to the inconsistent system $Ax = b$ if A and b are given by:

(a) $\begin{bmatrix} 1 & 1 \\ 1 & -1 \\ 1 & 0 \end{bmatrix}$ and $\begin{bmatrix} 1 \\ 0 \\ 3 \end{bmatrix}$ (b) $\begin{bmatrix} 1 & 1 \\ -1 & 1 \\ 0 & 2 \end{bmatrix}$ and $\begin{bmatrix} -3 \\ 2 \\ 5 \end{bmatrix}$

(c) $\begin{bmatrix} 1 & 2 \\ -1 & 3 \\ 1 & 1 \end{bmatrix}$ and $\begin{bmatrix} -1 \\ 0 \\ 2 \end{bmatrix}$ (d) $\begin{bmatrix} 1 & -4 \\ -1 & 1 \\ 1 & 5 \end{bmatrix}$ and $\begin{bmatrix} 1 \\ 1 \\ 1 \end{bmatrix}$

(e) $\begin{bmatrix} 4 & 1 \\ 2 & -2 \\ 2 & 2 \\ 1 & -4 \end{bmatrix}$ and $\begin{bmatrix} 0 \\ 1 \\ 0 \\ 1 \end{bmatrix}$ (f) $\begin{bmatrix} -2 & 1 \\ 4 & -1 \\ 1 & 2 \\ 2 & -2 \end{bmatrix}$ and $\begin{bmatrix} 1 \\ 2 \\ 1 \\ 1 \end{bmatrix}$

(g) $\begin{bmatrix} -2 & -2 & 1 \\ -1 & 4 & -2 \\ 4 & 1 & 2 \\ 2 & -2 & -4 \end{bmatrix}$ and $\begin{bmatrix} 1 \\ 0 \\ 0 \\ 0 \end{bmatrix}$

(h) $\begin{bmatrix} 1 & 0 & 0 \\ 0 & 1 & 0 \\ 0 & 0 & 1 \\ 0 & 0 & 1 \end{bmatrix}$ and $\begin{bmatrix} 3 \\ 2 \\ 1 \\ 0 \end{bmatrix}$ (i) $\begin{bmatrix} 1 & -1 & 1 \\ 1 & 1 & -1 \\ 0 & 2 & 1 \\ 0 & 0 & 3 \end{bmatrix}$ and $\begin{bmatrix} 1 \\ 1 \\ 1 \\ 4 \end{bmatrix}$

*2. The vectors $(1, 1, 1, 1)$, $(1, -1, 1, -1)$, and $(1, 2, -1, -2)$ are orthogonal. Express $(6, 5, 0, -7)$ as a linear combination of these vectors.

3. Suppose that we wish to find the best straight-line fit to the data $(t_1, y_1), \ldots, (t_m, y_m)$. Show that if the t-coordinates of the data are symmetric about 0, then the columns of the matrix A are orthogonal.

*4. Find the best straight-line fit to the data

t	-5	-4	-2	2	4	5
y	7	5	1	0	-2	-5

Carefully plot the data and draw the straight line (see Problem 3).

5. Let $w_1, \ldots, w_m$ be an orthonormal basis for R^m. If u and v are in R^m, show that
*(a) $\langle u, v \rangle = \langle u, w_1 \rangle \langle v, w_1 \rangle + \cdots + \langle u, w_m \rangle \langle v, w_m \rangle$
(b) $\| u \|^2 = \langle u, w_1 \rangle^2 + \cdots + \langle u, w_m \rangle^2$

6. Let $w_1, \ldots, w_n$ be an orthonormal basis for a subspace V of R^m, and let A be the matrix whose columns are these orthonormal vectors.

*(a) Show that the projection matrix for V reduces to $P = AA^T$.

(b) Show that $P = w_1w_1^T + \cdots + w_nw_n^T$.

(c) Show that $w_iw_i^T$ is a projection matrix for the projection onto the one-dimensional subspace spanned by w_i.

7. Suppose that $w_1, \ldots, w_k$ are orthonormal vectors in R^m and that for every u in R^m

$$\|u\|^2 = \langle u, w_1 \rangle^2 + \cdots + \langle u, w_k \rangle^2 \quad \text{for} \quad \ldots$$

Show that $w_1, \ldots, w_k$ form an orthonormal basis for ... Let ... be the subspace of R^m spanned by $w_1, \ldots, w_k$ and ... the projection matrix for V. Show that $Pu = u$ for all u in R^m by showing that $\|Pu - u\|^2 = 0$ for all u in R^m.)

3.8 THE GRAM–SCHMIDT PROCESS

In Section 3.7 we saw some of the advantages that result from using an orthogonal basis for a vector space. Such bases substantially simplify the algebraic computations involved in finding the representation of a vector in terms of the basis and the projection of a vector onto a subspace. Such a basis also corresponds to our mental image of the geometry of space. In view of these remarks it is certainly worthwhile to investigate the following two questions:

1. Does a vector space always have an orthogonal basis?
2. If so, is there a reasonable way to construct an orthogonal basis?

The student should realize that an affirmative answer to the second question implies an affirmative answer to the first. Hence we will show that the answer to both questions is affirmative if we exhibit a procedure for constructing an orthogonal basis for a vector space.

Suppose that V is a vector space. We construct an orthogonal basis for V as follows. Choose any nonzero vector v_1 in V. If V has dimension greater than 1, there is a vector u_2 in V that is not in the subspace spanned by v_1. In general, u_2 will not be orthogonal to v_1. Let v_2 be the projection of u_2 onto the orthogonal complement of the subspace spanned by v_1. Then v_1, v_2 are nonzero orthogonal (and hence independent) vectors. Since

$$u_2 = v_2 + \frac{\langle v_1, u_2 \rangle}{\langle v_1, v_1 \rangle} v_1,$$

the vector u_2 is a linear combination of v_1 and v_2. Similarly, v_2 is a linear combination of v_1 and u_2. Thus v_1 and v_2 span the same space as v_1 and u_2. Now either v_1 and v_2 form a basis for V or there is a vector u_3 in V that is not in the subspace spanned by v_1 and v_2. In the latter case let v_3 be the projection of u_3 onto the orthogonal complement of the subspace of V

spanned by v_1 and v_2.

$$v_3 = u_3 - \frac{\langle v_1, u_3 \rangle}{\langle v_1, v_1 \rangle} v_1 - \frac{\langle v_2, u_3 \rangle}{\langle v_2, v_2 \rangle} v_2.$$

Then v_1, v_2, and v_3 are nonzero orthogonal vectors that span the same subspace as v_1, v_2, u_3. Again, either v_1, v_2, and v_3 form a basis for V (in which case we are done) or else there is a vector u_4 in V that is not in the subspace spanned by v_1, v_2, and v_3. In the latter case we let v_4 be the orthogonal complement of the subspace of V

$$v_4 = u_4 - \frac{\langle v_1, u_4 \rangle}{\langle v_1, v_1 \rangle} v_1 - \frac{\langle v_2, u_4 \rangle}{\langle v_2, v_2 \rangle} v_2 - \frac{\langle v_3, u_4 \rangle}{\langle v_3, v_3 \rangle} v_3$$

Then v_1, v_2, v_3, and v_4 are nonzero orthogonal (and hence independent) vectors that span the same subspace as v_1, v_2, v_3, and u_4. Repeating this procedure must eventually produce an orthogonal basis for V.

Theorem 3.5 (*Gram–Schmidt*) *Every subspace V of R^m has an orthogonal basis, and hence an orthonormal basis.*

GRAM–SCHMIDT PROCESS

Given

An n-dimensional subspace V of R^m.

Goal

To produce an orthogonal basis for V.

Procedure

1. Choose any nonzero vector v_1 in V.
2. Choose any vector u_2 in V that is not in the subspace spanned by v_1 and let v_2 be its projection onto the orthogonal complement of this subspace:

$$v_2 = u_2 - \frac{\langle v_1, u_2 \rangle}{\langle v_1, v_1 \rangle} v_1.$$

$\quad\vdots$

n. Choose any vector u_n in V that is not in the subspace spanned by $v_1, \ldots, v_{n-1}$ and let v_n be its projection onto the orthogonal complement of this subspace:

$$v_n = u_n - \frac{\langle v_1, u_n \rangle}{\langle v_1, v_1 \rangle} v_1 - \cdots - \frac{\langle v_{n-1}, u_n \rangle}{\langle v_{n-1}, v_{n-1} \rangle} v_{n-1}.$$

The orthogonal basis for V is $v_1, \ldots, v_n$.

EXAMPLE 1   Let us find an orthogonal basis for R^3 that contains the vector $(1, 1, 1)$. Following the Gram–Schmidt process, we let $v_1 = (1, 1, 1)$ and choose a vector u_2 in R^3 that is not a multiple of v_1, say $u_2 = (1, 0, 0)$. Then

$$v_2 = (1, 0, 0) - \frac{1}{3}(1, 1, 1) = \frac{1}{3}(2, -1, -1).$$

Since any multiple of v_2 is also orthogonal to v_1, we let $v_2 = (2, -1, -1)$. As a final step, we must find a vector u_3 not in the subspace spanned by $(1, 1, 1)$ and $(2, -1, -1)$. Letting A be the matrix whose columns are these vectors and applying Gaussian elimination the augmented matrix $[A : u_3]$, we find that $u_3 = (a, b, c)$ will not be in this subspace if we choose $b \neq c$. Therefore, $u_3 = (1, -1, 1)$ is not in the subspace and we have

$$v_3 = (1, -1, 1) - \frac{1}{3}(1, 1, 1) - \frac{2}{6}(2, -1, -1)$$
$$= (0, -1, 1).$$

Thus $(1, 1, 1)$, $(2, -1, -1)$, and $(0, -1, 1)$ form an orthogonal basis for R^3 containing the vector $(1, 1, 1)$.

EXERCISE 1   Find an orthogonal basis for R^3 containing the vector $(1, 2, 3)$.

The Gram–Schmidt process can also be used to replace any set of independent vectors with a set of orthogonal vectors that span the same space. However, in this case the procedure is somewhat simpler. At the ith stage of the process we do not have to find a vector that is not in the subspace spanned by $v_1, \ldots, v_{i-1}$. The ith basis vector u_i is such a vector. For example, suppose that $u_1, \ldots, u_k$ are k linearly independent vectors in R^n. Let $v_1 = u_1$ and v_2 be the projection of u_2 onto the orthogonal complement of the subspace spanned by u_1.

$$v_2 = u_2 - \frac{\langle v_1, u_2 \rangle}{\langle v_1, v_1 \rangle} v_1$$

Then v_1, v_2 are orthogonal vectors that span the same space as u_1, u_2. Now let v_3 be the projection of u_3 onto the orthogonal complement of the subspace spanned by v_1, v_2.

$$v_3 = u_3 - \frac{\langle v_1, u_3 \rangle}{\langle v_1, v_1 \rangle} v_1 - \frac{\langle v_2, u_3 \rangle}{\langle v_2, v_2 \rangle} v_2$$

Then v_1, v_2, v_3 are orthogonal vectors that span the same space as u_1, u_2, u_3. In general, having found $v_1, \ldots, v_i$, let v_{i+1} be the projection of u_{i+1} onto the orthogonal complement of the subspace spanned by $v_1, \ldots, v_i$.

$$v_{i+1} = u_{i+1} - \frac{\langle v_1, u_{i+1} \rangle}{\langle v_1, v_1 \rangle} v_1 - \cdots - \frac{\langle v_i, u_{i+1} \rangle}{\langle v_i, v_i \rangle} v_i$$

Then $v_1, \ldots, v_{i+1}$ are orthogonal vectors that span the same space as $u_1, \ldots, u_{i+1}$. This proves the following result.

Result 1   Let $u_1, \ldots, u_k$ be independent vectors in R^n. Then there is an orthogonal set of vectors $v_1, \ldots, v_k$ with the property that for each i, $1 \leq i \leq k$, $v_1, \ldots, v_i$ and $u_1, \ldots, u_i$ span the same space.

EXAMPLE 2   The three vectors $(1, 1, 0)$, $(2, 1, 0)$, and $(1, 1, 1)$ form a basis for R^3. If we apply the Gram–Schmidt process to these vectors, we obtain

$$v_1 = (1, 1, 0),$$

$$v_2 = (2, 1, 0) - \frac{3}{2}(1, 1, 0) = \left(\frac{1}{2}, -\frac{1}{2}, 0\right),$$

$$v_3 = (1, 1, 1) - \frac{2}{2}(1, 1, 0) - \frac{0}{1/2}\left(\frac{1}{2}, -\frac{1}{2}, 0\right)$$

$$= (0, 0, 1).$$

EXERCISE 2   Find an orthogonal basis for the subspace V spanned by $(1, -1, 3)$, $(0, 1, 2)$, and $(-1, 2, -1)$.

PROBLEMS 3.8

*1. Find an orthogonal basis of R^4 containing the vectors $(1, 1, 1, 0)$ and $(-1, 1, 0, 1)$.

2. Find an orthogonal basis for the subspace V spanned by the given vectors.
  (a) $(2, 1, 3)$ and $(1, 2, 0)$
  *(b) $(1, 1, 1)$, $(0, 1, 1)$, and $(0, 0, 1)$
  (c) $(1, 1, 0, 0)$, $(1, -2, 1, 0)$, and $(0, 1, 1, 1)$
  *(d) $(-1, 2, 0, 2)$, $(2, -4, 1, -4)$, and $(-1, 3, 1, 1)$
  (e) $(1, 1, 2, 1)$, $(-1, 1, 0, 1)$, and $(2, 1, 1, 0)$

3. Find an orthogonal basis for the solution space of each of the following systems.

  *(a) $x_1 - x_2 + x_3 + 2x_4 = 0$

  (b) $x_1 + 2x_2 - x_3 = 0$
  $x_2 + x_3 = 0$
  $x_1 + x_2 - 2x_3 = 0$

*4. Apply Gram–Schmidt to change $(1, -1, 2)$, $(1, 1, 1)$, and $(2, -1, 1)$ to an orthogonal set of vectors spanning the same subspace of R^3.

5. Show that if the Gram–Schmidt process is applied to the vectors u_1, u_2, u_3, and u_4, where u_4 is a linear combination of u_1, u_2, and u_3, then $v_4 = 0$.

6. Describe what happens when the Gram–Schmidt process is applied to a dependent set of vectors.

7. Let A be an $m \times n$ matrix with independent columns $u_1, u_2, \ldots, u_n$. Apply the Gram–Schmidt process to these vectors to obtain the orthonormal vectors $w_1, w_2, \ldots, w_n$, where

$$v_1 = u_1,$$

$$\vdots$$

$$v_n = u_n - \frac{\langle v_1, u_n \rangle}{\langle v_1, v_1 \rangle} v_1 - \cdots - \frac{\langle v_{n-1}, u_n \rangle}{\langle v_{n-1}, v_{n-1} \rangle} v_{n-1},$$

and

$$w_i = \frac{v_i}{\|v_i\|}, \qquad i = 1, \ldots, n.$$

Let Q be the matrix whose columns are $w_1, \ldots, w_n$.

(a) Show that

$$u_1 = \langle w_1, u_1 \rangle w_1,$$

$$u_2 = \langle w_1, u_2 \rangle w_1 + \langle w_2, u_2 \rangle w_2,$$

$$\vdots =$$

$$u_n = \langle w_1, u_n \rangle w_1 + \cdots + \langle w_n, u_n \rangle w_n.$$

(*Hint:* Apply the Gram–Schmidt process, normalizing each v as it is obtained.)

(b) Find a triangular matrix R such that $A = QR$. This decomposition of the matrix A is called its *QR decomposition*.

(c) Show that R is nonsingular.

(d) Show that if A is square, then Q is an orthogonal matrix.

(e) Show that if A is square, then the normal equations become $R\bar{x} = Q^T b$.

(f) Suppose that $m > n$, so that A is not square. Extend the orthonormal vectors $w_1, \ldots, w_n$ to an orthonormal basis of R^m, and let Q_1 be the matrix whose columns are this orthonormal basis. Thus Q_1 is an orthogonal matrix. Can the matrix R be replaced by a matrix R_1 so that $A = Q_1 R_1$?

8. For each part of Problem 2, let A be the matrix whose columns are the given vectors. Find the QR decomposition of each of the matrices.

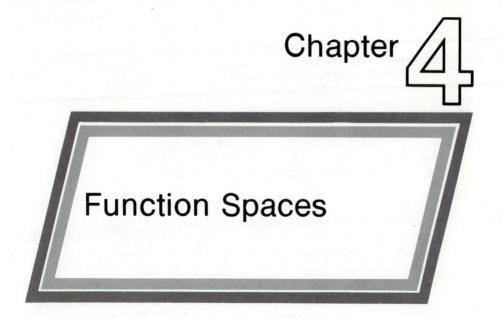

Chapter 4

Function Spaces

In Chapter 2 we defined a vector space as a collection of vectors in R^m that is closed under addition and scalar multiplication. In this chapter we see how to view collections of real-valued functions as vector spaces. This will show that the basic ideas of linear algebra can be applied to objects that are not in R^m.

4.1   FUNCTION SPACES

We begin by defining the sum of two functions, multiplication of a function by a scalar, and the zero function. Let f and g be two functions defined on an interval I and let α be a scalar. Then we define

$$(f + g)(x) = f(x) + g(x) \qquad \text{for all } x \text{ in } I,$$
$$(\alpha f)(x) = \alpha f(x) \qquad \text{for all } x \text{ in } I.$$

Thus the sum $f + g$ of the functions f and g is defined to be the function whose value at any point x in I is given by $f(x) + g(x)$. Similarly, the scalar multiple αf of the function f is the function whose value at any point x in I is given by $\alpha f(x)$. Finally, the zero function, denoted by the symbol 0, is the function whose value at any point x in I is the number 0; that is, $0(x) = 0$ for all x in I.

Definition   A **function space** is a collection of real-valued functions defined on an interval I that is closed under addition and scalar multiplication of

163

functions. (We are assuming that all function spaces are nonempty, that is, that they contain at least one function.)

EXAMPLE 1   We know from calculus that if f and g are continuous functions on an interval I and α is a scalar, then $f + g$ and αf are continuous functions on I. Thus the collection of all functions continuous on an interval I is a function space. When $I = (-\infty, \infty)$ we denote this function space by $C(R)$. When I is any other interval, we denote it by $C(I)$.

All the examples of function spaces in this chapter are *subspaces* of $C(I)$ for some interval I. That is, they are subsets of $C(I)$ which are themselves function spaces. The reader should observe that when I_1 and I_2 are different intervals, the function spaces $C(I_1)$ and $C(I_2)$ are different. (Why?)

EXAMPLE 2   The collection $C^n(I)$, $n \geq 1$, of all functions on an interval I which have n continuous derivatives is a function space. For any n, it is a subspace of $C(I)$. Notice that these spaces are not equal to $C(I)$. For example, if we consider the function $f(x) = |x|$ on the interval $I = (-1, 1)$, then f is in $C(I)$, but since f is not differentiable at $x = 0$, f is not in $C^1(I)$. Thus $C^1(I)$ is a **proper** subspace of $C(I)$, that is, a subspace of $C(I)$ that is not equal to $C(I)$. It is also true that $C^2(I)$ is a proper subspace of $C^1(I)$, and so on.

EXERCISE 1   Let $I = [0, 1]$. Determine which of the following collections of functions is a function space. If the collection fails to be a function space, list the conditions that are not satisfied.

(a)   The collection of all functions f in $C(I)$ such that $f(1/2) = 0$.
(b)   The collection of all functions f in $C(I)$ such that $f(1) = 2$.

EXAMPLE 3   Recall that a *polynomial of degree n* is a function of the form

$$p(x) = a_n x^n + a_{n-1}x^{n-1} + \cdots + a_1 x + a_0 \qquad (a_n \neq 0).$$

The numbers $a_0, a_1, \ldots, a_n$ are called the *coefficients* of $p(x)$. The zero polynomial is not assigned a degree by this definition. We declare it to have degree 0 together with all the other constant polynomials. Since the sum of two polynomials is a polynomial and the product of a scalar and a polynomial is a polynomial, the collection $P(I)$ of all polynomials defined on an interval I is a function space.

EXERCISE 2   Is the collection $P_n(R)$ of all polynomials of degree $\leq n$, where n is a fixed nonnegative integer, a function space?

EXERCISE 3   Is the collection of all polynomials defined on R with degree exactly 5, together with the zero polynomial, a function space?

EXAMPLE 4   A *homogeneous linear differential equation* is an equation of the form

$$y^{(n)} + p_{n-1}(x)y^{(n-1)} + \cdots + p_0(x)y = 0,$$

where $p_0, \ldots, p_{n-1}$ are continuous functions and $y^{(k)}$ denotes the kth derivative of y. [The symbols y', y'', and y''' are usually used to denote $y^{(1)}$, $y^{(2)}$, and $y^{(3)}$.] The number n is called the *order* of the equation. $y' + 2xy = 0$ and $y'' + 2x^2y' + y = 0$ are examples of homogeneous linear differential equations of order 1 and 2, respectively. By a *solution* to a differential equation we mean a function $y = y(x)$ which when substituted into the equation makes it an identity. For example, $y = \sin x$ is a solution to the second-order equation $y'' + y = 0$. So is $y = \cos x$. So are $y = \alpha \sin x$ and $y = \sin x + \cos x$. (Check this!) In fact, the set of all solutions to this differential equation is a function space called the *solution space*. For suppose that y_1 and y_2 are both solutions and α is a scalar. Then

$$(y_1 + y_2)'' + (y_1 + y_2) = (y_1'' + y_1) + (y_2'' + y_2) = 0 \qquad \text{and}$$
$$(\alpha y_1)'' + \alpha y_1 = \alpha(y_1'' + y_1) = \alpha 0 = 0.$$

Thus $y_1 + y_2$ and αy_1 are solutions, as asserted.

EXERCISE 4   Show that the set of all solutions to a homogeneous linear differential equation of order 3 is a function space.

EXAMPLE 5   Consider the first-order differential equation $y' - \lambda y = 0$. A solution to this equation is a function $y = y(x)$ whose derivative is λ times y. The function $y = Ce^{\lambda x}$ has this property for any constant C. Indeed, if

$$y = Ce^{\lambda x},$$

then

$$y' = \lambda Ce^{\lambda x} = \lambda y.$$

These exponential functions are the only nonzero functions with this property, and hence they are the only solutions to the differential equation. To prove this, suppose that y is a solution of $y' - \lambda y = 0$. We assert that $ye^{-\lambda x}$ is a constant. Let

$$z = ye^{-\lambda x}.$$

Then

$$\begin{aligned} z' &= y'e^{-\lambda x} - \lambda ye^{-\lambda x} \\ &= \lambda ye^{-\lambda x} - \lambda ye^{-\lambda x} \\ &= 0. \end{aligned}$$

Since the derivative of z is the zero function, z is a constant function. To determine the constant, we observe that $z(0) = y(0)$. Denoting $y(0)$ by C, it follows that $z(x) = C$ for all x and hence that $y = Ce^{\lambda x}$.

PROBLEMS 4.1

*1. Determine whether the collection of all functions f in $C(R)$ that satisfy the stated condition is a function space. If the collection fails to be a function space, list those conditions that are not satisfied.
(a) $f(4) = 0$
(b) $f(4) = 1$
(c) $f(4) = f(1)$
(d) $2f(1) + f(-1) = 0$
(e) f is an *odd* function; that is, $f(-x) = -f(x)$ for all x
(f) f is an *even* function; that is, $f(x) = f(-x)$ for all x
(g) f is a nonnegative function; that is, $f(x) \geq 0$ for all x

*2. Determine whether the collection of all polynomials in $C(R)$ that satisfy the stated condition is a function space. If the collection fails to be a function space, list those conditions that are not satisfied.
(a) All polynomials with integer coefficients
(b) All polynomials with zero constant term
(c) All polynomials with nonzero constant term
(d) All polynomials with leading coefficient 1
(e) All polynomials containing only odd powers of x
(f) All polynomials containing only even powers of x

3. Show that for any interval I, $P(I)$ is a subspace of $C(I)$.

4. Show that $P_n(I)$ is a subspace of $P(I)$ for any n (see Exercise 2).

*5. Let $I = [0, 1]$. Determine which of the following collections of functions are function spaces. If the collection fails to be a function space, list the conditions that are not satisfied.
(a) The collection of all functions f in $C(I)$ such that $f(0) = f(1)$
(b) The collection of all functions f in $C^2(I)$ such that $f''(1/2) = 0$
(c) The collection of all functions f in $C^2(I)$ such that $f''(1/2) = 1$

*6. For each of the following differential equations, determine whether the set of its solutions is a function space. If it fails to be a function space, list those conditions that are not satisfied.
(a) $y'' + 2y' + y = 0$
(b) $y'' + xy' + 7y = 0$
(c) $y''' + xy'' + x^2 y' + x^3 y = 0$, $y(1) = 0$
(d) $y''' + xy'' + x^2 y' + x^3 y = 0$, $y(1) = 2$

4.2 CONCEPTS OF LINEAR ALGEBRA IN FUNCTION SPACES

Thus far we have seen that function spaces are defined in the same way vector spaces are defined. They are both collections of objects, called vectors, which are closed under the operations of addition and scalar multiplication. The only difference is that in one case the vectors are

n-tuples of real numbers and in the other case the vectors are functions. Hence it is not surprising that the concepts of linear combination, spanning, linear independence, dimension, and basis in a function space are defined exactly the same way in which they are defined in R^m.

For example, if $f_1, \ldots, f_n$ and f are functions defined on an interval I, then f is a linear combination of $f_1, \ldots, f_n$ if there exist scalars $\alpha_1, \ldots, \alpha_n$ such that

$$f = \alpha_1 f_1 + \cdots + \alpha_n f_n. \qquad (1)$$

We must be careful to understand exactly what equation (1) means. Two functions that are defined on the same interval are *equal* if they produce the same value at every point in the interval. In other words, two functions are equal if they are the same function (even though they may be written differently). Thus equation (1) means that

$$f(x) = \alpha_1 f_1(x) + \cdots + \alpha_n f_n(x) \qquad \text{for } all \ x \text{ in } I.$$

EXAMPLE 1   Suppose that $f(x) = x^2 - 6$, $f_1(x) = (x-1)^2$, $f_2(x) = x - 1$, and $f_3(x) = 1$ for all x. Then $f = f_1 + 2f_2 - 5f_3$; that is, $f(x) = f_1(x) + 2f_2(x) - 5f_3(x)$ for all x.

EXAMPLE 2   Let $f(x) = 1$, $g(x) = \sin^2 x$, and $h(x) = \cos^2 x$ for all x. Then $f = g + h$.

As another example, consider the definition of linear independence in the function space setting. Let $f_1, \ldots, f_n$ be functions defined on an interval I. They are called **linearly independent** if the only solution of the equation

$$\alpha_1 f_1 + \cdots + \alpha_n f_n = 0 \qquad \text{is} \qquad \alpha_1 = \cdots = \alpha_n = 0.$$

The equation $\alpha_1 f_1 + \cdots + \alpha_n f_n = 0$ is a special case of equation (1). Thus the equation $\alpha_1 f_1 + \cdots + \alpha_n f_n = 0$ (the zero function) means that $\alpha_1 f_1(x) + \cdots + \alpha_n f_n(x) = 0$ (the number zero) for all x in I.

The importance of the phrase "for all x in I" cannot be overemphasized. The function $f(x) = x^2 - 1$ and the zero function have the same value when x equals 1 or -1, but they are certainly not the same function. Two functions may have the same value at one or many points in the interval I without being equal. To be equal (to be the same function), they must have the same value at *every* point of I.

EXAMPLE 3   Consider the polynomials 1, x, and x^2 as vectors in $C(R)$. We assert that these polynomials are independent. For suppose that

$$0 = \alpha_0 + \alpha_1 x + \alpha_2 x^2 \qquad \text{for all } x. \qquad (2)$$

If we choose three distinct numbers in I, say -1, 0, and 1, and set $x = -1$,

0, and 1, we obtain three equations in three unknowns.

$$\begin{aligned} \alpha_0 - \alpha_1 + \alpha_2 &= 0 \\ \alpha_0 &= 0 \\ \alpha_0 + \alpha_1 + \alpha_2 &= 0 \end{aligned} \quad \text{or} \quad \begin{bmatrix} 1 & -1 & 1 \\ 1 & 0 & 0 \\ 1 & 1 & 1 \end{bmatrix} \begin{bmatrix} \alpha_0 \\ \alpha_1 \\ \alpha_2 \end{bmatrix} = \begin{bmatrix} 0 \\ 0 \\ 0 \end{bmatrix}$$

The only solution of this system is $\alpha_0 = \alpha_1 = \alpha_2 = 0$. Thus the functions 1, x, and x^2 are independent.

There is another useful technique for showing that a collection of functions is independent. Setting $x = 0$ in (2), we obtain $\alpha_0 = 0$. Differentiating both sides of (2) we obtain $0 = \alpha_1 + 2\alpha_2 x$ for all x in I, and setting $x = 0$ in this equation gives $\alpha_1 = 0$. Differentiating again, we obtain $2\alpha_2 = 0$ or $\alpha_2 = 0$. Thus $\alpha_0 = \alpha_1 = \alpha_2 = 0$.

This technique gives us another occasion to stress the importance of understanding equation (2). If two functions are equal, they must have the same derivative. The equation

$$0 = \alpha_0 + \alpha_1 x + \alpha_2 x^2 \qquad \text{for all } x \text{ in } I$$

states that the function $\alpha_0 + \alpha_1 x + \alpha_2 x^2$ and the zero function are equal *as functions*. Thus it is legitimate to differentiate both sides to obtain $0 = \alpha_1 + 2\alpha_2 x$ for all x in I. Contrast this with the following situation. We cannot differentiate both sides of the equation $x^2 - 1 = 0$. For if we do, we obtain $2x = 0$ and hence $x = 0$, an absurdity. The reason we cannot differentiate both sides of the equation is that it is not an identity in x; it does not assert that the function $x^2 - 1$ and the zero function are *equal functions*.

EXERCISE 1 Show that the polynomials 1, x, x^2, and x^3 are independent in $C(R)$. If you are confident that you could also show that 1, x, x^2, x^3, and x^4 are independent, stop. Otherwise, do it.

It should be clear that the same reasoning could be used to show that for any n the polynomials $1, x, \ldots, x^n$ are independent in $C(R)$.

EXAMPLE 4 The functions e^x and e^{2x} are independent in $C(R)$. For suppose that

$$\alpha_1 e^x + \alpha_2 e^{2x} = 0 \qquad \text{for all } x. \tag{3}$$

Setting $x = 0$, we obtain $\alpha_1 + \alpha_2 = 0$. Differentiating both sides of (3), we obtain $\alpha_1 e^x + 2\alpha_2 e^{2x} = 0$ for all x and setting $x = 0$ in this equation gives $\alpha_1 + 2\alpha_2 = 0$. Thus

$$\begin{aligned} \alpha_1 + \alpha_2 &= 0 \\ \alpha_1 + 2\alpha_2 &= 0 \end{aligned} \quad \text{or} \quad \begin{bmatrix} 1 & 1 \\ 1 & 2 \end{bmatrix} \begin{bmatrix} \alpha_1 \\ \alpha_2 \end{bmatrix} = \begin{bmatrix} 0 \\ 0 \end{bmatrix}.$$

The only solution of this system is $\alpha_1 = \alpha_2 = 0$.

EXERCISE 2 Show that e^x, e^{2x}, and e^{3x} are independent in $C(R)$.

Example 4 and Exercise 2 suggest that if $a_1, a_2, \ldots, a_n$ are distinct numbers, then $e^{a_1 x}, e^{a_2 x}, \ldots, e^{a_n x}$ are independent in $C(R)$. This is true (see Problem 6).

EXAMPLE 5   The method for showing the independence of functions based on differentiation works even if 0 is not in the interval. To prove that the functions 1, x, and x^2 are independent on $I = [1, 4]$, we proceed as follows. Suppose that

$$\alpha_0 + \alpha_1 x + \alpha_2 x^2 = 0 \qquad \text{for all } x \text{ in } I. \tag{4}$$

Setting $x = 3$ we obtain the equation $\alpha_0 + 3\alpha_1 + 9\alpha_2 = 0$. Again, differentiating both sides of (4) gives $\alpha_1 + 2\alpha_2 x = 0$ for all x in I and setting $x = 3$ we obtain $\alpha_1 + 6\alpha_2 = 0$. A final differentiation gives $2\alpha_2 = 0$. Thus we have the system of equations

$$
\begin{aligned}
\alpha_0 + 3\alpha_1 + 9\alpha_2 &= 0 \\
\alpha_1 + 6\alpha_2 &= 0 \\
2\alpha_2 &= 0
\end{aligned}
\qquad \text{or} \qquad
\begin{bmatrix} 1 & 3 & 9 \\ 0 & 1 & 6 \\ 0 & 0 & 2 \end{bmatrix}
\begin{bmatrix} \alpha_0 \\ \alpha_1 \\ \alpha_2 \end{bmatrix}
=
\begin{bmatrix} 0 \\ 0 \\ 0 \end{bmatrix}.
$$

Hence $\alpha_0 = \alpha_1 = \alpha_2 = 0$.

EXAMPLE 6   The polynomials 1, x, and x^2 clearly span the space $P_2(R)$ of polynomials of degree ≤ 2. In Example 3 we showed that they are independent. Hence they form a basis and dim $P_2(R) = 3$.

EXERCISE 3   Show that 1, x, x^2, and x^3 form a basis for $P_3(R)$. If you are confident that you could also show that 1, x, x^2, x^3, and x^4 form a basis for $P_4(R)$, stop. Otherwise, do it.

Example 6 and Exercise 3 should make it clear that for any n, the polynomials 1, $x, \ldots, x^n$ form a basis for $P_n(R)$. Hence dim $P_n(R) = n + 1$ (see Problem 7).

Now consider the space $P(R)$ of all polynomials. For every n, $P_n(R)$ is a subspace of $P(R)$ of dimension $n + 1$. This implies that the dimension of $P(R)$ (if it is defined) would have to be greater than or equal to $n + 1$ for every n. Clearly, no finite number can satisfy this condition. Hence we say that the dimension of $P(R)$ is infinite.

We can also prove that the dimension of $P(R)$ is not finite by showing that no finite set of polynomials can possibly span the collection of all polynomials. To prove this, we note that any finite set of polynomials must contain a polynomial of largest degree, say k. It follows that no polynomial of degree larger than k can be a linear combination of these polynomials. Hence $P(R)$ cannot have a finite basis, that is, a basis with a finite number of elements.

Definition   If a function space V has a basis consisting of a finite number of functions, we say that V is **finite-dimensional**. If a function space V does not have a basis consisting of a finite number of functions, we say that V is **infinite-dimensional**.

EXERCISE 4   Show that $C(R)$ is infinite-dimensional.

PROBLEMS 4.2

*1. Determine which of the following collections of polynomials are independent in $P(R)$.
(a) $1, 1 - t, 1 + t$ (b) $t, t^2 + 1, t^2 - 1$
(c) $1, 2t, 1 + t, 3t^2, 4t^3$ (d) $1 + t, 1 + t + t^2, 1 + t + t^2 + t^3$

*2. Determine which of the following collections of functions are independent in $C(R)$.
(a) $\sin x, \cos x, \sin 2x$ (b) $\sin x, \cos x, x$
(c) $e^x, \sin x$ (d) $\sin x, \sin x + \cos x, \sin x - \cos x$

3. Show that $1, 1 + t, 1 + t^2$, and $1 + t^3$ form a basis for $P_3(R)$.

*4. Find a basis for $P_0(R)$.

*5. For each of the following, consider the subspace V of $C(R)$ spanned by the given collection of functions. Find a basis for V and the dimension of V.
(a) $1, 1 - t, 1 + t$ (b) $t, t^2 + 1, t^2 - 1$
(c) $\sin x, \cos x, \sin 2x$ (d) $\sin x, \cos x, x$
(e) $e^x, \sin x$

6. Show that if a_1, a_2, a_3 are distinct numbers, then $e^{a_1 x}$, $e^{a_2 x}$, and $e^{a_3 x}$ are independent in $C(R)$.

7. Show that for any n, $1, x, \ldots, x^n$ form a basis for $P_n(R)$.

4.3   INNER PRODUCTS IN FUNCTION SPACES

In this section we extend the concept of an inner product to function spaces. The properties of the inner product in R^m which were necessary to define length and angle and to prove the Cauchy–Schwarz inequality are the following:

(a) $\langle x, y \rangle = \langle y, x \rangle$
(b) $\langle x, y + z \rangle = \langle x, y \rangle + \langle x, z \rangle$
(c) $\langle \alpha x, y \rangle = \alpha \langle x, y \rangle$
(d) $\langle x, x \rangle \geq 0$, and $\langle x, x \rangle = 0$ if and only if $x = 0$.

All of these properties are consequences of the definition of the inner product in R^m (see Theorem 3.1). Hence when attempting to define an inner product in function spaces we should demand that these properties be satisfied. The reason is that if these properties are satisfied, all the results of Chapter 3 will carry over to this new setting with no change.

Definition Let I be a closed and bounded interval, say $I = [a, b]$, and let f and g be two functions in $C(I)$. The **inner product** of f and g, denoted by $\langle f, g \rangle$, is defined by the equation

$$\langle f, g \rangle = \int_a^b f(x)g(x)\,dx. \tag{1}$$

There is a way to view this definition which shows that it is closely related to the inner product in R^m. In R^m the inner product $\langle x, y \rangle$ is defined to be the sum of products of corresponding components. If we think of the values of a function as the "components of a vector" and of the integral as a "sum," then

$$\langle f, g \rangle = \int_a^b f(x)g(x)\,dx$$

can also be thought of as a sum of products of corresponding components.

Properties (a)–(d) above follow from elementary properties of the integral (the student should verify this). Also, notice that we have not defined an inner product on $C(I)$ when I is an interval of infinite length or when I is an open interval (a, b). Why?

The remainder of this section is devoted to defining length and orthogonality in function spaces. In the next section we discuss projections, least-squares approximations, and the Gram–Schmidt process in the function space setting.

Length and Orthogonality

Let $I = [a, b]$. The **norm** (or **length**) of a function f in $C(I)$ is

$$\| f \| = \langle f, f \rangle^{1/2} = \left[\int_a^b f(x)^2\,dx \right]^{1/2}.$$

The **distance** between two functions f and g in $C(I)$ is

$$\| f - g \| = \left[\int_a^b (f(x) - g(x))^2\,dx \right]^{1/2}.$$

Two functions f and g in $C(I)$ are **orthogonal** if

$$\langle f, g \rangle = \int_a^b f(x)g(x)\,dx = 0.$$

The reader should observe that if we think of the values of functions as their components and think of integrals as sums, then the definitions of length, distance, and orthogonality are similar to those given in R^m. Also observe that the norm of a function is not its arc length, and that orthogonality of functions has nothing whatsoever to do with their graphs intersecting orthogonally.

EXAMPLE 1 Let $I = [0, 2\pi]$. Then

$$\|\sin x\|^2 = \int_0^{2\pi} \sin^2 x \, dx = \frac{x}{2} - \frac{\sin 2x}{4} \Big|_0^{2\pi} = \pi.$$

The distance between the two functions $\sin x$ and $\cos x$ can be computed from the relation

$$
\begin{aligned}
\|\sin x - \cos x\|^2 &= \int_0^{2\pi} (\sin x - \cos x)^2 \, dx \\
&= \int_0^{2\pi} (\sin^2 x - 2 \sin x \cos x + \cos^2 x) \, dx \\
&= \int_0^{2\pi} (1 - 2 \sin x \cos x) \, dx \\
&= (x - \sin^2 x) \big|_0^{2\pi} \\
&= 2\pi.
\end{aligned}
$$

The functions $\sin x$ and $\cos x$ are orthogonal since

$$\langle \sin x, \cos x \rangle = \int_0^{2\pi} \sin x \cos x \, dx = \frac{\sin^2 x}{2} \Big|_0^{2\pi} = 0.$$

EXERCISE 1   Compute the length of $\sin x$ and the distance between $\sin x$ and $\cos x$ if $I = [0, \pi/2]$. Also, test the orthogonality of $\sin x$ and $\cos x$.

EXERCISE 2   Interpret the norm of a function and the distance between two functions in terms of area. Also, show that if $I = [-1, 1]$, the functions x and x^2 are orthogonal in $C(I)$. Then draw the graphs of these functions and observe that the graphs do not intersect orthogonally.

EXAMPLE 2   If $I = [0, 2\pi]$, then the $2n + 1$ functions

$$1, \sin x, \cos x, \ldots, \sin nx, \cos nx$$

are orthogonal in $C(I)$. In fact,

$$\int_0^{2\pi} \sin mx \sin nx \, dx = \begin{cases} 0 & \text{if } m \neq n \\ \pi & \text{if } m = n \end{cases}$$

$$\int_0^{2\pi} \cos mx \cos nx \, dx = \begin{cases} 0 & \text{if } m \neq n \\ \pi & \text{if } m = n \end{cases}$$

$$\int_0^{2\pi} \sin mx \cos nx \, dx = 0 \qquad \text{for all positive integers } m \text{ and } n.$$

The norm of the constant function 1 is $\sqrt{2\pi}$ and the norm of each of the other functions is $\sqrt{\pi}$. We ask the reader to verify these facts in Problem 10.

PROBLEMS 4.3

1. Let $I = [0, 1]$. Compute the norm of each of the following functions in $C(I)$.

   *(a) 1   (b) x (c) $x^2 + 2$ *(d) $\dfrac{1}{\sqrt{x^2 + 1}}$

2. Let $I = [0, 1]$. Find each of the following.

   *(a) $\langle 1, x^2 \rangle$ (b) $\langle x, \sqrt{x} \rangle$ *(c) $\left\langle x, \dfrac{1}{1 + x^2} \right\rangle$

3. Let $I = [0, 1]$. Find two nonzero polynomials of degree ≤ 2 that are orthogonal to:

   *(a) x (b) $x + 1$ *(c) x^2 (d) x^3

*4. Show that the polynomials 1 and $x - \frac{1}{2}$ are orthogonal in $C(I)$, $I = [0, 1]$. Then find a nonzero polynomial of degree ≤ 2 that is orthogonal to both of these polynomials.

5. Let $I = [0, 1]$. Show that the only polynomial of degree ≤ 2 that is orthogonal to every polynomial of degree ≤ 2 is the zero polynomial. Can you generalize this result?

6. (a) Show that if f_1, f_2, and f_3 are nonzero orthogonal functions in $C(I)$, $I = [a, b]$, then they are linearly independent.
   (b) If $f_1, \ldots, f_n$ are nonzero orthogonal functions, are they independent? Why?

*7. Let $I = [-1, 1]$. Find a basis for the orthogonal complement of the subspace of $P_3(I)$ that is spanned by:
   (a) x (b) x^2
   (c) 1 and x (d) x and x^2

*8. For each of the subspaces and their orthogonal complements given in Problem 7, express the function x^3 as a sum of a vector in the subspace and a vector in the orthogonal complement. Do the same for $1 + x^2$ and $1 + x^3$.

9. It is important to observe that the definition of the inner product (1) depends on the interval I. Let $I_1 = [0, 1]$ and $I_2 = [0, 2]$. Then both x and x^2 belong to $C(I_1)$ and $C(I_2)$. However, show that their inner product when they are viewed as elements of the function space $C(I_1)$ is different from their inner product when they are viewed as elements of the function space $C(I_2)$. Hence we should expect that those properties which are defined in terms of the inner product will depend on the interval I.

10. Use the identities

$$\sin a \sin b = \frac{1}{2}[\cos(a - b) - \cos(a + b)]$$

$$\cos a \cos b = \frac{1}{2}[\cos(a - b) + \cos(a + b)]$$

$$\sin a \cos b = \frac{1}{2}[\sin(a - b) + \sin(a + b)]$$

$$\sin^2 mx = \frac{1}{2}(1 - \cos 2mx)$$

$$\cos^2 mx = \frac{1}{2}(1 + \cos 2mx)$$

to prove the assertions made in Example 2.

11. Show that the results of Example 2 remain valid if we replace the interval $I = [0, 2\pi]$ by the interval $I = [-\pi, \pi]$. Will they remain valid for any interval of length 2π?

12. A function f is orthogonal to a subspace V of $C(I)$ if f is orthogonal to every function in V.
    (a) Let $I = [-1, 1]$. Show that the function x is orthogonal to the subspace of $P(I)$ spanned by $1, x^2, x^4$.
    (b) Show that a function f is orthogonal to a finite-dimensional subspace V of $C(I)$ if and only if f is orthogonal to every function in a basis for V.

4.4 PROJECTIONS IN FUNCTION SPACES

Having extended the notion of an inner product to function spaces, we now extend the other concepts and results of Chapter 3 to function spaces.

Projections

The projection of a function f in $C(I)$ onto a k-dimensional subspace V of $C(I)$ is defined to be that function h in V such that $f - h$ is orthogonal to V. As in the case of R^m, the formula for calculating h is simplest when we have an orthogonal basis $g_1, \ldots, g_k$ for the subspace V. (As in R^m, we can always find an orthogonal basis for the subspace using the Gram–Schmidt process which we discuss below.) In this case, the projection h is simply the sum of the projections of f onto the orthogonal basis vectors $g_1, \ldots, g_k$.

$$h = \frac{\langle f, g_1 \rangle}{\langle g_1, g_1 \rangle} g_1 + \cdots + \frac{\langle f, g_k \rangle}{\langle g_k, g_k \rangle} g_k \tag{1}$$

EXERCISE 1 Verify that if h is given by this formula, then h is in V and $f - h$ is orthogonal to V.

EXAMPLE 1 The projection of x onto $\cos x$ in $C(I)$, $I = [0, \pi]$, is the function

$$h(x) = \frac{\langle x, \cos x \rangle}{\langle \cos x, \cos x \rangle} \cos x.$$

Since

$$\langle x, \cos x \rangle = \int_0^\pi x \cos x \, dx = x \sin x + \cos x \Big|_0^\pi = -2,$$

$$\langle \cos x, \cos x \rangle = \int_0^\pi \cos^2 x \, dx = \frac{x}{2} + \frac{\sin 2x}{4} \Big|_0^\pi = \frac{\pi}{2},$$

we obtain

$$h(x) = \frac{-4}{\pi} \cos x.$$

EXERCISE 2 Find the projection of x onto $\sin x$ in $C(I)$, where $I = [0, \pi]$.

EXAMPLE 2 Let $I = [0, 2\pi]$. By equation (1), the projection of x onto the subspace V of $C(I)$ spanned by the three orthogonal functions 1, $\sin x$, $\cos x$ is

$$h(x) = \frac{\langle x, 1 \rangle}{\langle 1, 1 \rangle} 1 + \frac{\langle x, \sin x \rangle}{\langle \sin x, \sin x \rangle} \sin x + \frac{\langle x, \cos x \rangle}{\langle \cos x, \cos x \rangle} \cos x.$$

Since

$$\langle 1, 1 \rangle = 2\pi,$$
$$\langle x, 1 \rangle = \int_0^{2\pi} x \, dx = 2\pi^2,$$
$$\langle x, \sin x \rangle = \int_0^{2\pi} x \sin x \, dx = -x \cos x + \sin x \Big|_0^{2\pi} = -2\pi,$$
$$\langle x, \cos x \rangle = \int_0^{2\pi} x \cos x \, dx = x \sin x + \cos x \Big|_0^{2\pi} = 0,$$
$$\langle \sin x, \sin x \rangle = \langle \cos x, \cos x \rangle = \pi,$$

we have

$$h(x) = \pi - 2 \sin x.$$

Now let us find the projection of x onto the subspace V_1 of $C(I), I = [0, 2\pi]$, spanned by the four orthogonal functions 1, $\sin x$, $\cos x$, and $\sin 2x$. The projection is given by

$$h_1(x) = \frac{\langle x, 1 \rangle}{\langle 1, 1 \rangle} 1 + \frac{\langle x, \sin x \rangle}{\langle \sin x, \sin x \rangle} \sin x + \frac{\langle x, \cos x \rangle}{\langle \cos x, \cos x \rangle} \cos x$$
$$+ \frac{\langle x, \sin 2x \rangle}{\langle \sin 2x, \sin 2x \rangle} \sin 2x,$$

and since the first three terms have already been computed, we need only compute the last term.

$$\langle x, \sin 2x \rangle = \int_0^{2\pi} x \sin 2x \, dx$$
$$= -\frac{x \cos 2x}{2} + \frac{\sin 2x}{4} \Big|_0^{2\pi} = -\pi$$
$$\langle \sin 2x, \sin 2x \rangle = \int_0^{2\pi} \sin^2 2x \, dx = \pi$$

Thus

$$h_1(x) = \pi - 2 \sin x - \sin 2x.$$

EXERCISE 3   Let $I = [0, 1]$. Use the result of Problem 4 in Section 4.3 to find the projection of x^2 onto $P_1(I)$.

The Gram–Schmidt Process

If V is a finite-dimensional function space and $f_1, \ldots, f_k$ is a basis for V, we can construct an orthogonal basis $g_1, \ldots, g_k$ for V using the **Gram–Schmidt** process. The procedure is exactly the same as it was in R^m. We begin by letting $g_1 = f_1$. Then let g_2 be the projection of f_2 onto the orthogonal complement of the subspace spanned by g_1.

$$g_2 = f_2 - \frac{\langle f_2, g_1 \rangle}{\langle g_1, g_1 \rangle} g_1$$

In general, having constructed $g_1, \ldots, g_i$ $(1 \le i < k)$, let g_{i+1} be the projection of f_{i+1} onto the orthogonal complement of the subspace spanned by the orthogonal vectors $g_1, \ldots, g_i$. Thus

$$g_{i+1} = f_{i+1} - \frac{\langle f_{i+1}, g_1 \rangle}{\langle g_1, g_1 \rangle} g_1 - \cdots - \frac{\langle f_{i+1}, g_i \rangle}{\langle g_i, g_i \rangle} g_i.$$

EXAMPLE 3   Let $I = [-1, 1]$. The function space $P_3(I)$ has 1, x, x^2, and x^3 as a basis. Let us apply the Gram–Schmidt orthogonalization process to find an orthogonal basis. Let $g_1(x) = 1$. Then since $\langle x, 1 \rangle = 0$ and $\langle 1, 1 \rangle = 2$, we have $g_2(x) = x - (0/2) \cdot 1 = x$. Similarly, since $\langle x^2, 1 \rangle = \frac{2}{3}$, $\langle x^2, x \rangle = 0$, and $\langle x, x \rangle = \frac{2}{3}$, we obtain

$$
\begin{aligned}
g_3(x) &= x^2 - \frac{\langle x^2, 1 \rangle}{\langle 1, 1 \rangle} 1 - \frac{\langle x^2, x \rangle}{\langle x, x \rangle} x \\
&= x^2 - \frac{\frac{2}{3}}{2} - \frac{0}{\frac{2}{3}} x \\
&= x^2 - \tfrac{1}{3}.
\end{aligned}
$$

Since $\langle x^3, 1 \rangle = 0$, $\langle x^3, x \rangle = \frac{2}{5}$, $\langle x^3, x^2 - \frac{1}{3} \rangle = 0$, and $\langle x^2 - \frac{1}{3}, x^2 - \frac{1}{3} \rangle = \frac{8}{45}$, we obtain

$$
\begin{aligned}
g_4(x) &= x^3 - \frac{\langle x^3, 1 \rangle}{\langle 1, 1 \rangle} 1 - \frac{\langle x^3, x \rangle}{\langle x, x \rangle} x - \frac{\langle x^3, x^2 - \frac{1}{3} \rangle}{\langle x^2 - \frac{1}{3}, x^2 - \frac{1}{3} \rangle} \left(x^2 - \tfrac{1}{3} \right) \\
&= x^3 - \tfrac{3}{5} x.
\end{aligned}
$$

Thus the polynomials 1, x, $x^2 - \frac{1}{3}$, and $x^3 - \frac{3}{5}x$ form an orthogonal basis for the space of polynomials of degree ≤ 3 on $[-1, 1]$. If these polynomials are adjusted so that $g_1(1) = g_2(1) = g_3(1) = g_4(1) = 1$, then they are called the *Legendre polynomials*. The first four Legendre polynomials are 1, x, $\frac{3}{2}(x^2 - \frac{1}{3})$, $\frac{5}{2}(x^3 - \frac{3}{5}x)$.

Least-Squares Approximations

In Chapter 3 we saw that the projection of a vector b onto a subspace gives the vector in the subspace which is closest to b. Thus we can think of the projection p of b onto a subspace as the best possible approximation to b by vectors in the subspace. It is this last interpretation of the projection that is most useful in the function space setting. Given a function f in $C(I)$ that is not in a finite-dimensional subspace V of $C(I)$, the projection h of f onto V gives the best possible approximation to f by functions in V in the sense that $\| f - h \|$ is as small as possible. It is called the **least-squares approxima-tion** to f by a function in V.

In many of the applications of this result, the subspace V is one of the following subspaces of $C(I)$:

1. The polynomials of degree $\leq n$ for some n.
2. The *trigonometric polynomials* of degree $\leq n$; that is, V is the subspace of $C(I)$ spanned by $1, \sin x, \cos x, \sin 2x, \cos 2x, \ldots, \sin nx, \cos nx$ for some n.

The main reason for the interest in approximating a function by a poly-nomial is that the function may be complicated and difficult to compute. Polynomials, on the other hand, are simple, both conceptually and compu-tationally. The reason for the interest in approximating a function by a trigonometric polynomial is that these approximations are important in any branch of physics or engineering that deals with oscillatory phenomena, heat conduction, or potential theory.

EXAMPLE 4   Consider the problem of finding the polynomial of degree ≤ 1 which is nearest $\sin \pi x$ on the interval $[-1, 1]$. The solution is the projection of $\sin \pi x$ onto the subspace $P_1(I)$ of $C(I)$, where $I = [-1, 1]$. Since 1 and x form an orthogonal basis for P_1, the projection is

$$h(x) = \frac{\langle \sin \pi x, 1 \rangle}{\langle 1, 1 \rangle} 1 + \frac{\langle \sin \pi x, x \rangle}{\langle x, x \rangle} x.$$

Since

$$\langle \sin \pi x, 1 \rangle = \int_{-1}^{1} \sin \pi x \, dx = 0,$$

$$\langle 1, 1 \rangle = 2,$$

$$\langle \sin \pi x, x \rangle = \int_{-1}^{1} x \sin \pi x \, dx$$

$$= -\frac{1}{\pi} x \cos \pi x + \frac{1}{\pi^2} \sin \pi x \Big|_{-1}^{1}$$

$$= \frac{2}{\pi},$$

$$\langle x, x \rangle = \int_{-1}^{1} x^2 \, dx = \frac{2}{3},$$

we obtain

$$h(x) = \frac{2/\pi}{2/3} x = \frac{3}{\pi} x.$$

EXAMPLE 5   Consider now the problem of finding the least-squares approximation to $\sin \pi x$ on $I = [-1, 1]$ by a polynomial of degree ≤ 2. In Example 3
we found that the functions 1, x, and $x^2 - \frac{1}{3}$ form an orthogonal basis for
$P_2(I)$. Hence the least-squares approximation is

$$h_1(x) = \frac{\langle \sin \pi x, 1 \rangle}{\langle 1, 1 \rangle} 1 + \frac{\langle \sin \pi x, x \rangle}{\langle x, x \rangle} x + \frac{\langle \sin \pi x, x^2 - \frac{1}{3} \rangle}{\langle x^2 - \frac{1}{3}, x^2 - \frac{1}{3} \rangle} \left(x^2 - \frac{1}{3} \right).$$

Notice that the first two terms have already been computed in Example 4.
Since $\langle \sin \pi x, x^2 - \frac{1}{3} \rangle = 0$,

$$h_1(x) = \frac{3}{\pi} x = h(x).$$

EXERCISE 4   Find the polynomial of degree ≤ 3 which is nearest $\sin \pi x$ on the
interval $[-1, 1]$. Having done this, how would you find the least-squares
approximation to $\sin \pi x$ on $[-1, 1]$ by a polynomial of degree ≤ 5?

EXAMPLE 6   Let us find the least-squares approximation to the function x^2 on
$[-\pi, \pi]$ by a trigonometric polynomial of degree ≤ 1. We have seen that
the trigonometric functions

$$1, \sin x, \cos x$$

are orthogonal. Hence the projection is

$$h = \frac{\langle x^2, 1 \rangle}{\langle 1, 1 \rangle} 1 + \frac{\langle x^2, \sin x \rangle}{\langle \sin x, \sin x \rangle} \sin x + \frac{\langle x^2, \cos x \rangle}{\langle \cos x, \cos x \rangle} \cos x$$

$$= \frac{\langle x^2, 1 \rangle}{2\pi} 1 + \frac{\langle x^2, \sin x \rangle}{\pi} \sin x + \frac{\langle x^2, \cos x \rangle}{\pi} \cos x.$$

Since

$$\langle x^2, 1 \rangle = \int_{-\pi}^{\pi} x^2 \, dx = \frac{2\pi^3}{3},$$

$$\langle x^2, \sin x \rangle = \int_{-\pi}^{\pi} x^2 \sin x \, dx = 0,$$

$$\langle x^2, \cos x \rangle = \int_{-\pi}^{\pi} x^2 \cos x \, dx = -4\pi,$$

we obtain

$$h(x) = \frac{\pi^2}{3} - 4\pi \cos x.$$

PROBLEMS 4.4

1. Let $I = [0, 1]$. Find the projection of:
   (a) $1 + x$ onto x^2
   *(b) $x + x^2$ onto x
   (c) $\sin \pi x$ onto 1

*2. Let $I = [0, 2\pi]$. Find the projection of x^2 onto the subspace of $C(I)$ spanned by 1, $\sin x$, and $\cos x$. Do the same for $\sin^2 x$ and $\sin x \cos x$.

*3. Let $I = [0, 2\pi]$. Find the projections of x, of x^2, and of $\sin x$ onto the subspace of $C(I)$ spanned by 1 and $\sin 2x$.

4. Find an orthogonal basis for the subspace of $C(I)$ spanned by:
   *(a) 1, x, and x^2 when $I = [0, 1]$ (b) e^x and e^{-x} when $I = [0, 1]$
   *(c) x and e^x when $I = [0, 1]$ (d) x and $\sin x$ when $I = [0, 2\pi]$
   *(e) x, x^2, and x^3 when $I = [-1, 1]$

5. Let $I = [-1, 1]$. Find all polynomials in $P_2(I)$ that are orthogonal to:
   *(a) $x + 1$ (b) x and $1 + x^2$

*6. Let $I = [0, 2\pi]$. Find the function in the subspace of $C(I)$ spanned by 1 and $\cos 2x$ that is nearest x. Do the same for $\sin x$.

*7. Find the polynomial of degree ≤ 2 that is nearest:
   (a) $\sin x$ on the interval $[-\pi, \pi]$ (b) $\sqrt[3]{x}$ on the interval $[-1, 1]$
   (c) x^3 on the interval $[-1, 1]$ (d) e^x on the interval $[0, 1]$
   (e) $\cos x$ on the interval $[-\pi, \pi]$

8. Let $g_1, \ldots, g_k$ be a basis for a subspace V of $C(I)$, $I = [a, b]$. The least-squares approximation to a function f in $C(I)$ by a function in V is that function h in V such that

$$\|h - f\|^2 = \int_a^b (h(x) - f(x))^2 \, dx$$

is as small as possible. To find h we must find scalars $\alpha_1, \ldots, \alpha_k$ such that

$$h = \alpha_1 g_1 + \cdots + \alpha_k g_k$$

and

$$e = \int_a^b [\alpha_1 g_1(x) + \cdots + \alpha_k g_k(x) - f(x)]^2 \, dx$$

is as small as possible. If e is viewed as a function of $\alpha_1, \ldots, \alpha_k$, then e will be minimized only if the equations

$$\frac{\partial e}{\partial \alpha_j} = 0, \quad j = 1, \ldots, k, \tag{1}$$

are satisfied.
   (a) Show that for $j = 1, \ldots, k$,

$$\frac{\partial e}{\partial \alpha_j} = 2 \int_a^b g_j(x)[\alpha_1 g_1(x) + \cdots + \alpha_k g_k(x) - f(x)] \, dx.$$

   (b) Using the inner product notation, show that equations (1) can be written as

$$\alpha_1 \langle g_j, g_1 \rangle + \cdots + \alpha_k \langle g_j, g_k \rangle = \langle g_j, f \rangle, \quad j = 1, \ldots, k.$$

(c) Show that these last equations can be written in matrix notation as

$$
\begin{bmatrix} \langle g_1, g_1 \rangle & \cdots & \langle g_1, g_k \rangle \\ \vdots & & \vdots \\ \langle g_k, g_1 \rangle & \cdots & \langle g_k, g_k \rangle \end{bmatrix} \begin{bmatrix} \alpha_1 \\ \vdots \\ \alpha_k \end{bmatrix} = \begin{bmatrix} \langle g_1, f \rangle \\ \vdots \\ \langle g_k, f \rangle \end{bmatrix}.
$$

(d) Show that if $g_1, \ldots, g_k$ are thought of as the "columns" of a matrix A and f is thought of as a "column vector" b, then this last equation can be written as

$$
\begin{bmatrix} -g_1- \\ \vdots \\ -g_k- \end{bmatrix} \begin{bmatrix} | & & | \\ g_1 & \cdots & g_k \\ | & & | \end{bmatrix} \begin{bmatrix} \alpha_1 \\ \vdots \\ \alpha_k \end{bmatrix} = \begin{bmatrix} -g_1- \\ \vdots \\ -g_k- \end{bmatrix} \begin{bmatrix} | \\ f \\ | \end{bmatrix}
$$

or

$$
A^T A \bar{x} = A^T b.
$$

*9. Clearly, the function $\sin x$ is not in $P_2(I)$, $I = [-\pi, \pi]$. Thus the equation

$$\sin x = a + bx + cx^2 \qquad \text{for all } x \text{ in } I$$

is certainly inconsistent. As above, if we view the functions 1, x, and x^2 as the columns of a matrix A and $\sin x$ as a column vector, this equation becomes

$$
\begin{bmatrix} | & | & | \\ 1 & x & x^2 \\ | & | & | \end{bmatrix} \begin{bmatrix} a \\ b \\ c \end{bmatrix} = \begin{bmatrix} | \\ \sin x \\ | \end{bmatrix}.
$$

Find a least-squares solution to this problem.

10. Interpret Problems 6 and 7 as inconsistent systems.

11. Investigate the difference between a least-squares approximation of $\sin x$ by a polynomial of degree ≤ 2 and the second-degree Taylor polynomial approximation of $\sin x$ about $x = 0$.

Chapter 5

Linear Transformations

In the first three chapters we used matrices mainly to represent the coefficients of a linear system. We regarded the equation $y = Ax$ as an equation to be solved for x. Whenever we were given a vector y, we found the x's. In this chapter we look at the equation $y = Ax$ from a different point of view. We start with a vector x and use the equation to compute a vector y. From this perspective the equation $y = Ax$ defines a function.

5.1 LINEAR TRANSFORMATIONS

The reader is no doubt familiar with ordinary functions such as the function f defined by the equation $f(x) = x^2$. This function transforms a real number into a real number, namely its square. For instance, the number 2 is transformed into the number 4 [i.e., $f(2) = 4$]. Next, we study functions that transform vectors into vectors.

If A is an $m \times n$ matrix, then the equation

$$T(x) = Ax$$

defines a function T that transforms a vector x in R^n into the vector Ax in R^m. Thus T is a function from R^n into $\dot{R}^m$. In general, if V and W are vector spaces, a function or **transformation** T from V into W is a rule that associates with every vector x in V a unique vector in W that is denoted by $T(x)$. If x is a vector in V, then $T(x)$ is called the **image** of x under the transformation T. We also say that T **maps** or **transforms** x into $T(x)$. For

181

example, if T is the transformation from R^3 into R^2 defined by the equation

$$T(x_1, x_2, x_3) = (x_1 + x_2, x_2 + x_3),$$

then T maps the vector $(1, 1, 1)$ onto the vector $T(1, 1, 1) = (2, 2)$. Similarly, T maps $(3, 2, 0)$ onto $(5, 2)$. In order to indicate that a transformation T maps vectors in a vector space V into a vector space W, we write $T : V \rightarrow W$ and say that T maps V into W. Two transformations $T : V \rightarrow W$ and $S : V \rightarrow W$ are equal if they have have the same effect on all vectors in V. That is, $T = S$ if $T(x) = S(x)$ for all vectors x in V.

Every $m \times n$ matrix A defines a transformation $T : R^n \rightarrow R^m$ via the equation $T(x) = Ax$. Transformations that are defined in this way are called **matrix transformations**.

EXAMPLE 1   If $A = \begin{bmatrix} 1 & 1 & 2 \\ 0 & 2 & 3 \end{bmatrix}$, then the matrix transformation $T : R^3 \rightarrow R^2$ defined by $T(x) = Ax$ maps the vector $(1, 1, 1)$ onto the vector $T(1, 1, 1) = (4, 5)$ because

$$\begin{bmatrix} 1 & 1 & 2 \\ 0 & 2 & 3 \end{bmatrix} \begin{bmatrix} 1 \\ 1 \\ 1 \end{bmatrix} = \begin{bmatrix} 4 \\ 5 \end{bmatrix}.$$

Matrix transformations have two important properties. If A is an $m \times n$ matrix and $T : R^n \rightarrow R^m$ is defined by $T(x) = Ax$, then for any two vectors x and y in R^n and any scalar α,

$$T(x + y) = A(x + y) = Ax + Ay = T(x) + T(y),$$
$$T(\alpha x) = A(\alpha x) = \alpha Ax = \alpha T(x).$$

Transformations that, like matrix transformations, have these two properties are the most important transformations in linear algebra.

Definition   Let V and W be vector spaces. A transformation $T : V \rightarrow W$ is called **linear** if for all vectors x and y in V and for all scalars α,

1. $T(x + y) = T(x) + T(y)$
2. $T(\alpha x) = \alpha T(x)$

A linear transformation $T : V \rightarrow W$ preserves vector addition and scalar multiplication. If x and y are two vectors in V, then first adding x and y and then applying T to their sum gives the same result as first applying T to x and y separately and then adding (see Figure 5.1). Similarly, first multiplying x by a scalar α and then applying T results in the same vector as multiplying the image of x under T by α.

If v_1 and v_2 are vectors in V and α_1, α_2 are scalars, then

$$\begin{aligned} T(\alpha_1 v_1 + \alpha_2 v_2) &= T(\alpha_1 v_1) + T(\alpha_2 v_2) \quad &\text{(since } T \text{ preserves addition)} \\ &= \alpha_1 T(v_1) + \alpha_2 T(v_2) \quad &\text{(since } T \text{ preserves} \\ & & \text{multiplication by scalars).} \end{aligned}$$

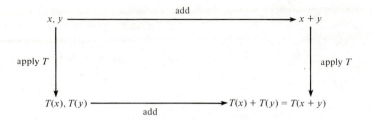

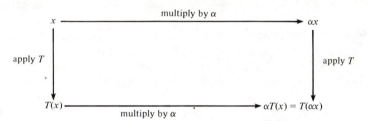

Figure 5.1

Clearly, this result extends to any linear combination of vectors in V. If $v_1, \ldots, v_n$ are vectors in V and $\alpha_1, \ldots, \alpha_n$ are scalars, then

$$T(\alpha_1 v_1 + \cdots + \alpha_n v_n) = \alpha_1 T(v_1) + \cdots + \alpha_n T(v_n).$$

Thus T preserves linear combinations. This is the reason for the importance of linear transformations in linear algebra. Since linear transformations preserve linear combinations, they are the natural transformations to study in linear algebra.

EXERCISE 1   Let $T : V \to W$ be a linear transformation. Show that for all vectors x and y in V,

$$T(-x) = -T(x) \quad \text{and} \quad T(x - y) = T(x) - T(y).$$

EXAMPLE 2   Consider the transformation $T : R^2 \to R^3$ defined by $T(x_1, x_2) = (x_1 + x_2, 3x_2, 2x_1 - x_2)$. We will prove that T is linear. Let $x = (x_1, x_2)$ and $y = (y_1, y_2)$ be any two vectors in R^2 and let α be any scalar. Then

$$
\begin{aligned}
T(x + y) &= T(x_1 + y_1, x_2 + y_2) \\
&= (x_1 + y_1 + x_2 + y_2, 3(x_2 + y_2), 2(x_1 + y_1) - (x_2 + y_2)) \\
&= (x_1 + x_2, 3x_2, 2x_1 - x_2) + (y_1 + y_2, 3y_2, 2y_1 - y_2) \\
&= T(x) + T(y), \\
T(\alpha x) &= T(\alpha x_1, \alpha x_2) \\
&= (\alpha x_1 + \alpha x_2, 3\alpha x_2, 2\alpha x_1 - \alpha x_2) \\
&= \alpha(x_1 + x_2, 3x_2, 2x_1 - x_2) \\
&= \alpha T(x).
\end{aligned}
$$

Thus T is linear.

Since all matrix transformations are linear, we can show that a transformation $T : R^n \to R^m$ is linear by finding an $m \times n$ matrix A with the property that $T(x) = Ax$ for all x in R^n. For example, we could have proved that the transformation T in Example 1 is linear by noting that $T(x) = Ax$, where

$$A = \begin{bmatrix} 1 & 1 \\ 0 & 3 \\ 2 & -1 \end{bmatrix}.$$

EXERCISE 2 Verify that the transformation $T : R^2 \to R^2$ defined by $T(x_1, x_2) = (2x_1 + 3x_2, - x_1)$ is linear.

EXAMPLE 3 The transformation $T : R^2 \to R^2$ defined by $T(x_1, x_2) = (x_1 x_2, x_2)$ is not linear because T does not preserve vector addition. For instance, if $x = (2, 3)$ and $y = (3, 2)$, then $T(x + y) = T(5, 5) = (25, 5)$ but $T(x) + T(y) = (6, 3) + (6, 2) = (12, 5) \neq (25, 5)$.

EXERCISE 3 Show that the transformation T in Example 3 does not preserve scalar multiplication.

EXERCISE 4 Determine which of the following transformations are linear.

(a) $T : R^2 \to R^2$ defined by $T(x_1, x_2) = (x_1^2, x_2 + x_1)$
(b) $T : R^1 \to R^2$ defined by $T(x_1) = (2x_1, 3x_1)$
(c) $T : R^3 \to R^4$ defined by $T(x_1, x_2, x_3) = (x_1, x_2, x_3, x_1 + x_3)$
(d) $T : R^2 \to R^1$ defined by $T(x_1, x_2) = x_1 + \sqrt{x_2}$

EXAMPLE 4 Let V be any vector space. The transformation $I : V \to V$ defined by $I(x) = x$ maps every vector in V onto itself. I is called the **identity transformation** in V. The transformation $0 : V \to V$ defined by $0(x) = 0$ maps every vector in V onto the zero vector and is called the **zero transformation** in V. We leave it to the reader to check that these two transformations are linear.

EXAMPLE 5 Let V be a vector space and let α be a fixed scalar. We leave it as an exercise to verify that the transformation $T : V \to V$ defined by $T(x) = \alpha x$ is linear. If $\alpha > 1$, T is called a *dilation*. If $0 < \alpha < 1$, then T is called a *contraction*. Geometrically, a dilation stretches the vectors in V by a factor of α while a contraction contracts the vectors by a factor α (see Figure 5.2).

EXAMPLE 6 Let $T : R^2 \to R^2$ be the matrix transformation defined by $T(x) = Ax$, where $A = \begin{bmatrix} 1 & 0 \\ 0 & 2 \end{bmatrix}$. Since

$$\begin{bmatrix} 1 & 0 \\ 0 & 2 \end{bmatrix}\begin{bmatrix} x_1 \\ x_2 \end{bmatrix} = \begin{bmatrix} x_1 \\ 2x_2 \end{bmatrix},$$

we see that T stretches the y-coordinates of vectors in R^2 (see Figure 5.3).

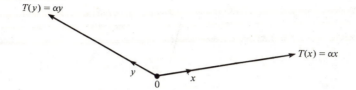

$T(y) = \alpha y$

$T(x) = \alpha x$

y

x

0

Figure 5.2

Every $m \times n$ matrix A gives rise to a linear transformation $T : R^n \to R^m$ defined by $T(x) = Ax$. We will now show that every linear transformation from R^n into R^m is in fact a matrix transformation. In other words, given any linear transformation $T : R^n \to R^m$, it is always possible to find a matrix A such that $T(x) = Ax$ for all x in R^n. To see this, let $e_1, e_2, \ldots, e_n$ be the standard basis for R^n. Each $T(e_j)$ is a vector in R^m. Let A be the $m \times n$ matrix whose jth column is the vector $T(e_j)$. If $x = (x_1, x_2, \ldots, x_n)$ is a vector in R^n, then $x = x_1 e_1 + x_2 e_2 + \cdots + x_n e_n$. Since T is linear, we have

$$T(x) = T(x_1 e_1 + x_2 e_2 + \cdots + x_n e_n)$$
$$= x_1 T(e_1) + x_2 T(e_2) + \cdots + x_n T(e_n) = Ax.$$

For example, the transformation $T : R^2 \to R^3$ which is defined by $T(x_1, x_2) = (x_1 + x_2, 3x_2, 2x_1 - x_2)$ is linear (see Example 1). Since $T(1, 0) = (1, 0, 2)$ and $T(0, 1) = (1, 3, -1)$, we have

$$T(x_1, x_2) = T(x_1 e_1 + x_2 e_2) = x_1 T(e_1) + x_2 T(e_2)$$
$$= x_1 \begin{bmatrix} 1 \\ 0 \\ 2 \end{bmatrix} + x_2 \begin{bmatrix} 1 \\ 3 \\ -1 \end{bmatrix} = \begin{bmatrix} 1 & 1 \\ 0 & 3 \\ 2 & -1 \end{bmatrix} \begin{bmatrix} x_1 \\ x_2 \end{bmatrix}.$$

Definition   Let $e_1, e_2, \ldots, e_n$ be the standard basis for R^n. If $T : R^n \to R^m$ is a linear transformation, then the $m \times n$ matrix whose jth column, $j = 1, \ldots, n$, is the vector $T(e_j)$ is called the **standard matrix** for T.

Result 1   If $T : R^n \to R^m$ is a linear transformation and A is the standard matrix for T, then $T(x) = Ax$ for all vectors x in R^n.

This result says that all linear transformations from R^n into R^m are matrix transformations. The power of this result is that it enables us to reduce questions about linear transformations to questions about matrices.

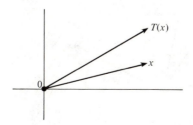

$T(x)$

x

0

Figure 5.3

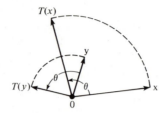

Figure 5.4

If A is an $m \times n$ matrix and $T : R^n \to R^m$ is the matrix transformation defined by $T(x) = Ax$, then the matrix A is of course the standard matrix for T.

EXAMPLE 7   The transformation $T : R^3 \to R^1$ which is defined by $T(x_1, x_2, x_3) = x_1 - x_2 + 2x_3$ is linear. Since $T(e_1) = 1$, $T(e_2) = -1$, $T(e_3) = 2$, the standard matrix for T is the 1×3 matrix $[1 \quad -1 \quad 2]$ and

$$T(x_1, x_2, x_3) = [1 \quad -1 \quad 2] \begin{bmatrix} x_1 \\ x_1 \\ x_3 \end{bmatrix}.$$

EXERCISE 5   Let the linear transformation $T : R^2 \to R^2$ be defined by $T(x_1, x_2) = (2x_1, 3x_2 - x_1)$. Find the standard matrix for T and check that multiplication by this matrix gives $T(x)$ for all vectors x in R^2.

EXAMPLE 8   The standard matrix for the identity transformation $I : R^n \to R^n$ is the $n \times n$ identity matrix. The standard matrix for the zero transformation $0 : R^n \to R^n$ is then $n \times n$ zero matrix.

EXAMPLE 9   Let θ be a fixed angle and let $T : R^2 \to R^2$ be the transformation that rotates every vector in R^2 counterclockwise through the angle θ (Figure 5.4). T is called the *rotation* transformation of R^2 through the angle θ. If x and y are vectors in R^2 and α is a scalar, then $T(x + y) = T(x) + T(y)$ and $T(\alpha x) = \alpha T(x)$ (see Figure 5.5). Thus T is a linear transformation. Clearly, T preserves length and angles; that is, if x and y are vectors in R^2, then

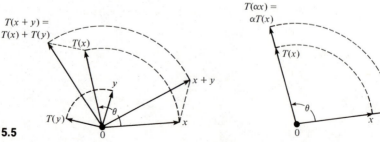

Figure 5.5

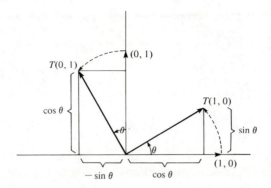

Figure 5.6

$\|T(x)\| = \|x\|$ and the angle between x and y is equal to the angle between $T(x)$ and $T(y)$. Since $T(1,0) = (\cos\theta, \sin\theta)$ and $T(0,1) = (-\sin\theta, \cos\theta)$ (see Figure 5.6), the standard matrix for T is

$$A = \begin{bmatrix} \cos\theta & -\sin\theta \\ \sin\theta & \cos\theta \end{bmatrix}.$$

For instance, if $\theta = 60°$, then the standard matrix for T is $A = \dfrac{1}{2}\begin{bmatrix} 1 & -\sqrt{3} \\ \sqrt{3} & 1 \end{bmatrix}$. Hence rotating the vector $x = (2,4)$ through $60°$ results in the vector $Ax = (1 - 2\sqrt{3}, \sqrt{3} - 2)$.

EXERCISE 6 Let V be the straight line in R^2 spanned by a vector v. Let $T : R^2 \to R^2$ be the transformation that projects every vector in R^2 onto V. Recall that if x is a vector in R^2, then the projection of x onto V is given by $T(x) = \dfrac{\langle x, v \rangle}{\langle v, v \rangle} v$. Use this formula to show that T is linear.

EXAMPLE 10 Let V be a subspace of R^n and let $T : R^n \to R^n$ be the transformation that projects all vectors in R^n onto V (see Figure 5.7). T is called the *projection* transformation of R^n onto V. Let A be a matrix whose columns form a basis for V. Then $P = A(A^TA)^{-1}A^T$ is the projection matrix for the subspace V (see Example 6 in Section 3.5). If x is any vector in R^n, then Px is the projection of x onto V and $T(x) = Px$. Hence T is linear and P is the standard matrix for T.

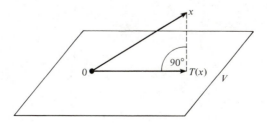

Figure 5.7

EXAMPLE 11   Let V be a straight line in R^2 passing through the origin. Let T : $R^2 \rightarrow R^2$ be the transformation that reflects every vector in R^2 across the line V (see Figure 5.8). Note that if the vector p is the projection of x onto the line V (see Figure 5.8), then $T(x) = x + 2(p - x) = 2p - x$. T is a linear transformation (see Figure 5.9). Clearly, T preserves length and angles. In general, if V is a subspace of R^n, we define a transformation $T : R^n \rightarrow R^n$ by the equation

$$T(x) = 2p - x,$$

where p is the projection of x onto the subspace V (see Figure 5.10). T is called the *reflection* transformation of R^n across V and $T(x)$ is called the *reflection* of the vector x across the subspace V. If P is the projection matrix for V, then

$$T(x) = 2Px - x = (2P - I)x.$$

Hence T is linear and $2P - I$ is the standard matrix for T. This matrix is called the *reflection matrix* for V.

EXERCISE 7   Let V be the straight line in R^2 passing through the origin with slope m. Then V is spanned by the vector $(1, m)$. Verify that

$$\frac{1}{1 + m^2} \begin{bmatrix} 1 - m^2 & 2m \\ 2m & m^2 - 1 \end{bmatrix}$$

is the reflection matrix for V.

So far our examples of linear transformations have all been transformations from R^n into R^m. We conclude this section with two examples involving function spaces (Chapter 4 is required for these examples). Recall that if I is an interval, then $C(I)$ denotes the vector space of all real-valued functions that are continuous on I and $C^1(I)$ stands for the subspace of all real-valued functions that have a continuous first derivative on I.

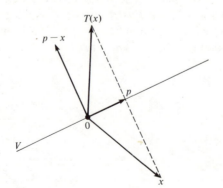

Figure 5.8

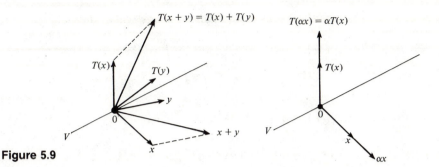

Figure 5.9

EXAMPLE 12   Let I be an interval. If f is a function in $C^1(I)$, then its derivative f' is continuous on I and hence belongs to $C(I)$. Let $D : C^1(I) \to C(I)$ be the transformation that maps f onto its derivative, that is, $D(f) = f'$. For instance, if $f(x) = x^3$, then $D(f)(x) = 3x^2$. Similarly, if $g(x) = \sin x$, then $D(g)(x) = \cos x$. D is a linear transformation because

$$D(f + g) = (f + g)' = f' + g' = D(f) + D(g),$$
$$D(\alpha f) = (\alpha f)' = \alpha f' = \alpha D(f).$$

The linear transformation D is often called the *differential operator*.

EXAMPLE 13   Let $I = [a, b]$ be a closed interval. Let $T : C(I) \to R^1$ be the transformation that maps a function f in $C(I)$ to its definite integral from a to b,

$$T(f) = \int_a^b f(x)\, dx.$$

For instance, if $I = [0, 1]$, $f(x) = 2x$, and $g(x) = 3x^2 - 1$, then

$$T(f) = \int_0^1 2x\, dx = x^2 \Big|_0^1 = 1,$$

$$T(g) = \int_0^1 (3x^2 - 1)\, dx = (x^3 - x) \Big|_0^1 = 0.$$

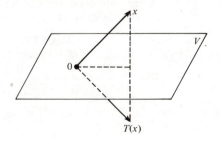

Figure 5.10

If f and g are any functions in $C(I)$ and α is any scalar, then

$$T(f+g) = \int_a^b (f+g)(x)\, dx = \int_a^b (f(x) + g(x))\, dx$$

$$= \int_a^b f(x)\, dx + \int_a^b g(x)\, dx = T(f) + T(g),$$

$$T(\alpha f) = \int_a^b (\alpha f)(x)\, dx = \int_a^b \alpha f(x)\, dx$$

$$= \alpha \int_a^b f(x)\, dx = \alpha T(f).$$

Thus T is a linear transformation.

PROBLEMS 5.1

*1. Which of the following transformations are linear? If the transformation is linear, find its standard matrix.
   (a) $T : R^3 \to R^2$ defined by $T(x_1, x_2, x_3) = (x_2, x_1)$
   (b) $T : R^2 \to R^2$ defined by $T(x_1, x_2) = (x_2, x_1 x_2)$
   (c) $T : R^1 \to R^3$ defined by $T(x_1) = (x_1, 0, 1)$
   (d) $T : R^2 \to R^1$ defined by $T(x_1, x_2) = 2x_1 + 3x_2$
   (e) $T : R^n \to R^1$ defined by $T(x) = \|x\|$

*2. In each part find the standard matrix for the given linear transformation.
   (a) $T : R^3 \to R^2$ defined by $T(x_1, x_2, x_3) = (2x_1 + x_2, x_2 - x_3)$
   (b) $T : R^4 \to R^1$ defined by $T(x_1, x_2, x_3, x_4) = x_1 + 2x_2 - x_3 + 5x_4$
   (c) $T : R^2 \to R^4$ defined by $T(x_1, x_2) = (0, x_2, x_1, x_1 + x_2)$

3. Let α be a scalar. Find the standard matrix for the linear transformation T : $R^n \to R^n$ defined by $T(x) = \alpha x$.

*4. Find the components of the vector obtained by rotating the vector $(1, 2)$ through an angle of:
   (a) $30°$ (b) $150°$ (c) $45°$ (d) $-45°$

*5. Find the components of the vector obtained by reflecting the vector $(5, 2)$ across the straight line spanned by the vector $(3, 9)$.

*6. Let $T : R^3 \to R^3$ be the projection transformation of R^3 onto the plane spanned by the vectors $(1, 0, 0)$ and $(0, 1, 1)$. Find the standard matrix for T.

*7. Find the standard matrix for the reflection transformation T of R^3 across the plane spanned by $(1, 0, 0)$ and $(0, 1, 1)$. Compute $T(1, 2, 3)$ and $T(2, -1, 2)$.

*8. Let T be the linear transformation in Problem 7. For which vectors x in R^3 is $T(x) = x$?

9. Let T be the rotation transformation of R^2 through the angle θ. Show that the standard matrix for T is orthogonal.

*10. Let V be a subspace of R^n and let T be the reflection transformation of R^n across V. Let A be the standard matrix for T.
   (a) Show that A is a symmetric orthogonal matrix.
   (b) Show that for all vectors x and y in R^n, $\|T(x)\| = \|x\|$ and $\langle T(x), T(y)\rangle = \langle x, y\rangle$.

11. Let L be the straight line in R^2 defined by the equation $ax + by = 0$.
   (a) Show that the projection matrix for L is

$$\frac{1}{a^2 + b^2}\begin{bmatrix} b^2 & -ab \\ -ab & a^2 \end{bmatrix}.$$

   (b) Show that the reflection matrix for L is

$$\frac{1}{a^2 + b^2}\begin{bmatrix} b^2 - a^2 & -2ab \\ -2ab & a^2 - b^2 \end{bmatrix}.$$

12. Show that conditions (1) and (2) in the definition of a linear transformation can be replaced by the single condition that

$$T(\alpha x + \beta y) = \alpha T(x) + \beta T(y)$$

holds for all vectors x and y and scalars α and β.

13. Let y be an m-vector. Show that the transformation $T : R^m \to R^1$ defined by $T(x) = \langle x, y \rangle$ is a linear transformation. What is the standard matrix for T?

14. Let $v_1, v_2, \ldots, v_n$ be vectors in a vector space V. Show that the transformation $T : R^n \to V$ defined by $T(x_1, x_2, \ldots, x_n) = x_1 v_1 + x_2 v_2 + \cdots + x_n v_n$ is linear.

15. (a) Let $T : V \to W$ and $S : V \to W$ be two linear transformations and let α be a scalar. We define two transformations $T + S : V \to W$ and $\alpha T : V \to W$ by the equations

$$(T + S)(x) = T(x) + S(x),$$

$$(\alpha T)(x) = \alpha T(x).$$

   Show that $T + S$ and αT are linear.
   (b) Let $T : R^n \to R^m$ and $S : R^n \to R^m$ be linear transformations. Let A and B be the standard matrices for T and S, respectively. Show that $A + B$ is the standard matrix for $T + S$. Also show that αA is the standard matrix for αT, where α is any scalar.

*16. (Chapter 4 required) Which of the following transformations are linear?
   (a) $T : C^2(R) \to C(R)$ defined by $T(f) = f'' + 2f'$
   (b) $T : C(R) \to R^1$ defined by $T(f) = f(0)$
   (c) $T : P(R) \to R^1$ defined by $T(f) = $ degree of f

17. Let $T : U \to V$ and $S : V \to W$ be two linear transformations.
   (a) Show that the transformation $ST : U \to W$ defined by $(ST)(x) = S(T(x))$ is linear.
   (b) Show that if $U = R^n$, $V = R^m$, and $W = R^k$, and A and B are the standard matrices for T and S respectively, then BA is the standard matrix for ST.

5.2   PROPERTIES OF LINEAR TRANSFORMATIONS

The nature of a linear transformation $T : V \to W$ cannot be fully understood without studying how T transforms subsets of V and, in particular, subspaces of V. For example, a rotation of R^2 through an angle θ carries each straight line onto a straight line and carries the whole plane onto itself. The dimension of a subspace of R^2 does not change under rotation. In contrast to this, projecting R^2 onto, say, the x-axis collapses the whole plane onto a one-dimensional subspace. All one-dimensional subspaces of R^2 with the exception of the y-axis are carried onto the x-axis. The y-axis is projected into the origin.

Let $T : V \to W$ be a linear transformation and let U be a set of vectors in V. Then T transforms U into a set of vectors in W, namely the set Z of all images (under T) of the vectors in U. This subset Z of W is called the **image of U under T** and we say that T maps U **onto** Z. When convenient we will denote the image of U under T by $T(U)$.

EXAMPLE 1   Let $T : R^2 \to R^2$ be the reflection transformation of R^2 across the x-axis. Then T maps every straight line passing through the origin onto a straight line passing through the origin. Note that T maps the x-axis onto itself. If L is any line segment in R^2, then T maps L onto a line segment of the same length (see Figure 5.11). Since every vector in R^2 is the reflection across the x-axis of some vector, we see that the image of R^2 under T is R^2.

EXERCISE 1   Let T be as in Example 1. Describe the image under T of the square with vertices $(1, 1)$, $(2, 1)$, $(2, 2)$, and $(1, 2)$. What can you say in general about the images under T of a rectangle, a parallelogram, a circle?

The reflection transformation in Example 1 maps each subspace of R^2 onto a subspace of R^2 of the same dimension. It is true in general that a linear transformation $T : V \to W$ maps every subspace of V onto a subspace of W. But the dimension of the image of a subspace U may be less than the dimension of U.

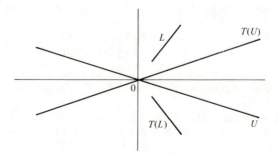

Figure 5.11

To prove this, let $u_1, u_2, \ldots, u_k$ be vectors that span U. Then every vector u in U is a linear combination

$$u = \alpha_1 u_1 + \alpha_2 u_2 + \cdots + \alpha_k u_k$$

of the u_i's. Since

$$
\begin{aligned}
T(u) &= T(\alpha_1 u_1 + \alpha_2 u_2 + \cdots + \alpha_k u_k) \\
&= \alpha_1 T(u_1) + \alpha_2 T(u_2) + \cdots + \alpha_k T(u_k),
\end{aligned}
$$

we see that the image of U under T is the set of all linear combinations of the vectors $T(u_1), T(u_2), \ldots, T(u_k)$. In particular, if the u_i's form a basis for U (i.e., $\dim U = k$), then $\dim T(U) \le k$ because $T(U)$ is spanned by k vectors. We have proved the following result.

Result 1 Let $T : V \to W$ be a linear transformation and let U be a subspace of V. Then T maps U onto a subspace of W of dimension less than or equal to the dimension of U. Moreover, if the vectors $u_1, u_2, \ldots, u_k$ span U, then the vectors $T(u_1), T(u_2), \ldots, T(u_k)$ span the image of U.

In particular, T maps the whole space V onto a subspace of W. If U is a subspace of V, then T maps U onto a subspace of W. In fact, since U is contained in V, the image of U must be a subspace of the image of V (which is possibly smaller than W). It is easy to determine the image of the trivial subspace $\{0\}$ of V. It is the trivial subspace of W because a linear transformation T must map the zero vector onto the zero vector. Indeed, since $T(0) = T(0 + 0) = T(0) + T(0)$, subtracting $T(0)$ from both sides of the equation yields $0 = T(0)$.

Result 1 leads to a practical method for determining the images of subspaces under a linear transformation mapping R^n into R^m.

Result 2 Let $T : R^n \to R^m$ be a linear transformation and let A be the standard matrix for T.

(a) The column space of A is the image of R^n under T.
(b) If U is a subspace of R^n spanned by the vectors $u_1, u_2, \ldots, u_k$ and B is the matrix with columns $u_1, u_2, \ldots, u_k$, then the column space of AB is the image of U under T.

Proof The columns of A are the images $T(e_j)$ of the standard basis vectors $e_1, e_2, \ldots, e_n$ in R^n. Since the e_j's span R^n, it follows that the columns of A span the image of R^n under T. Thus $C(A) = T(R^n)$, proving (a). The proof of part (b) is similar. The columns of the matrix AB are the vectors $Au_1, Au_2, \ldots, Au_k$. But $Au_j = T(u_j), j = 1, \ldots, k$. Thus the columns of AB span the image of U under T.

In particular, if T is a linear transformation from R^n into R^m, then the dimension of $T(R^n)$ is equal to the rank of the standard matrix for T.

EXAMPLE 2   Let $T : R^3 \to R^3$ be the linear transformation defined by $T(x) = Ax$, where

$$A = \begin{bmatrix} 1 & 1 & 3 \\ 0 & 1 & 1 \\ 2 & 0 & 4 \end{bmatrix}.$$

Reducing A to an echelon matrix shows that the first two columns of A form a basis for the column space of A. Thus T maps R^3 onto the plane spanned by the vectors $(1, 0, 2)$ and $(1, 1, 0)$. T maps the plane U spanned by $(1, 0, 0)$ and $(0, 1, 0)$ onto the column space of the matrix

$$\begin{bmatrix} 1 & 1 & 3 \\ 0 & 1 & 1 \\ 2 & 0 & 4 \end{bmatrix} \begin{bmatrix} 1 & 0 \\ 0 & 1 \\ 0 & 0 \end{bmatrix} = \begin{bmatrix} 1 & 1 \\ 0 & 1 \\ 2 & 0 \end{bmatrix}.$$

The two columns of this matrix are independent. Hence T maps U onto a plane. This plane is a subspace of the plane $T(R^3)$ and hence $T(U) = T(R^2)$.

EXERCISE 2   Let T be as in Example 2. Show that T maps the plane spanned by $(1, 0, 0)$ and $(5, 2, -2)$ onto a straight line.

EXERCISE 3   Let $T : R^3 \to R^2$ be the linear transformation defined by $T(x_1, x_2, x_3) = (x_1 + x_2 + x_3, x_2 + x_3)$.

(a) Find a basis for the image of R^3 under T. Is $T(R^3) = R^2$?
(b) Describe the image of the plane spanned by the vectors $(0, 1, 0)$ and $(0, 0, 1)$.

We know that a linear transformation $T : V \to W$ cannot map a subspace U of V onto a space of dimension higher than the dimension of U. But the dimension of the image of U under T may be less than the dimension of U. The following result indicates when these dimensions are equal.

Result 3   Let $T : V \to W$ be a linear transformation and let U be a subspace of V. Let $u_1, u_2, \ldots, u_k$ be a basis for U. Then the following conditions are equivalent.

1. $T(u) \neq 0$ for all nonzero vectors u in U.
2. The vectors $T(u_1), T(u_2), \ldots, T(u_k)$ form a basis for $T(U)$.
3. $\dim T(U) = \dim U$.

Proof   The dimension of U is k. First we show that conditions 1 and 2 are equivalent. Suppose that condition 1 holds. If $\alpha_1, \alpha_2, \ldots, \alpha_k$ are scalars such that

$$0 = \alpha_1 T(u_1) + \alpha_2 T(u_2) + \cdots + \alpha_k T(u_k),$$

then $0 = T(\alpha_1 u_1 + \alpha_2 u_2 + \cdots + \alpha_k u_k).$ Hence $\alpha_1 u_1 + \alpha_2 u_2 + \cdots +$

$\alpha_k u_k = 0$ because T maps nonzero vectors in U into nonzero vectors. Since the u_i's are independent, we conclude that $\alpha_1 = \alpha_2 = \cdots = \alpha_k = 0$. This shows that the vectors $T(u_1), T(u_2), \ldots, T(u_k)$ are independent. Since these vectors span $T(U)$ (Result 1), they form a basis for $T(U)$. Conversely, suppose that $T(u_1), T(u_2), \ldots, T(u_k)$ is a basis for $T(U)$. Let u be any nonzero vector in U. Then u is a linear combination of the u_i's, say, $u = \alpha_1 u_1 + \alpha_2 u_2 + \cdots + \alpha_k u_k$. Since $u \neq 0$, at least one of the scalars α_i is not zero. Now $T(u) = \alpha_1 T(u_1) + \alpha_2 T(u_2) + \cdots + \alpha_k T(u_k)$. Since one of the α_i's is not zero and since the vectors $T(u_j)$ are independent, it follows that $T(u) \neq 0$. This shows that conditions 1 and 2 are equivalent.

Next we prove that conditions 2 and 3 are equivalent. If condition 2 holds, then $\dim T(U) = k$. Conversely, if $\dim T(U) = k$, then $T(u_1)$, $T(u_2), \ldots, T(u_k)$ is a basis for $T(U)$ because k vectors spanning a k-dimensional space must be a basis for the space (Theorem 2.4). We have now shown that conditions 1 and 2 are equivalent and that conditions 2 and 3 are equivalent. This means that all three conditions are equivalent.

EXAMPLE 3   Let T be the projection transformation of R^3 onto a plane W. Let U be a plane in R^3 spanned by the vectors u_1 and u_2. If u is a vector in U, then $T(u) = 0$ if and only if u lies in the orthogonal complement of W. Thus by Result 3 $\dim T(U) = 2$ if and only if the plane U and the straight line $W^\perp$ have no vectors in common besides the zero vector. If this is the case, then $T(U) = W$ because $T(U)$ is a two-dimensional subspace of the plane W.

Let $T : V \to W$ be a linear transformation. Result 3 implies that if there is a nonzero vector u in V such that $T(u) = 0$, then T maps every subspace containing u onto a subspace of smaller dimension. This suggests that we should take a close look at the set N of all vectors in V that are mapped by T onto the zero vector. This set is a subspace of V. To prove this, let u and v be any vectors in N and let α be any scalar. Then

$$T(u + v) = T(u) + T(v) = 0 + 0 = 0,$$
$$T(\alpha u) = \alpha T(u) = \alpha 0 = 0.$$

Hence $u + v$ and αu are in N.

Definition   Let $T : V \to W$ be a linear transformation. Then the subspace of V consisting of all vectors that T maps onto 0 is called the **null space** (or **kernel**) of T. We denote the null space of T by $N(T)$.

In this terminology, Result 3 says that T maps a subspace U onto a subspace of the same dimension as U if and only if the zero vector is the only vector that is both in U and in the null space of T.

The relationship between the null space of a linear transformation $T : R^n \to R^m$ and the null space of the standard matrix A for T is very simple. Since $T(x) = Ax$ for all x in R^n, we see that $N(T) = N(A)$.

EXAMPLE 4   If T is the rotation transformation of R^2 through an angle θ, then $N(T) = \{0\}$. T cannot rotate a nonzero vector onto the zero vector.

EXAMPLE 5   Let U be a k-dimensional subspace of R^n, $k < n$, and let T be the projection transformation of R^n onto U. Then $T(v) = 0$ if and only if v is orthogonal to U. Hence the null space of T is the orthogonal complement of the subspace U. The nullity of T is $n - k$.

EXAMPLE 6   The standard matrix for the linear transformation $T : R^3 \to R^2$ defined by $T(x_1, x_2, x_3) = (x_1 - x_3, x_2 + x_3)$ is $A = \begin{bmatrix} 1 & 0 & -1 \\ 0 & 1 & 1 \end{bmatrix}$. Thus $N(T) = N(A)$. The vector $(1, -1, 1)$ forms a basis for the null space of A. Hence $N(T)$ is the straight line in R^3 spanned by this vector.

Definition   A linear transformation $T : V \to W$ is called **one-to-one** if T maps any two distinct vectors in V onto distinct vectors in W. That is, T is one-to-one if whenever u and v are vectors in V and $u \neq v$, then $T(u) \neq T(v)$.

The transformation in Example 6 is not one-to-one. Neither is the projection transformation in Example 5. There are certainly many different vectors in R^n which have the same projection onto the subspace U. On the positive side we note that every rotation transformation of R^2 is one-to-one. Rotating two different vectors through the same angle surely cannot result in the same vector. The following result shows that a linear transformation $T : V \to W$ is one-to-one if and only if T maps all nonzero vectors onto nonzero vectors.

Result 4   A linear transformation $T : V \to W$ is one-to-one if and only if $N(T) = \{0\}$.

Proof   First, assume that T is one-to-one. If $v \neq 0$ is a vector in V, then $T(v) \neq T(0) = 0$. Hence $N(T)$ contains no nonzero vectors. Conversely, assume that $N(T) = \{0\}$ and let u and v be two distinct vectors in V. Then $u - v \neq 0$ and hence $u - v$ is not in the null space of T. We conclude that $T(u) - T(v) = T(u - v) \neq 0$ [i.e., $T(u) \neq T(v)$]. This shows that T is one-to-one.

EXAMPLE 7   Let us prove that the reflection transformation T of R^n across a subspace V of R^n is one-to-one. Let P be the projection matrix for V. Then $T(x) = 2Px - x$ for all x in R^n (see Example 11 in Section 5.1). Suppose that u is a vector in the null space of T. Then $0 = T(u) = 2Pu - u$, or $u = 2Pu$. Since Pu, the projection of u onto V, is in V, it follows that u is in V. But this means that $Pu = u$. Therefore, $u = 2u$ and hence $u = 0$. We conclude that $N(T) = \{0\}$.

EXERCISE 4   Show that the linear transformation $T : R^2 \to R^3$ defined by $T(x_1, x_2) = (x_1 + x_2, x_1 - x_2, x_1 + 2x_2)$ is one-to-one.

Let $T : V \to W$ be a one-to-one linear transformation. Then T maps all nonzero vectors in V onto nonzero vectors. Hence, if U is any subspace of V and $u_1, u_2, \ldots, u_k$ is a basis for U, then $T(u_1), T(u_2), \ldots, T(u_k)$ is a basis for the image of U under T (Result 4) and $\dim T(U) = \dim U$. Thus linear transformations that are one-to-one preserve bases and dimensions.

EXAMPLE 8   The linear transformation $T : R^2 \to R^3$ defined by $T(x) = Ax$, where

$$A = \begin{bmatrix} 1 & 1 \\ 0 & 1 \\ 1 & 1 \end{bmatrix},$$

is one-to-one because $N(T) = N(A) = \{0\}$. Hence T maps every one-dimensional subspace of R^2 onto a one-dimensional subspace or R^3. More precisely, if the vector u spans the straight line U in R^2, then $T(u)$ spans the straight line $T(U)$ in R^3. T maps R^2 onto the plane in R^3 spanned by $T(1, 0) = (1, 0, 1)$ and $T(0, 1) = (1, 1, 1)$.

We conclude this section with a result about the relationship between the dimension of the null space of a linear transformation $T : V \to W$ and the dimension of the image of V under T.

Result 5   Let V be a vector space of dimension n and let $T : V \to W$ be a linear transformation. Then

$$\dim T(V) + \dim N(T) = n.$$

A proof of this result is outlined in Problem 5. If $V = R^n$ and $W = R^m$, we can derive Result 5 from the rank-plus-nullity theorem for matrices. Let A be the standard matrix for T. A is an $m \times n$ matrix, $T(V)$ is the column space of A, and $N(T)$ is the null space of A; therefore,

$$\dim T(V) + \dim N(T) = \text{rank } A + \text{nullity} A = n.$$

PROBLEMS 5.2

*1. Let $T : R^3 \to R^3$ be the linear transformation defined by $T(x_1, x_2, x_3) = (x_1 + x_2, 2x_1 + 3x_2 + x_3, x_2 + x_3)$. For each of the following subspaces, find a basis for its image under T.
   (a) R^3
   (b) The plane spanned by $(1, 1, 0)$ and $(0, 0, 1)$
   (c) The plane spanned by $(0, 1, 1)$ and $(1, 1, 1)$
   (d) The line spanned by $(5, 1, 3)$

*2. Repeat Problem 1 with the linear transformation $T : R^3 \to R^2$ defined by $T(x_1, x_2, x_3) = (x_1 + x_3, x_2 + x_3)$.

*3. Determine which of the following linear transformations are one-to-one.
  (a) $T : R^2 \to R^2$ defined by $T(x_1, x_2) = (x_1 + x_2, 2x_1 + 2x_2)$
  (b) $T : R^2 \to R^3$ defined by $T(x_1, x_2) = (x_1, x_1 + x_2, x_1 - x_2)$
  (c) $T : R^3 \to R^2$ defined by $T(x_1, x_2, x_3) = (x_1 - x_2, x_3)$
  (d) $T : R^3 \to R^3$ defined by $T(x_1, x_2, x_3) = (x_3, x_1, x_2)$
  (e) $T : R^2 \to R^3$ defined by $T(x_1, x_2) = (x_1, x_1, x_1)$

*4. Let A be a nonsingular matrix and let $T : R^n \to R^n$ be defined by $T(x) = Ax$. Show that T is one-to-one and maps R^n onto R^n.

5. Let $T : V \to W$ be a linear transformation and let n be the dimension of V. Show that $\dim T(V) + \dim N(T) = n$. [*Hint:* Let $u_1, u_2, \ldots, u_r$ be a basis for $N(T)$. Extend this basis to a basis $u_1, u_2, \ldots, u_n$ of V. Then show that $T(u_{r+1}), T(u_{r+2}), \ldots, T(u_n)$ is a basis for $T(V)$.]

6. Let V be a vector space of dimension n. Show that a linear transformation $T : V \to V$ is one-to-one if and only if T maps V onto V.

7. Let V be a vector space of dimension n and let $T : V \to W$ be a linear transformation. Show that if $\dim W < n$, then T is not one-to-one.

8. Let $T : R^2 \to R^2$ be a linear transformation and let P be the parallelogram determined by two nonzero vectors u and v in R^2. Show that T maps P onto a parallelogram or onto a line segment (a degenerate parallelogram). (*Hint:* Every vector in P is of the form $\alpha u + \beta v$, where $0 \le \alpha \le 1$ and $0 \le \beta \le 1$.)

9. A linear transformation $T : R^n \to R^n$ is called an *isometry* if $\langle T(x), T(y) \rangle = \langle x, y \rangle$ for all vectors x and y in R^n. Let T be an isometry.
  (a) Show that $\|T(x)\| = \|x\|$ for every vector x in R^n.
  (b) Show that the standard matrix for T is orthogonal and that $T(e_1)$, $T(e_2), \ldots, T(e_n)$ is an orthonormal basis for R^n.
  (c) Show that T is one-to-one and that $T(R^n) = R^n$.

10. Show that if A is an orthogonal $n \times n$ matrix and $T : R^n \to R^n$ is defined by $T(x) = Ax$, then T is an isometry.

5.3   MATRICES OF LINEAR TRANSFORMATIONS

We saw in Section 5.1 that every linear transformation $T : R^n \to R^m$ is a matrix transformation; $T(x) = Ax$, where A is the standard matrix for T. In this section we show that if V and W are any vector spaces of dimension n and m, respectively, then every linear transformation $T : V \to W$ can be represented by an $m \times n$ matrix. To do this we need the concept of a coordinate vector relative to a basis.

Let V be an n-dimensional vector space and let $v_1, v_2, \ldots, v_n$ be a basis for V. We saw in Section 2.4 (Result 1) that every vector v in V can be expressed as a linear combination

$$v = x_1 v_1 + x_2 v_2 + \cdots + x_n v_n$$

in a unique way. The n-vector $(x_1, x_2, \ldots, x_n)$ is called the **coordinate vector** of v with respect (or relative) to the basis $v_1, v_2, \ldots, v_n$. The scalar x_i is called the ith **coordinate** of v with respect to the basis $v_1, v_2, \ldots, v_n$. Note that the coordinate vector of v_j with respect to the basis $v_1, v_2, \ldots, v_n$ is e_j, the jth vector in the standard basis for R^n.

EXAMPLE 1   If $b = (b_1, b_2, \ldots, b_n)$ is a vector in R^n, then $b = b_1 e_1 + b_2 e_2 + \cdots + b_n e_n$, where $e_1, e_2, \ldots, e_n$ are the standard basis vectors in R^n. Consequently, the coordinates of b with respect to the standard basis for R^n are just the components of b and the coordinate vector of b with respect to the standard basis is b itself.

EXAMPLE 2   The vectors $v_1 = (1, 2)$ and $v_2 = (3, 5)$ form a basis for R^2. To find the coordinate vector of $v = (3, 4)$ with respect to this basis we have to find scalars x_1 and x_2 such that $v = x_1 v_1 + x_2 v_2$. The matrix form of this equation is

$$\begin{bmatrix} 1 & 3 \\ 2 & 5 \end{bmatrix} \begin{bmatrix} x_1 \\ x_2 \end{bmatrix} = \begin{bmatrix} 3 \\ 4 \end{bmatrix}.$$

The only solution is $x_1 = -3$, $x_2 = 2$. Thus $v = -3v_1 + 2v_2$ and the coordinate vector of v with respect to the basis v_1, v_2 is $(-3, 2)$.

In general, if $v_1, v_2, \ldots, v_n$ is a basis for a subspace V of R^m and v is a vector in V, then in order to find the coordinate vector of v with respect to the given basis, we have to solve a linear system. The equation

$$x_1 v_1 + x_2 v_2 + \cdots + x_n v_n = v$$

is equivalent to the system $Ax = v$, where A is the matrix with columns $v_1, v_2, \ldots, v_n$. The rank of A is n and hence $Ax = v$ has a unique solution. This unique solution is the coordinate vector of v relative to the basis $v_1, v_2, \ldots, v_n$.

EXERCISE 1   Find the coordinate vector of $v = (1, 5, 9)$ with respect to the basis:

(a) $(1, 0, 1)$, $(1, 1, 2)$, $(0, 2, 3)$ of R^3
(b) $(1, 0, 0)$, $(1, 1, 0)$, $(1, 1, 1)$ of R^3

EXAMPLE 3   (Chapter 4 required) We saw in Chapter 4 that the functions $1, x, x^2, \ldots, x^n$ form a basis for the space $P_n(R)$ of all polynomials of degree less than or equal to n. The dimension of $P_n(R)$ is $n + 1$. If

$$f(x) = a_0 + a_1 x + a_2 x^2 + \cdots + a_n x^n$$

is a polynomial in this space, then the coordinate vector of $f(x)$ with respect to the basis $1, x, x^2, \ldots, x^n$ is the vector $(a_0, a_1, a_2, \ldots, a_n)$ in R^{n+1}.

Coordinate vectors allow us to represent the vectors in any n-dimensional space V by vectors in R^n. Once we have chosen a basis $v_1, v_2, \ldots, v_n$ for V, every vector in V corresponds to a unique vector in R^n, namely its coordinate vector with respect to the basis $v_1, v_2, \ldots, v_n$. Conversely, every n-vector $x = (x_1, x_2, \ldots, x_n)$ corresponds to a unique vector in V, namely the vector $v = x_1 v_1 + x_2 v_2 + \cdots + x_n v_n$. When we pass from the vectors in V to their coordinate vectors, we pass from V to the familiar n-tuples of real numbers. This transformation from a vector to its coordinates is linear. To prove this, let u and v be two vectors in V and let $x = (x_1, x_2, \ldots, x_n)$, $y = (y_1, y_2, \ldots, y_n)$ be their coordinate vectors relative to the basis $v_1, v_2, \ldots, v_n$. Then

$$u + v = x_1 v_1 + x_2 v_2 + \cdots + x_n v_n + y_1 v_1 + y_2 v_2 + \cdots + y_n v_n$$
$$= (x_1 + y_1)v_1 + (x_2 + y_2)v_2 + \cdots + (x_n + y_n)v_n.$$

Hence $x + y = (x_1 + y_1, x_2 + y_2, \ldots, x_n + y_n)$ is the coordinate vector of $u + v$ relative to the basis $v_1, v_2, \ldots, v_n$. Similarly, if α is a scalar, then

$$\alpha u = \alpha(x_1 v_1 + x_2 v_2 + \cdots + x_n v_n) = \alpha x_1 v_1 + \alpha x_2 v_2 + \cdots + \alpha x_n v_n$$

and hence $\alpha x = (\alpha x_1, \alpha x_2, \ldots, \alpha x_n)$ is the coordinate vector of αu with respect to the basis $v_1, v_2, \ldots, v_n$. In general, any linear combination of vectors in V corresponds to the same linear combination of their coordinate vectors. That is, if $u_1, u_2, \ldots, u_k$ are vectors in V and $z_1, z_2, \ldots, z_k$ are their coordinate vectors relative to the basis $v_1, v_2, \ldots, v_n$, then the coordinate vector of $u = \alpha_1 u_1 + \alpha_2 u_2 + \cdots + \alpha_k u_k$ relative to the basis $v_1, v_2, \ldots, v_n$ is $\alpha_1 z_1 + \alpha_2 z_2 + \cdots + \alpha_k z_k$.

Definition Let $T : V \to W$ be a linear transformation and suppose that $v_1, v_2, \ldots, v_n$ and $w_1, w_2, \ldots, w_m$ are bases for V and W, respectively. The $m \times n$ matrix whose jth column is the coordinate vector of $T(v_j)$ with respect to the basis $w_1, w_2, \ldots, w_m$ is called the **matrix for T** with respect to the chosen bases for V and W.

Next, we show that if we represent the vectors in V and W by their coordinate vectors relative to the chosen bases, then the matrix for T with respect to these bases represents the linear transformation T.

Result 1 Suppose that $v_1, v_2, \ldots, v_n$ and $w_1, w_2, \ldots, w_m$ are bases for the vector spaces V and W, respectively. Let $T : V \to W$ be a linear transformation and let A be the matrix for T with respect to the chosen bases. If v is a vector in V and x is the coordinate vector of v with respect to the basis $v_1, v_2, \ldots, v_n$, then Ax is the coordinate vector of $T(v)$ with respect to the basis $w_1, w_2, \ldots, w_m$.

Proof   Let v be a vector in V and let $x = (x_1, x_2, \ldots, x_n)$ be the coordinate vector of v with respect to the basis $v_1, v_2, \ldots, v_n$. Then

$$v = x_1 v_1 + x_2 v_2 + \cdots + x_n v_n$$

and hence

$$T(v) = x_1 T(v_1) + x_2 T(v_2) + \cdots + x_n T(v_n). \tag{1}$$

Let $a_1, a_2, \ldots, a_n$ be the columns of A. Since the coordinate vector of $T(v_j)$ with respect to the basis $w_1, w_2, \ldots, w_m$ is a_j, equation (1) shows that the coordinate vector of $T(v)$ with respect to the basis $w_1, w_2, \ldots, w_m$ is the m-vector

$$Ax = x_1 a_1 + x_2 a_2 + \cdots + x_n a_n.$$

EXAMPLE 4   Let $T : R^n \to R^m$ be a linear transformation. Let $e_1, e_2, \ldots, e_n$ be the standard basis for R^n. Then the coordinate vector of $T(e_j), j = 1, \ldots, n$, with respect to the standard basis for R^m is $T(e_j)$ itself. Thus the matrix of T with respect to the standard bases for R^n and R^m is the standard matrix for T as defined in Section 5.1.

EXAMPLE 5   Consider the linear transformation $T : R^3 \to R^2$ defined by $T(x_1, x_2, x_3) = (x_1 - x_2, x_2 + x_3)$. The standard matrix for T is $\begin{bmatrix} 1 & -1 & 0 \\ 0 & 1 & 1 \end{bmatrix}$. The vectors $v_1 = (2, 0, 1)$, $v_2 = (0, 2, 2)$, and $v_3 = (0, 2, 3)$ form a basis for R^3. The vectors $w_1 = (1, 2)$ and $w_2 = (0, 1)$ form a basis for R^2. In order to find the matrix for T with respect to the bases v_1, v_2, v_3 and w_1, w_2, we compute

$$T(v_1) = (2, 1) = 2w_1 - 3w_2,$$
$$T(v_2) = (-2, 4) = -2w_1 + 8w_2,$$
$$T(v_3) = (-2, 5) = -2w_1 + 9w_2.$$

Hence the coordinate vectors of $T(v_1)$, $T(v_2)$, and $T(v_3)$ with respect to the basis w_1, w_2 are $(2, -3)$, $(-2, 8)$, and $-2, 9)$, respectively. Thus the matrix for T with respect to the bases v_1, v_2, v_3 and w_1, w_2 is

$$A = \begin{bmatrix} 2 & -2 & -2 \\ -3 & 8 & 9 \end{bmatrix}.$$

Consider the vector $v = (-6, 6, 7)$. Since $v = -3v_1 - v_2 + 4v_3$, the coordinate vector of v with respect to the basis v_1, v_2, v_3 is $x = (-3, -1, 4)$. Now

$$\begin{bmatrix} 2 & -2 & -2 \\ -3 & 8 & 9 \end{bmatrix} \begin{bmatrix} -3 \\ -1 \\ 4 \end{bmatrix} = \begin{bmatrix} -12 \\ 37 \end{bmatrix}$$

and hence $(-12, 37)$ is the coordinate vector of $T(v)$ with respect to basis w_1, w_2. We have

$$T(v) = -12w_1 + 37w_2 = (-12, 13).$$

EXERCISE 2   Let $T : R^2 \to R^3$ be the linear transformation defined by $T(x_1, x_2)$ $= (x_1, x_2, x_1 + x_2)$. The vectors $v_1 = (1, 0)$ and $v_2 = (1, 1)$ form a basis for R^2. The vectors $w_1 = (1, 0, 0)$, $w_2 = (1, 1, 0)$, and $w_3 = (1, 1, 1)$ form a basis for R^3. Find the matrix of T with respect to these bases.

When T is a linear transformation of a vector space V into itself, we usually use only one basis to construct a matrix for T. That is, if $T : V \to V$ is a linear transformation and $v_1, v_2, \ldots, v_n$ is a basis for V, then the $n \times n$ matrix A whose jth column is the coordinate vector of $T(v_j)$ with respect to the basis $v_1, v_2, \ldots, v_n$ is called the matrix of T with respect to the basis $v_1, v_2, \ldots, v_n$. If v is any vector in V and x is the coordinate vector of v with respect to $v_1, v_2, \ldots, v_n$, then Ax is the coordinate vector of $T(v)$ with respect to $v_1, v_2, \ldots, v_n$.

EXAMPLE 6   Let V be a straight line in R^2 passing through the origin and let $T : R^2 \to R^2$ be the projection transformation of R^2 onto V. We will show that if we choose the right basis for R^2, then the matrix of T with respect to this basis is particularly simple. Choose any nonzero vector w_1 on the line V, and let w_2 be any nonzero vector in R^2 orthogonal to w_1. Then w_1 and w_2 are independent and hence form a basis for R^2. Since $T(w_1) = w_1 = 1w_1 + 0w_2$ and $T(w_2) = 0 = 0w_1 + 0w_2$, the coordinate vectors of $T(w_1)$ and $T(w_2)$ with respect to the basis w_1, w_2 are $(1, 0)$ and $(0, 0)$, respectively. Hence $\begin{bmatrix} 1 & 0 \\ 0 & 0 \end{bmatrix}$ is the matrix of T relative to the basis w_1, w_2 of R^2.

EXERCISE 3   Let V, w_1, and w_2 be as in Example 6 and let $T : R^2 \to R^2$ be the reflection transformation of R^2 across the straight line V. Find the matrix of T relative to the basis w_1, w_2.

EXAMPLE 7   (Chapter 4 required) The polynomials 1, x, and x^2 form a basis for the space $P_2(R)$ of all polynomials of degree less than or equal to 2. The differential operator $D : P_2(R) \to P_2(R)$ defined by $D(f) = f'$ is a linear transformation. Now

$$D(1) = 0 = 0 \cdot 1 + 0x + 0x^2,$$
$$D(x) = 1 = 1 \cdot 1 + 0x + 0x^2,$$
$$D(x^2) = 2x = 0 \cdot 1 + 2x + 0x^2.$$

Thus the coordinate vectors of $D(1)$, $D(x)$, and $D(x^2)$ with respect to the basis $1, x, x^2$ of $P_2(R)$ are $(0, 0, 0)$, $(1, 0, 0)$, and $(0, 2, 0)$, respectively. Consequently, the matrix of D with respect to the basis $1, x, x^2$ is

$$A = \begin{bmatrix} 0 & 1 & 0 \\ 0 & 0 & 2 \\ 0 & 0 & 0 \end{bmatrix}.$$

If $f(x) = a_0 + a_1 x + a_2 x^2$ is a polynomial in $P_2(R)$, then (a_0, a_1, a_2) is the coordinate vector of $f(x)$ with respect to the basis $1, x, x^2$ and

$$\begin{bmatrix} 0 & 1 & 0 \\ 0 & 0 & 2 \\ 0 & 0 & 0 \end{bmatrix} \begin{bmatrix} a_0 \\ a_1 \\ a_2 \end{bmatrix} = \begin{bmatrix} a_1 \\ 2a_2 \\ 0 \end{bmatrix}$$

is the coordinate vector of $f'(x) = a_1 + 2a_2 x$ with respect to the basis $1, x, x^2$ of $P_2(R)$.

PROBLEMS 5.3

*1. Find the coordinate vector of $(2, 1, 3)$ with respect to:
   (a) the basis $(1, 0, 1)$, $(1, 1, 0)$, $(1, 1, 1)$ of R^3
   (b) the basis $(0, 0, 1)$, $(0, 1, 1)$, $(1, 1, 1)$ of R^3

*2. Let $T : R^3 \to R^2$ be defined by $T(x_1, x_2, x_3) = (3x_1 + 2x_2, x_2 - 2x_3)$. Find the matrix for T with respect to the bases $(1, 0, 1)$, $(1, 2, 1)$, $(0, 1, 2)$ for R^3 and $(1, 2)$, $(4, 0)$ for R^2.

*3. Let T be the linear transformation in Problem 2. Find the matrix for T with respect to the bases $(0, 0, 1)$, $(0, 1, 1)$, $(1, 1, 1)$ for R^3 and $(1, 0)$, $(1, 1)$ for R^2.

4. Let $T : R^2 \to R^2$ be defined by $T(x_1, x_2) = (x_1 + x_2, 2x_2)$. Find the matrix for T with respect to the basis $(1, 0)$, $(1, 1)$ for R^2.

5. Let $v_1, v_2, \ldots, v_n$ be a basis for R^n. Let A be the matrix whose columns are the coordinate vectors of $e_1, e_2, \ldots, e_n$ (the standard basis) relative to the basis $v_1, v_2, \ldots, v_n$. Show that for every x in R^n, Ax is the coordinate vector of x relative to $v_1, v_2, \ldots, v_n$.

6. Let $v_1, v_2, \ldots, v_n$ be a basis for the vector space V and let A be the matrix for a linear transformation $T : V \to V$ with respect to this basis. Show that if A has rank n, then T is one-to-one.

*7. Let V be the line in R^2 spanned by the vector $(1, 2)$. Find a basis of R^2 such that the matrix of the reflection transformation of R^2 across V with respect to this basis is $\begin{bmatrix} 1 & 0 \\ 0 & -1 \end{bmatrix}$.

8. Let $T : R^n \to R^n$ be an isometry (see Problem 9 in Section 5.2).
   (a) Show that if $v_1, v_2, \ldots, v_n$ is an orthogonal (orthonormal) basis for R^n, then $T(v_1), T(v_2), \ldots, T(v_n)$ also is an orthogonal (orthonormal) basis for R^n.
   (b) Show that the matrix of T with respect to an orthonormal basis for R^n is orthogonal.

Chapter 6

Abstract Vector Spaces

The purpose of this brief chapter is to give the abstract definitions of a vector space and an inner product space. These abstract definitions specify those properties of *n*-vectors and inner products that are required to derive the results in Chapters 2 and 3. The advantage of this abstract approach is that if we know that a collection of objects has these properties, then all of the results about vector spaces and inner product spaces apply to that collection.

6.1 ABSTRACT VECTOR SPACES

In Chapter 2 we saw that the study of systems of linear equations led naturally to the concept of a vector space and the related notions of linear independence and dependence, spanning, basis, and dimension. Since *n*-vectors play an important role in the theory of linear equations, we were led to formulate these concepts for sets of *n*-vectors. However, the idea of a vector space need not be restricted to collections of *n*-vectors. It can be generalized so that it applies to many other sets of objects. We achieve this generalization by making the following abstract definition of a vector space.

Definition   A **vector space** is a collection V of objects, called **vectors**, together with two operations, called addition and scalar multiplication, which satisfy the following conditions. (In these conditions, x, y, and z are vectors in V and α and β are scalars.)

I. Conditions for addition
   (a) $x + y = y + x$
   (b) $x + (y + z) = (x + y) + z$
   (c) There is a unique vector 0 such that $x + 0 = x$ for all x in V.
   (d) For every vector x in V there is a unique vector $-x$ in V such that $x + (-x) = 0$.

II. Conditions for scalar multiplication
   (e) $\alpha(\beta x) = (\alpha\beta)x$
   (f) $1x = x$
   (g) $\alpha(x + y) = \alpha x + \alpha y$
   (h) $(\alpha + \beta)x = \alpha x + \beta x$

This definition does not specify the particular nature of the objects that are called vectors. All that has been specified is that there must be two operations defined on the collection of these objects and that these operations must satisfy certain rules. Comparing these rules with the properties of vectors given in Chapter 1 shows that R^n is a vector space with respect to the operations of adding n-vectors and multiplying an n-vector by a real number. The vectors in this vector space are n-vectors. Comparing these rules with the properties of $m \times n$ matrices given in Chapter 1 will show that the collection of all $m \times n$ matrices is a vector space with respect to the operations of addition of matrices and multiplying a matrix by a scalar. The vectors in this vector space are $m \times n$ matrices.

EXERCISE 1   (Chapter 4 required) Show that the set $C(I)$ of all functions that are continuous on an interval I is a vector space with respect to the usual operations of addition and scalar multiplication. The vectors in this vector space are continuous functions.

EXERCISE 2   (Chapter 5 required) In Chapter 5 (Problem 15 in Section 5.1) we defined addition and scalar multiplication of linear transformations. Show that the collection V of all linear transformations from R^n to R^m is a vector space with respect to these operations. The vectors in this vector space are linear transformations.

EXERCISE 3   The collection of all infinite sequences $(a_1, a_2, \dots)$ of real numbers can be thought of as an infinite-dimensional generalization of R^n. Thus it is natural to denote this collection by R^∞. Show that R^∞ is a vector space with respect to the obvious two operations. (The reader should define these operations.)

The abstract definition of a vector space clearly indicates that a vector space is a much more complicated notion than we have indicated in Chapter 2. Not only must the vector space be closed under the operations of addition and scalar multiplication, but also these operations must satisfy several rules. In the case of m-vectors, these rules follow easily from the definitions of addition of vectors, multiplication of a vector by a scalar, and the properties of real numbers. Hence when dealing with collections of m-vectors, the crucial condition is that the collection be closed under addition and scalar multiplication. In other situations the rules for vector operations may not be obvious and hence must be verified.

Why are the rules needed? They are needed because they specify the legal manipulations that can be performed when working (or computing) with vectors. All the theorems in Chapter 2 dealing with the concepts of linear independence, spanning, basis, and dimension can be proved using only these rules. However, we cannot always use the proofs that we gave in Chapter 2 for abstract vector spaces. The reason is that many of them are coordinate proofs. That is, many of the proofs in Chapter 2 depend on a choice of a basis for the vector space. This dependence is not noticed there because there is such a natural basis for R^n, namely the standard basis. It is in fact so natural that any n-vector $x = (x_1, \ldots, x_n)$ is its own coordinate vector with respect to the standard basis. It turns out that if V is a finite-dimensional vector space and if a basis of V is specified, and if all vectors in Chapter 2 are assumed to be the coordinate vectors of the vectors in V with respect to this basis, then all the proofs are valid. However, it is not always convenient to be required to first choose a basis for V in order to interpret some result about a set of vectors in V or to perform some calculations with the vectors in V. In addition, reasoning with the vectors themselves (instead of their coordinates with respect to some basis) often illuminates the structure of the proof more clearly. For these reasons it is desirable to give proofs of theorems about vector spaces which do not depend on the choice of a basis. Such proofs are called coordinate-free proofs. Some of the proofs that we have given in Chapter 2 are coordinate-free proofs. For example, in Section 2.2 we gave two proofs of the following result. Given a set of vectors $v_1, \ldots, v_n$ in R^m, the collection of all linear combinations of these vectors is a vector space. In the first proof, we formed the matrix having these vectors as its columns. By doing this we have identified the vectors with their coordinate vectors with respect to the standard basis. For this reason, the proof is a coordinate proof. However, the second proof shows that this result can be proved by manipulating the vectors themselves, without ever mentioning their coordinates. Thus the second proof is a coordinate-free proof.

Finally, let us reconcile the definition of a vector space given in Chapter 2 with the definition given above. We define a **subspace** of a vector space V to be a collection W of vectors in V which is itself a vector space with respect to the operations of addition and scalar multiplication defined on V. Thus a subspace of a vector space is a subset of the vector space which is closed under addition and scalar multiplication and satisfies all the defining conditions of a vector space. But this last requirement follows automatically from the fact that vectors in W are also in V and hence (since V is a vector space) satisfy the defining conditions of a vector space. This proves the following theorem.

Theorem 6.1   *A subset W of a vector space V is a subspace of V if and only if it is closed under the operations in V of addition and scalar multiplication.*

Since R^n is a vector space, it follows from this theorem that it is legitimate to say (as we did in Chapter 2) that a set of n-vectors is a vector space if and only if it is closed under addition and scalar multiplication of vectors.

EXERCISE 4   (Chapter 4 required) Show that Exercise 1 together with this result justifies the definition of a function space given in Chapter 4.

EXERCISE 5   Let v_0 be a fixed vector in R^n. Show that the set of all linear transformations from R^n into R^n with the property that $T(v_0) = 0$ is a vector space.

EXERCISE 6   Use Exercise 3 to show that the set of all infinite sequences of the form $(a_1, a_2, \ldots)$ with $a_2 = 0$ is a vector space.

PROBLEMS 6.1

*1. Consider the set of all infinite sequences that have only finitely many nonzero terms. A sequence $(a_1, a_2, \ldots)$ is in this collection if and only if there is an N such that $a_n = 0$ for all $n \geq N$. Is this set a vector space?

*2. Consider the set of all infinite sequences that have infinitely many terms equal to zero. A sequence $(a_1, a_2, \ldots)$ is in this set if and only if $a_n = 0$ for infinitely many n. [For example, $(1, 0, 1, 0, \ldots)$ is in this set.] Is this set a vector space?

3. (a) Show that the matrices $\begin{bmatrix} 1 & 0 \\ 0 & 0 \end{bmatrix}, \begin{bmatrix} 0 & 1 \\ 0 & 0 \end{bmatrix}, \begin{bmatrix} 0 & 0 \\ 1 & 0 \end{bmatrix},$ and $\begin{bmatrix} 0 & 0 \\ 0 & 1 \end{bmatrix}$ form a basis for the vector space of all 2×2 matrices.
   *(b) Let V be the vector space of all $m \times n$ matrices. Find a basis for V. What is the dimension of V?
   *(c) Is the set of all $m \times n$ echelon matrices a subspace of V?

*4. Let V be the vector space of all $n \times n$ matrices.
   (a) Show that the set W of all upper triangular matrices is a subspace of V. Find a basis for W and the dimension of W.
   (b) Show that the set U of all diagonal $n \times n$ matrices is a subspace of V. Find a basis for U and the dimension of U.
   (c) Show that the set of all $n \times n$ symmetric matrices is a subspace of V. What is its dimension?
   (d) Show that the set of all $n \times n$ skew-symmetric matrices is a subspace of V. What is its dimension?

*5. Let V be the vector space of all $n \times n$ matrices. Which of the following sets of matrices are subspaces of V?

(a) The matrices with trace equal to zero.

(b) The nonsingular matrices.

(c) The singular matrices.

(d) The matrices with all diagonal entries equal to zero.

(e) The orthogonal matrices.

6. Let A be a 3×3 matrix. Consider the 10 matrices $I, A, A^2, \ldots, A^9$.

(a) Show that these 10 matrices are linearly dependent.

(b) Use part (a) to show that there exists an integer $k \le 9$ and scalars $a_0, a_1, \ldots, a_{k-1}$ such that

$$A^k + a_{k-1}A^{k-1} + \cdots + a_1 A + a_0 I = 0.$$

(c) Generalize your argument to apply to the case where A is an $n \times n$ matrix.

7. Explain why R^∞, the vector space of all infinite sequences, cannot have a basis consisting of finitely many sequences (see Exercise 3).

8. Identify those proofs in Chapter 2 that are coordinate proofs.

9. (Calculus required) Show that the set of all convergent infinite sequences is a vector space.

10. (Calculus required) Show that the set of all infinite sequences which converge to zero is a vector space.

*11. (Calculus required) Is the set of all infinite sequences that converge to 1 a vector space?

6.2   INNER PRODUCT SPACES

In Chapter 4 we pointed out that all of the theorems and results about inner products that were derived in Chapter 3 depended only on those properties of an inner product given in Theorem 3.1. When we defined an inner product in function spaces, all these properties were satisfied. Because of this, all the results from Chapter 3 carried over to function spaces unchanged.

Our motivation for the abstract definition of an inner product is exactly the same. We define an inner product on an abstract vector space to be a function that satisfies those properties given in Theorem 3.1. Then all of the results and theorems about inner products that we derived in Chapter 3 will hold for a vector space with an (abstract) inner product.

Definition   An **inner product** on a vector space V is a function that assigns a real number $\langle u, v \rangle$ to every pair of vectors u and v in V in such a way that for all vectors u and v and scalars α,

(a) $\langle u, v \rangle = \langle v, u \rangle$

(b) $\langle u, v + w \rangle = \langle u, v \rangle + \langle u, w \rangle$
(c) $\langle \alpha u, v \rangle = \alpha \langle u, v \rangle$
(d) $\langle u, u \rangle \geq 0$, and $\langle u, u \rangle = 0$ if and only if $u = 0$

An **inner product space** is a vector space with an inner product.

EXERCISE 1   Let $\alpha_1, \ldots, \alpha_m$ be fixed positive numbers. Show that the formula

$$\langle u, v \rangle = \alpha_1 u_1 v_1 + \cdots + \alpha_m u_m v_m$$

defines an inner product on R^m.

EXERCISE 2   (Chapter 4 required) Let $h(x)$ be a nonnegative function in $C(I)$, $I = [a, b]$. Show that the formula

$$\langle f, g \rangle = \int_a^b f(x)g(x)h(x)\, dx$$

defines an inner product on $C(I)$.

EXERCISE 3   An $m \times m$ matrix A is called *positive definite* if $x^T A x > 0$ for all nonzero vectors x in R^m. Let A be a positive definite symmetric matrix. Show that the formula

$$\langle u, v \rangle = u^T A v$$

defines an inner product on R^m.

PROBLEMS 6.2

In each of the problems we assume that V is an inner product space.

1. The norm (or length) of a vector in V, denoted by $\| v \|$, is defined by the equation

$$\| v \| = \sqrt{\langle v, v \rangle}.$$

   (a) Show that for any vector v in V and scalar α,

   $$\| \alpha v \| = | \alpha | \cdot \| v \|.$$

   (b) Show that for any vector v in V, $\| v \| \geq 0$, and $\| v \| = 0$ if and only if $v = 0$.
   (c) State and prove the Cauchy–Schwarz inequality for vectors in V. (*Hint:* Use the proof given in Chapter 3.)
   (d) State and prove the triangle inequality for vectors in V. (*Hint:* Use the proof given in Chapter 3.)
   (e) How would you define the distance between two vectors in V?

2. Define two vectors u and v in V to be orthogonal if $\langle u, v \rangle = 0$.
   (a) Show that the zero vector in V is orthogonal to every vector in V.
   (b) Show that two vectors u and v in V are orthogonal if and only if $\| u + v \| = \| u - v \|$.
   (c) Show that if u and v are orthogonal vectors in V, then $\| u + v \|^2 = \| u \|^2 + \| v \|^2$.

3. Define a collection $v_1, \ldots, v_k$ of vectors in V to be orthogonal if $\langle v_i, v_j \rangle = 0$ whenever $i \neq j$. Show that if $v_1, \ldots, v_k$ are nonzero orthogonal vectors in V, then they are linearly independent. (*Hint:* Use the proof given in Chapter 3.)

4. Let S be a subset of V. Define a vector v in V to be orthogonal to the set S, written $v \perp S$, if v is orthogonal to every vector in S.
   (a) Show that the only vector in V that is orthogonal to V is the zero vector.
   (b) Show that if W is a subspace of V, then $v \perp W$ if and only if v is orthogonal to a spanning set for W.

5. Let S be a subset of V. Define the orthogonal complement of S, denoted by $S^\perp$, to be the set of all vectors in V that are orthogonal to S.
   (a) Show that $S^\perp$ is a subspace of V.
   (b) Let W be a subspace of V. Show that the only vector contained in both W and $W^\perp$ is the zero vector.
   (c) If V is n-dimensional and W is an r-dimensional subspace of V, show that the dimension of $W^\perp$ is $n - r$. (*Hint:* Use Problem 8 in Section 3.4.)
   (d) Show that if $w_1, \ldots, w_r$ is a basis for W and $w_{r+1}, \ldots, w_n$ is a basis for $W^\perp$, then $w_1, \ldots, w_r, w_{r+1}, \ldots, w_n$ is a basis for V. [*Hint:* See the proof of Theorem 3.2(d).]

6. Prove Theorem 3.3 when R^m is replaced by an abstract inner product space V.

7. (a) Define the projection of a vector b onto a subspace W of an abstract inner product space V.
   (b) Show that the projection of b onto W is the vector in W that is closest to b.

8. Show that the Gram–Schmidt process is valid in any finite-dimensional inner product space V.

Chapter 7

Determinants

In this chapter we define the determinant of a square matrix. The determinant function associates with every square matrix A a unique *number* (denoted det A) called the *determinant* of A. The determinant of a 2×2 matrix $\begin{bmatrix} a & b \\ c & d \end{bmatrix}$ is defined by the formula $ad - bc$. It is possible to similarly define the determinant of an $n \times n$ matrix by an explicit formula. However, for $n > 4$ this formula is quite complicated and useless for computation. Therefore, we choose to define determinants by means of their basic properties. These basic properties can be derived in a formal manner, but their derivation is quite messy and not very instructive. Our approach has the advantage that it places emphasis on those properties of determinants that are most useful for computational purposes. This will enable us to derive a reasonably simple procedure for computing determinants which is based on the Gaussian elimination process.

We are interested in determinants primarily as a tool to answer important questions that arise in Chapter 8. For this reason we concentrate on computational results and omit most proofs.

7.1 PROPERTIES OF THE DETERMINANT

We now list the fundamental properties of the determinant.

Property 1. If a scalar multiple of one row of a square matrix A is added to another row of A, then the determinant of the resulting matrix is equal to the determinant of A.

For example,

$$\det \begin{bmatrix} 1 & 2 & -3 \\ -4 & 3 & 5 \\ 4 & 8 & -5 \end{bmatrix} = \det \begin{bmatrix} 1 & 2 & -3 \\ -4 & 3 & 5 \\ 2 & 4 & 1 \end{bmatrix}.$$

Here the first matrix is the result of adding twice the first row of the second matrix to its third row.

Property 2. If any two rows of a square matrix A are interchanged, then the determinant of the resulting matrix is the negative of the determinant of A.

For example,

$$\det \begin{bmatrix} 1 & 2 & -3 \\ 2 & 4 & 1 \\ -4 & 3 & 5 \end{bmatrix} = (-1) \det \begin{bmatrix} 1 & 2 & -3 \\ -4 & 3 & 5 \\ 2 & 4 & 1 \end{bmatrix}.$$

Property 3. If any row of a square matrix A is multiplied by a scalar α, then the determinant of the resulting matrix is α times the determinant of A. Equivalently, if B is the matrix obtained from A by dividing a row of A by the constant α, then $\det A = \alpha \det B$.

For example,

$$\det \begin{bmatrix} 1 & 2 & -3 \\ -4 & 3 & 5 \\ 4 & 8 & 2 \end{bmatrix} = 2 \det \begin{bmatrix} 1 & 2 & -3 \\ -4 & 3 & 5 \\ 2 & 4 & 1 \end{bmatrix}.$$

Property 4. The determinant of the $n \times n$ identity matrix I is equal to 1.

For example, when $n = 3$ we have

$$\det \begin{bmatrix} 1 & 0 & 0 \\ 0 & 1 & 0 \\ 0 & 0 & 1 \end{bmatrix} = 1.$$

EXERCISE 1 Verify that each of these properties holds for the determinant of a 2×2 matrix by using the formula

$$\det \begin{bmatrix} a & b \\ c & d \end{bmatrix} = ad - bc.$$

The preceding four properties define the determinant function. Any property that can be derived from these is also a property of the determinant. We present below some of these additional properties and show how they can be used to compute the determinant of a matrix.

Suppose that a square matrix A has a zero row. Then multiplying this zero row by 2 does not change A; hence (by Property 3) det $A = 2$ det A. The only number that is equal to twice itself is the number 0. Thus:

Property 5.   If a square matrix has a zero row, then its determinant is equal to zero.

We leave it to the reader to deduce from Properties 1 and 5 that:

Property 6.   The determinant of a square matrix with two identical rows is equal to zero.

Next suppose that we are dealing with an upper triangular matrix that has no zeros on the main diagonal, say

$$A = \begin{bmatrix} 2 & 5 & 8 & 4 \\ 0 & -1 & 4 & 7 \\ 0 & 0 & 3 & -2 \\ 0 & 0 & 0 & 2 \end{bmatrix}.$$

We can make all entries above the main diagonal zero without changing either the determinant or the diagonal entries. We do this by a method similar to back substitution. We first subtract appropriate multiples of the last row from the preceding rows. Then we continue this process row by row until we are done. Thus (by Property 1)

$$\det A = \det \begin{bmatrix} 2 & 0 & 0 & 0 \\ 0 & -1 & 0 & 0 \\ 0 & 0 & 3 & 0 \\ 0 & 0 & 0 & 2 \end{bmatrix}.$$

Now we use Property 3 successively to "factor out" the diagonal entries. Since det $I = 1$ (Property 4),

$$\det A = 2 \cdot (-1) \cdot 3 \cdot 2 \cdot \det \begin{bmatrix} 1 & 0 & 0 & 0 \\ 0 & 1 & 0 & 0 \\ 0 & 0 & 1 & 0 \\ 0 & 0 & 0 & 1 \end{bmatrix} = -12.$$

This method works for any upper triangular matrix with no zero entries on the main diagonal. It follows that the determinant of an upper triangular matrix is equal to the product of the diagonal entries provided that these entries are nonzero. When one of the diagonal entries is zero, the back substitution procedure does not eliminate all entries above the main diagonal. But the last row with a zero diagonal entry will become a zero row and hence the determinant is zero. So in this case we also know that the determinant is equal to the product of the diagonal entries. For example,

$$
\det \begin{bmatrix} 5 & 1 & 2 & 1 \\ 0 & 0 & 3 & -3 \\ 0 & 0 & 4 & 7 \\ 0 & 0 & 0 & 2 \end{bmatrix} = \det \begin{bmatrix} 5 & 1 & 0 & 0 \\ 0 & 0 & 0 & 0 \\ 0 & 0 & 4 & 0 \\ 0 & 0 & 0 & 2 \end{bmatrix} = 0 = 5 \cdot 0 \cdot 4 \cdot 2.
$$

Property 7.   The determinant of any upper triangular square matrix A is equal to the product of the diagonal entries of A.

EXERCISE 2   Convince yourself that the determinant of a lower triangular square matrix (i.e., a matrix with all zeros above the main diagonal) is equal to the product of the diagonal entries.

We are now ready to develop a procedure to compute the determinant of an arbitrary square matrix.

EXAMPLE 1   Let

$$
A = \begin{bmatrix} 1 & 2 & -3 \\ 2 & 4 & 1 \\ -4 & 3 & 5 \end{bmatrix}.
$$

Subtracting multiples of the first row from the other rows gives (by Property 1)

$$
\det A = \det \begin{bmatrix} 1 & 2 & -3 \\ 0 & 0 & 7 \\ 0 & 11 & -7 \end{bmatrix}.
$$

Now an interchange of rows gives us a triangular matrix. Hence by Properties 2 and 7 we obtain

$$
\det A = -\det \begin{bmatrix} 1 & 2 & -3 \\ 0 & 11 & -7 \\ 0 & 0 & 7 \end{bmatrix} = -1 \cdot 11 \cdot 7 = -77.
$$

Example 1 illustrates how to proceed in general. We reduce the matrix to upper triangular form by forward elimination. If we add multiples of a row to another row, the determinant does not change. Of course, we have to keep track of how often we interchange rows because each such interchange switches the sign of the determinant. Finally, we read off the determinant from the triangular form.

In this reduction it is sometimes convenient (although not necessary) to multiply a row by a nonzero scalar. Property 3 tells us how to compensate for this. For example, if

$$A = \begin{bmatrix} 19 & -38 & 95 \\ 7 & 91 & 14 \\ 5 & -8 & 9 \end{bmatrix},$$

it is convenient to divide the first row by 19 (i.e., multiply by 1/19) and to divide the second row by 7. By Property 3 we have

$$\det A = 19 \cdot 7 \cdot \det \begin{bmatrix} 1 & -2 & 5 \\ 1 & 13 & 2 \\ 5 & -8 & 9 \end{bmatrix}.$$

In other words, when we have calculated the determinant of the new matrix we have to multiply our result by $19 \cdot 7 = 133$ to obtain the determinant of A.

EXAMPLE 2   Suppose that at some stage in the reduction of a square matrix A to triangular form we obtain the matrix

$$\begin{bmatrix} 1 & 3 & 3 & 6 \\ 0 & 0 & -4 & 9 \\ 0 & 0 & 5 & 8 \\ 0 & 0 & -3 & -2 \end{bmatrix}.$$

Further reduction will not change the zero entry in the second position on the diagonal. Thus the final triangular matrix will have a zero on the main diagonal. Hence we can conclude without finishing the reduction that $\det A = 0$.

EXAMPLE 3   Note that the fourth row of the matrix

$$A = \begin{bmatrix} 1 & 2 & -3 & 2 \\ 2 & 0 & -1 & 5 \\ 8 & -3 & 2 & 4 \\ 2 & 4 & -6 & 4 \end{bmatrix}$$

is a multiple of the first row. Thus we can obtain a zero row and it follows that $\det A = 0$. In general, if one row of a square matrix is a multiple of another row, then the determinant must be 0.

EXERCISE 3   Compute the determinants of the following matrices.

(a) $\begin{bmatrix} 2 & -3 \\ 6 & 5 \end{bmatrix}$ (b) $\begin{bmatrix} -1 & 2 & 9 \\ 3 & 0 & 4 \\ 5 & 1 & -2 \end{bmatrix}$ (c) $\begin{bmatrix} 25 & -10 & 55 \\ 3 & 15 & -9 \\ 7 & -2 & 6 \end{bmatrix}$

(d) $\begin{bmatrix} 1 & 0 & 3 & 0 \\ 2 & 2 & 0 & 4 \\ 3 & -1 & 1 & -2 \\ 4 & 1 & 3 & -1 \end{bmatrix}$ (e) $\begin{bmatrix} 2 & -1 & 0 & 1 \\ 4 & -2 & 5 & 1 \\ -2 & 1 & 3 & 2 \\ 8 & -4 & 9 & -5 \end{bmatrix}$

We defined the determinant of a 2×2 matrix by the formula

$$\det \begin{bmatrix} a & b \\ c & d \end{bmatrix} = ad - bc.$$

It turns out that this formula is a consequence of the properties of determinants given above. To see this, let us assume first that $a \neq 0$. Then

$$\det \begin{bmatrix} a & b \\ c & d \end{bmatrix} = a \det \begin{bmatrix} 1 & b/a \\ c & d \end{bmatrix}$$
$$= a \det \begin{bmatrix} 1 & b/a \\ 0 & d - cb/a \end{bmatrix}$$
$$= a \cdot 1 \cdot (d - cb/a)$$
$$= ad - bc.$$

If $a = 0$, then $\det \begin{bmatrix} 0 & b \\ c & d \end{bmatrix} = -\det \begin{bmatrix} c & d \\ 0 & b \end{bmatrix} = -bc = ad - bc$. We leave it as an exercise to verify a similar formula for the determinant of a 3×3 matrix.

$$\det \begin{bmatrix} a & b & c \\ d & e & f \\ g & h & i \end{bmatrix} = aei + bfg + cdh - gec - hfa - idb \qquad (1)$$

It is easy to remember this formula with the help of the following schematic device. Write the first two columns of the 3×3 matrix to the right of the matrix and compute the products as indicated by the arrows.

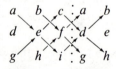

These six products are the products in (1). The products corresponding to the arrows pointing down have plus signs attached to them and the products corresponding to the arrows pointing up have negative signs attached to them. This schematic device works only for 3×3 determinants and cannot be used for higher-order determinants.

EXAMPLE 4   The determinant of the matrix $\begin{bmatrix} 2 & -1 & 4 \\ 3 & 1 & 2 \\ 5 & 2 & 3 \end{bmatrix}$ is

$$= 6 + (-10) + 24 - 20 - 8 - (-9) = 1.$$

We mention (without proof) two more properties of determinants.

Property 8. If A is a square matrix, then A and its transpose have the same determinant; $\det A = \det A^T$.

EXERCISE 4 Prove Property 8 for 2×2 matrices.

It follows that Properties 1–6 remain valid if the word row is replaced by the word column. Therefore, in reducing to triangular form we may add multiples of one column to another column, interchange columns, and multiply a column by a nonzero scalar. If a square matrix has a zero column or two identical columns, then its determinant is zero.

EXAMPLE 5 The third column of the matrix

$$A = \begin{bmatrix} -3 & 1 & -2 \\ 2 & 5 & 7 \\ 4 & 1 & 5 \end{bmatrix}$$

is the sum of its first and second columns. Thus subtracting these two columns from the third gives a zero column and hence $\det A = 0$.

The next property is a consequence of Properties 1–8. If we consider how involved matrix multiplication and the evaluation of determinants are, then this property is quite remarkable.

Property 9. If A and B are $n \times n$ matrices, then

$$\det(AB) = (\det A) \cdot (\det B).$$

An important consequence of this property is that

$$\det A^{-1} = \frac{1}{\det A} \tag{2}$$

for every nonsingular matrix A. To prove this, we note that

$$(\det A) \cdot (\det A^{-1}) = \det(AA^{-1}) = \det I = 1.$$

This shows us that $\det A \neq 0$. Dividing the equation by $\det A$ gives us (2).

We can use the value of the determinant of a square matrix A to determine whether A is singular or nonsingular. Suppose that we have reduced A to reduced echelon form E. Then E is upper triangular and $\det E = \alpha \cdot \det A$, where $\alpha \neq 0$ is a scalar. Recall that A is singular if and only if its rank is less than n (i.e., E has a zero row). But if E has a zero row, then $\det E = 0$; hence $\det A = 0$. Conversely, if $\det A = 0$, then $\det E = 0$ and one of the diagonal entries of E must be 0. Hence E has a zero row because of the echelon pattern. This proves the following important result.

Theorem 7.1 *A square matrix is singular if and only if its determinant is zero. Equivalently, a square matrix A is nonsingular if and only if $\det A \neq 0$.*

PROBLEMS 7.1

*1. Compute the determinants of the following matrices. Then use Theorem 7.1 to determine which of these matrices are singular.

(a) $\begin{bmatrix} -3 & 1 \\ 2 & 5 \end{bmatrix}$
(b) $\begin{bmatrix} 1 & 0 & 0 \\ -2 & 3 & 0 \\ 5 & 9 & 4 \end{bmatrix}$
(c) $\begin{bmatrix} 2 & 0 & 1 \\ -1 & 1 & 2 \\ 2 & -2 & -2 \end{bmatrix}$

(d) $\begin{bmatrix} 5 & 1 & 2 \\ 8 & -7 & 9 \\ 5 & 1 & 2 \end{bmatrix}$
(e) $\begin{bmatrix} 3 & 5 & -6 \\ 1 & 9 & -2 \\ -2 & 8 & 4 \end{bmatrix}$
(f) $\begin{bmatrix} \frac{1}{2} & \frac{1}{3} & \frac{1}{4} \\ \frac{1}{4} & -\frac{1}{6} & \frac{1}{4} \\ 0 & 2 & \frac{1}{5} \end{bmatrix}$

(g) $\begin{bmatrix} 0 & 1 & 2 & 3 \\ 1 & 0 & -1 & 2 \\ 3 & 4 & 0 & -1 \\ -1 & 2 & 2 & 0 \end{bmatrix}$
(h) $\begin{bmatrix} 3 & 1 & 0 & 0 \\ 4 & 2 & 0 & 0 \\ 0 & 0 & 1 & 5 \\ 0 & 0 & 2 & 1 \end{bmatrix}$

(i) $\begin{bmatrix} 1 & -3 & 0 & 0 & 0 \\ 2 & 1 & 5 & 0 & 0 \\ 0 & 2 & -2 & 1 & 0 \\ 0 & 0 & 1 & 3 & 2 \\ 0 & 0 & 0 & 4 & 1 \end{bmatrix}$

2. Show that the determinant of the matrix

$$\begin{bmatrix} a & b & c \\ d & e & f \\ g & h & i \end{bmatrix}$$

is equal to $aei + bfg + cdh - ceg - afh - bdi$.

3. Suppose that A is an $n \times n$ matrix and α is a scalar. Show that $\det(\alpha A) = \alpha^n \det A$. When is $\det(-A) = \det A$?

*4. Suppose that A, B, and C are 3×3 matrices such that $\det A = 2$, $\det B = -\frac{5}{3}$, and $\det C = 0$. Calculate the following determinants.
(a) $\det A^{-1}$ (b) $\det(3B)$ (c) $\det(ABC)$ (d) $\det(A^TB^{-1})$

5. Prove Property 9 for 2×2 matrices.

*6. Let A and B be $n \times n$ matrices. We know that in general $AB \neq BA$. However, show that AB and BA have the same determinant.

*7. Suppose that A is a skew-symmetric $n \times n$ matrix (skew-symmetric means that $A^T = -A$). Show that if n is odd, then $\det A = 0$. Give an example of a skew-symmetric 2×2 matrix with nonzero determinant.

8. Prove that if A is an orthogonal matrix, then $\det A$ is equal to 1 or -1. (*Hint:* If A is orthogonal, then $AA^T = I$.)

*9. Evaluate the determinants of the following matrices.

(a) $\begin{bmatrix} 0 & a \\ b & c \end{bmatrix}$
(b) $\begin{bmatrix} 0 & 0 & a \\ 0 & b & c \\ d & e & f \end{bmatrix}$
(c) $\begin{bmatrix} 0 & 0 & 0 & a \\ 0 & 0 & b & c \\ 0 & d & e & f \\ g & h & i & j \end{bmatrix}$

*10. What can you say about the determinant of the matrix

$$A = \begin{bmatrix} 0 & 0 & \cdots & 0 & a_{1n} \\ 0 & 0 & \cdots & a_{2,n-1} & a_{2n} \\ \vdots & & \ddots & & \vdots \\ 0 & a_{n-1,2} & \cdots & a_{n-1,n-1} & a_{n-1,n} \\ a_{n1} & a_{n2} & \cdots & a_{n,n-1} & a_{nn} \end{bmatrix}?$$

7.2   EXPANSION BY COFACTORS

Although the Gaussian elimination process provides us with a very efficient method for computing determinants, it is sometimes useful to use an alternative method. In this section we develop such a method, the expansion of a determinant by cofactors.

We mentioned in Section 7.1 a formula for the determinant of a 3×3 matrix A:

$$\det \begin{bmatrix} a_{11} & a_{12} & a_{13} \\ a_{21} & a_{22} & a_{23} \\ a_{31} & a_{32} & a_{33} \end{bmatrix} = a_{11}a_{22}a_{33} + a_{12}a_{23}a_{31} + a_{13}a_{21}a_{32}$$
$$- a_{13}a_{22}a_{31} - a_{11}a_{23}a_{32} - a_{12}a_{21}a_{33}.$$

Rearranging the terms in this formula and factoring out the quantities a_{11}, a_{12}, and a_{13} leads to

$$\det A = a_{11}(a_{22}a_{33} - a_{23}a_{32})$$
$$- a_{12}(a_{21}a_{33} - a_{23}a_{31})$$
$$+ a_{13}(a_{21}a_{32} - a_{22}a_{31}).$$

Hence

$$\det A = a_{11}A_{11} + a_{12}A_{12} + a_{13}A_{13}, \tag{1}$$

where

$$A_{11} = \det \begin{bmatrix} a_{22} & a_{23} \\ a_{32} & a_{33} \end{bmatrix}, \quad A_{12} = -\det \begin{bmatrix} a_{21} & a_{23} \\ a_{31} & a_{33} \end{bmatrix}, \quad A_{13} = \det \begin{bmatrix} a_{21} & a_{22} \\ a_{31} & a_{32} \end{bmatrix}.$$

The quantity A_{11}, which is called the cofactor of a_{11}, is the determinant of the matrix obtained from A by omitting its first row and first column. Similarly, A_{12}, the cofactor of a_{12}, is the negative of the determinant of the matrix obtained from A by omitting its first row and second column. Finally, A_{13} is the determinant of the matrix obtained from A by omitting its first row and third column.

In general, the evaluation of the determinant of an $n \times n$ matrix A can be reduced to the evaluation of n determinants of $(n-1) \times (n-1)$ matrices.

Let

$$A = \begin{bmatrix} a_{11} & a_{12} & \cdots & a_{1n} \\ a_{21} & a_{22} & \cdots & a_{2n} \\ \vdots & \vdots & & \vdots \\ a_{n1} & a_{n2} & \cdots & a_{nn} \end{bmatrix}.$$

The **cofactor** of an entry a_{ij} is defined to be the number

$$A_{ij} = (-1)^{i+j} D_{ij},$$

where D_{ij} is the determinant of the $(n-1) \times (n-1)$ matrix obtained from A by omitting the ith row and jth column of A. The sign $(-1)^{i+j}$ is $+1$ or -1, depending on whether $i+j$ is even or odd. If we replace the entries a_{ij} of A by the corresponding sign $(-1)^{i+j}$, we obtain a checkerboard pattern. The patterns for $n = 3$ and $n = 4$ are as follows.

$$\begin{bmatrix} + & - & + \\ - & + & - \\ + & - & + \end{bmatrix} \qquad \begin{bmatrix} + & - & + & - \\ - & + & - & + \\ + & - & + & - \\ - & + & - & + \end{bmatrix}$$

Note that the pattern always begins with a plus sign in the upper left corner and that the signs alternate.

EXAMPLE 1   Consider the matrix

$$A = \begin{bmatrix} 1 & -2 & 4 \\ 2 & -7 & 9 \\ -1 & 8 & -1 \end{bmatrix}.$$

The cofactor of the entry $a_{23} = 9$ is

$$A_{23} = (-1)^{2+3} \det \begin{bmatrix} 1 & -2 \\ -1 & 8 \end{bmatrix} = -(8-2) = -6.$$

The cofactor of the entry $a_{31} = -1$ is

$$A_{31} = (-1)^{3+1} \det \begin{bmatrix} -2 & 4 \\ -7 & 9 \end{bmatrix} = +(-18 + 28) = 10.$$

You should verify that the other cofactors are $A_{11} = -65$, $A_{12} = -7$, $A_{13} = 9$, $A_{21} = 30$, $A_{22} = 3$, $A_{32} = -1$, and $A_{33} = -3$.

The following result, whose proof we omit, gives us a method for computing determinants by means of cofactors.

Result 1   Let A be an $n \times n$ matrix. Then the sum of the products of the entries in any row (or column) of A and their corresponding cofactors is equal to the determinant of A. In symbols, for any i and any j,

$$\det A = a_{i1} A_{i1} + a_{i2} A_{i2} + \cdots + a_{in} A_{in}, \qquad (2)$$
$$\det A = a_{1j} A_{1j} + a_{2j} A_{2j} + \cdots + a_{nj} A_{nj}. \qquad (3)$$

Formula (2) is called the **cofactor expansion** of the determinant of A along the ith row. Formula (3) is the cofactor expansion down the jth column. Thus equation (1) is the cofactor expansion of the determinant of the 3×3 matrix A along the first row. If A is a 4×4 matrix, then $\det A = a_{21}A_{21} + a_{22}A_{22} + a_{23}A_{23} + a_{24}A_{24}$ is the cofactor expansion of $\det A$ along the second row and $\det A = a_{13}A_{13} + a_{23}A_{23} + a_{33}A_{33} + a_{43}A_{43}$ is the cofactor expansion down the third column.

EXAMPLE 2   The cofactors of the matrix

$$A = \begin{bmatrix} 1 & -2 & 4 \\ 2 & -7 & 9 \\ -1 & 8 & -1 \end{bmatrix}$$

were calculated in Example 1. Expanding the determinant of A along the third row gives

$$\begin{aligned} \det A &= (-1) \cdot A_{31} + 8 \cdot A_{32} + (-1) \cdot A_{33} \\ &= (-1) \cdot 10 + 8 \cdot (-1) + (-1) \cdot (-3) = -15. \end{aligned}$$

Expanding down the first column gives the same value,

$$\begin{aligned} \det A &= 1 \cdot A_{11} + 2 \cdot A_{21} + (-1) \cdot A_{31} \\ &= 1 \cdot (-65) + 2 \cdot 30 + (-1) \cdot 10 = -15. \end{aligned}$$

EXAMPLE 3   Any 1×1 matrix $[a]$ is upper triangular and a is its only diagonal entry. Hence $\det[a] = a$. In particular, if

$$A = \begin{bmatrix} a_{11} & a_{12} \\ a_{21} & a_{22} \end{bmatrix},$$

then the cofactor expansion along the first row yields the familiar formula $\det A = a_{11}A_{11} + a_{12}A_{12} = a_{11}a_{22} - a_{12}a_{21}$.

EXERCISE 1   Let

$$A = \begin{bmatrix} 1 & 2 & 3 \\ -2 & 0 & 4 \\ 5 & -7 & 0 \end{bmatrix}.$$

Calculate all nine cofactors of A. Then expand the determinant of A along each row and down each column and check that all expansions give the same value. Finally, check the result by reducing A to upper triangular form.

When we compute a determinant using cofactors, then for obvious reasons it is usually more efficient to choose the row or column of the matrix that has the most zero entries.

EXAMPLE 4 Let

$$A = \begin{bmatrix} 2 & 1 & 0 & 5 \\ -2 & 0 & -1 & 1 \\ 3 & 4 & 3 & 0 \\ 1 & -2 & 0 & 6 \end{bmatrix}.$$

Expanding the determinant of A down the third column gives

$$\det A = 0 \cdot A_{13} + (-1) \cdot A_{23} + 3 \cdot A_{33} + 0 \cdot A_{43} = -A_{23} + 3 \cdot A_{33},$$

where

$$A_{23} = -\det \begin{bmatrix} 2 & 1 & 5 \\ 3 & 4 & 0 \\ 1 & -2 & 6 \end{bmatrix} \quad \text{and} \quad A_{33} = +\det \begin{bmatrix} 2 & 1 & 5 \\ -2 & 0 & 1 \\ 1 & -2 & 6 \end{bmatrix}.$$

The expansion of A_{23} down the third column gives

$$A_{23} = -5 \cdot \det \begin{bmatrix} 3 & 4 \\ 1 & -2 \end{bmatrix} - 6 \cdot \det \begin{bmatrix} 2 & 1 \\ 3 & 4 \end{bmatrix} = -5 \cdot (-10) - 6 \cdot 5 = 20.$$

The expansion of A_{33} along the second row gives

$$A_{33} = 2 \cdot \det \begin{bmatrix} 1 & 5 \\ -2 & 6 \end{bmatrix} - 1 \cdot \det \begin{bmatrix} 2 & 1 \\ 1 & -2 \end{bmatrix} = 2 \cdot 16 - (-5) = 37.$$

Thus $\det A = -20 + 3 \cdot 37 = 91$.

Sometimes we can combine elementary row operations with expansion by cofactors and significantly reduce the computation required. We illustrate the procedure for the matrix in Example 4.

EXAMPLE 5 Let A be the matrix of Example 4. The determinant of A is equal to the determinant of the matrix

$$B = \begin{bmatrix} 2 & 1 & 0 & 5 \\ -2 & 0 & -1 & 1 \\ -3 & 4 & 0 & 3 \\ 1 & -2 & 0 & 6 \end{bmatrix}$$

obtained from A by adding three times the second row to the third row. Expanding $\det B$ down the third column gives

$$\det A = \det B = -(-1) \det \begin{bmatrix} 2 & 1 & 5 \\ -3 & 4 & 3 \\ 1 & -2 & 6 \end{bmatrix}.$$

Using row operations on the 3×3 matrix (adding suitable multiples of the third row to each of the others) gives

$$\det A = \det \begin{bmatrix} 0 & 5 & -7 \\ 0 & -2 & 21 \\ 1 & -2 & 6 \end{bmatrix}.$$

Now expanding down the first column yields

$$\det A = \det \begin{bmatrix} 5 & -7 \\ -2 & 21 \end{bmatrix} = 91.$$

We now have two methods for computing the determinant of an $n \times n$ matrix: reduction to triangular form and cofactor expansion. The reduction to triangular form is usually much faster than the cofactor expansion unless $n = 2$ or $n = 3$. For example, the computation of the determinant of a 5×5 matrix by cofactors involves in general more than 200 multiplications, whereas the reduction to triangular form can be done with fewer than 50 multiplications. The difference between the two methods increases dramatically as the size of the matrix increases. For a 10×10 matrix the reduction to triangular form involves fewer than 350 multiplications, but the cofactor method involves more than 6 million multiplications! No computer in the world could ever compute the determinant of a 100×100 matrix by cofactor expansion because this would involve about 10^{158} multiplications. However, a modern computer could calculate such a determinant by reduction to triangular form in a few seconds because fewer than 350,000 multiplications are needed.

We conclude this section with an example illustrating the typical situation that arises in Chapter 8.

EXAMPLE 6 Consider the matrix

$$A = \begin{bmatrix} \lambda & -1 & 0 \\ 0 & \lambda - 3 & -2 \\ 2 & 1 & \lambda + 2 \end{bmatrix},$$

where λ is an unknown quantity. Let us calculate the determinant of A. In this case it will not be a number but an expression in terms of λ. Expanding down the first column, we obtain

$$\begin{aligned} \det A &= \lambda \cdot \det \begin{bmatrix} \lambda - 3 & -2 \\ 1 & \lambda + 2 \end{bmatrix} + 2 \cdot \det \begin{bmatrix} -1 & 0 \\ \lambda - 3 & -2 \end{bmatrix} \\ &= \lambda[(\lambda - 3)(\lambda + 2) + 2] + 2 \cdot 2 \\ &= \lambda(\lambda^2 - \lambda - 4) + 4 \\ &= \lambda^3 - \lambda^2 - 4\lambda + 4 \\ &= (\lambda - 1)(\lambda - 2)(\lambda + 2). \end{aligned}$$

It follows that the matrix A is singular ($\det A = 0$) if and only if $\lambda = 1, 2,$ or -2.

PROBLEMS 7.2

*1. Calculate the determinants of the following matrices.

(a) $\begin{bmatrix} -2 & 2 & 9 \\ 3 & -4 & 1 \\ 5 & 0 & 8 \end{bmatrix}$ (b) $\begin{bmatrix} 0 & 2 & -3 \\ -2 & 0 & 3 \\ 2 & -3 & 0 \end{bmatrix}$ (c) $\begin{bmatrix} \frac{1}{3} & 0 & \frac{1}{3} \\ \frac{1}{2} & 1 & \frac{1}{2} \\ -1 & \frac{1}{3} & \frac{2}{3} \end{bmatrix}$

(d) $\begin{bmatrix} 1 & -1 & 1 & 0 \\ -1 & 1 & 0 & 1 \\ 1 & 0 & 1 & -1 \\ 0 & 1 & -1 & 1 \end{bmatrix}$ (e) $\begin{bmatrix} 1 & 2 & 3 \\ 1 & 1 & 2 \\ 2 & 3 & 1 \end{bmatrix}$ (f) $\begin{bmatrix} 7 & 0 & 2 & 1 \\ 2 & -2 & 1 & 0 \\ 0 & 1 & 0 & -2 \\ 5 & 1 & -1 & 3 \end{bmatrix}$

2. Verify that the determinant of the matrix $\begin{bmatrix} a - \lambda & b \\ c & d - \lambda \end{bmatrix}$ is $\lambda^2 - (a + d)\lambda + ad - bc$.

*3. Determine all scalars λ for which the matrix $\begin{bmatrix} 3 - \lambda & 1 \\ -6 & -4 - \lambda \end{bmatrix}$ is singular.

*4. Repeat Problem 3 with the matrices

(a) $\begin{bmatrix} 1 - \lambda & -2 \\ 1 & -1 - \lambda \end{bmatrix}$ (b) $\begin{bmatrix} -\lambda & 0 & -1 \\ -2 & 2 - \lambda & -3 \\ -1 & 2 & -2 - \lambda \end{bmatrix}$

5. Verify that if a, b, c, and d are integers, then the determinant of the matrix $\begin{bmatrix} a & b \\ c & d \end{bmatrix}$ is an integer.

6. Suppose that A is an $n \times n$ matrix and all entries of A are integers. Explain why det A also is an integer.

7. Verify that

$$\det \begin{bmatrix} 1 & x & x^2 \\ 1 & y & y^2 \\ 1 & z & z^2 \end{bmatrix} = (x - y)(y - z)(z - x).$$

8. Let d_n be the determinant of the $n \times n$ matrix

$$D_n = \begin{bmatrix} x & y & 0 & 0 & 0 & 0 & \cdots & 0 \\ y & x & y & 0 & 0 & 0 & \cdots & 0 \\ 0 & y & x & y & 0 & 0 & \cdots & 0 \\ \vdots & & & & & & & \vdots \\ 0 & \cdots & & 0 & 0 & 0 & y & x & y \\ 0 & \cdots & & 0 & 0 & 0 & 0 & y & x \end{bmatrix}.$$

Show that $d_n = xd_{n-1} - y^2 d_{n-2}$.

9. (a) Show that the determinants of the matrices

$\begin{bmatrix} 1 & x \\ x & 1 \end{bmatrix}$, $\begin{bmatrix} 1 & x & x^2 \\ x^2 & 1 & x \\ x & x^2 & 1 \end{bmatrix}$, and $\begin{bmatrix} 1 & x & x^2 & x^3 \\ x^3 & 1 & x & x^2 \\ x^2 & x^3 & 1 & x \\ x & x^2 & x^3 & 1 \end{bmatrix}$

are $1 - x^2$, $(1 - x^3)^2$, and $(1 - x^4)^3$, respectively.

(b) On the basis of part (a), give a formula for the determinant of the matrix

$$\begin{bmatrix} 1 & x & x^2 & \cdots & x^n \\ x^n & 1 & x & \cdots & x^{n-1'} \\ \vdots & & & & \vdots \\ x & x^2 & x^3 & \cdots & 1 \end{bmatrix}.$$

7.3   THE CROSS PRODUCT

In this section we define the cross product of two vectors in 3-space and then use this product to give a geometric interpretation of 2×2 and 3×3 determinants.

Definition   For any two vectors $a = (a_1, a_2, a_3)$ and $b = (b_1, b_2, b_3)$ in R^3, their **cross product** $a \times b$ is defined to be the vector

$$a \times b = (a_2 b_3 - a_3 b_2, \, a_3 b_1 - a_1 b_3, \, a_1 b_2 - a_2 b_1).$$

Note that the components of $a \times b$ are 2×2 determinants,

$$a \times b = \left(\det \begin{bmatrix} a_2 & a_3 \\ b_2 & b_3 \end{bmatrix}, \, -\det \begin{bmatrix} a_1 & a_3 \\ b_1 & b_3 \end{bmatrix}, \det \begin{bmatrix} a_1 & a_2 \\ b_1 & b_2 \end{bmatrix} \right). \tag{1}$$

This formula makes it somewhat easier to remember the definition of the cross product.

EXAMPLE 1   If $a = (2, 3, -1)$ and $b = (-2, 1, 5)$, then

$$a \times b = \left(\det \begin{bmatrix} 3 & -1 \\ 1 & 5 \end{bmatrix}, \, -\det \begin{bmatrix} 2 & -1 \\ -2 & 5 \end{bmatrix}, \det \begin{bmatrix} 2 & 3 \\ -2 & 1 \end{bmatrix} \right)$$
$$= (16, -8, 8).$$

EXERCISE 1   Compute $a \times b$ and $c \times d$ where $a = (1, 2, 3)$, $b = (-2, 4, 1)$, $c = (0, -3, 3)$, and $d = (-1, 0, 1)$.

If $a = (a_1, a_2, a_3)$ and $b = (b_1, b_2, b_3)$ are two collinear vectors in R^3, say $b = \alpha a$, then

$$\det \begin{bmatrix} a_i & a_j \\ b_i & b_j \end{bmatrix} = \det \begin{bmatrix} a_i & a_j \\ \alpha a_i & \alpha a_j \end{bmatrix} = \alpha \det \begin{bmatrix} a_i & a_j \\ a_i & a_j \end{bmatrix} = 0$$

and hence $a \times b = 0$. In particular, $a \times a = 0$.

The cross product, the inner product, and 3×3 determinants are related by the formula

$$\det \begin{bmatrix} a_1 & a_2 & a_3 \\ b_1 & b_2 & b_3 \\ c_1 & c_2 & c_3 \end{bmatrix} = \langle a \times b, c \rangle, \tag{2}$$

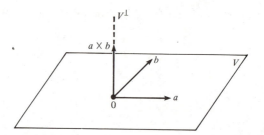

Figure 7.1

where $a = (a_1, a_2, a_3)$, $b = (b_1, b_2, b_3)$, and $c = (c_1, c_2, c_3)$. Indeed, expanding the determinant by cofactors along the third row gives

$$c_1 \det \begin{bmatrix} a_2 & a_3 \\ b_2 & b_3 \end{bmatrix} - c_2 \det \begin{bmatrix} a_1 & a_3 \\ b_1 & b_3 \end{bmatrix} + c_3 \det \begin{bmatrix} a_1 & a_2 \\ b_1 & b_2 \end{bmatrix},$$

which by (1) is equal to $\langle c, a \times b \rangle = \langle a \times b, c \rangle$.

Since the determinant of a square matrix with two identical rows is equal to zero, it follows from (2) that $\langle a \times b, a \rangle = 0$ and $\langle a \times b, b \rangle = 0$. Consequently, the vector $a \times b$ is orthogonal to both a and b. Suppose that a and b are not collinear. Then we can find a vector $c = (c_1, c_2, c_3)$ in R^3 such that a, b, c form a basis for R^3. Hence the matrix

$$\begin{bmatrix} a_1 & a_2 & a_3 \\ b_1 & b_2 & b_3 \\ c_1 & c_2 & c_3 \end{bmatrix}$$

is nonsingular and therefore by (2) we have $\langle a \times b, c \rangle \neq 0$. This implies that $a \times b \neq 0$. The vectors a and b span a plane in R^3 and the orthogonal complement of this plane is a line through the origin. The vector $a \times b \neq 0$, being orthogonal to a and b, is on this line and hence spans this line (see Figure 7.1).

We summarize this discussion in the following result.

Result 1   Let a and b be two vectors in R^3. Then their cross product $a \times b$ is orthogonal to both a and b. If a and b are not collinear, then $a \times b \neq 0$ and a vector c in R^3 is orthogonal to both a and b if and only if c is a scalar multiple of $a \times b$.

EXAMPLE 2   The vectors $a = (1, 2, 0)$ and $b = (-1, 3, 1)$ are not collinear and hence span a plane V in R^3. The orthogonal complement of V is the straight line spanned by the vector $a \times b = (2, -1, 5)$.

Now we wish to compute the length of the cross product of two vectors $a = (a_1, a_2, a_3)$ and $b = (b_1, b_2, b_3)$. We have

$$\|a \times b\|^2 = (a_2 b_3 - a_3 b_2)^2 + (a_3 b_1 - a_1 b_3)^2 + (a_1 b_2 - a_2 b_1)^2.$$

A straightforward algebraic manipulation (which the reader should perform)

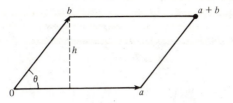

Figure 7.2

shows that this quantity is equal to

$$\left(a_1^2 + a_2^2 + a_3^2\right)\left(b_1^2 + b_2^2 + b_3^2\right) - \left(a_1b_1 + a_2b_2 + a_3b_3\right)^2,$$

which is simply

$$\|a\|^2\|b\|^2 - \langle a, b\rangle^2.$$

Since $\langle a, b\rangle = \|a\|\|b\| \cos\theta$, where θ is the angle between a and b, it follows that

$$\begin{aligned}
\|a \times b\|^2 &= \|a\|^2\|b\|^2 - \|a\|^2\|b\|^2\cos^2\theta \\
&= \|a\|^2\|b\|^2(1 - \cos^2\theta) \\
&= \|a\|^2\|b\|^2\sin^2\theta.
\end{aligned}$$

Taking square roots yields

$$\|a \times b\| = \|a\| \cdot \|b\| \sin\theta.$$

This formula has a nice geometric interpretation. Consider the parallelogram determined by a and b, that is, the parallelogram with vertices 0, a, b, and $a + b$ (see Figure 7.2). The area of the parallelogram is equal to the base $\|a\|$ times the height $h = \|b\| \sin\theta$. Hence the area is equal to $\|a\| \cdot \|b\| \sin\theta = \|a \times b\|$. This proves:

Result 2 If a and b are two vectors in R^3 and θ is the angle between a and b, then $\|a \times b\| = \|a\| \cdot \|b\| \sin\theta$ is equal to the area of the parallelogram determined by a and b.

EXAMPLE 3   The area of the parallelogram determined by the vectors $a = (1, 1, 1)$ and $b = (2, 0, 3)$ is $\|a \times b\| = \|(3, -1, -2)\| = \sqrt{14}$.

We conclude this section with two examples that give geometric interpretations for 2×2 and 3×3 determinants.

EXAMPLE 4   Let $A = \begin{vmatrix} a_1 & a_2 \\ b_1 & b_2 \end{vmatrix}$. We will show that the absolute value of det A is equal to the area of the parallelogram determined by $a = (a_1, a_2)$ and $b = (b_1, b_2)$ (see Figure 7.3).

First note that this area is equal to the area of the parallelogram determined by the vectors $\bar{a} = (a_1, a_2, 0)$ and $\bar{b} = (b_1, b_2, 0)$ in R^3. Hence,

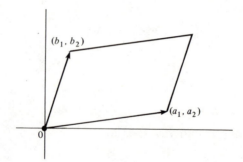

Figure 7.3

by Result 2, the area is equal to

$$\|\bar{a} \times \bar{b}\| = \|(0, 0, a_1 b_2 - a_2 b_1)\| = \left| \det \begin{bmatrix} a_1 & a_2 \\ b_1 & b_2 \end{bmatrix} \right|.$$

EXAMPLE 5 Let

$$A = \begin{bmatrix} a_1 & a_2 & a_3 \\ b_1 & b_2 & b_3 \\ c_1 & c_2 & c_3 \end{bmatrix}.$$

We will show that the absolute value of det A is equal to the volume of the parallelopiped determined by $a = (a_1, a_2, a_3)$, $b = (b_1, b_2, b_3)$, and $c = (c_1, c_2, c_3)$ (see Figure 7.4). The volume is equal to the area of the base times the height h. By Result 2 the area of the base is $\|a \times b\|$. Since $a \times b$ is perpendicular to a and b, the height h is equal to the length of the projection of c onto $a \times b$. Hence

$$h = \left\| \frac{\langle a \times b, c \rangle}{\langle a \times b, a \times b \rangle} a \times b \right\|$$

$$= \frac{|\langle a \times b, c \rangle|}{\|a \times b\|^2} \|a \times b\| = \frac{|\langle a \times b, c \rangle|}{\|a \times b\|}.$$

Thus the volume is

$$\|a \times b\| \cdot \frac{|\langle a \times b, c \rangle|}{\|a \times b\|} = |\langle a \times b, c \rangle|$$

and $|\langle a \times b, c \rangle|$ is equal to $|\det A|$ by equation (2).

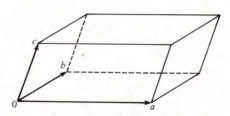

Figure 7.4

PROBLEMS 7.3

*1. For the given vectors a and b, find a vector orthogonal to a and b and find the area of the parallelogram determined by a and b.
(a) $a = (1, -1, 0)$, $b = (3, 2, 3)$ (b) $a = (-1, 2, 3)$, $b = (5, 1, 2)$
(c) $a = (2, 6, -1)$, $b = (1, 1, 2)$

*2. Find the area of the parallelogram determined by the vectors:
(a) $(1, 2)$, $(-2, 5)$ (b) $(2, 3)$, $(3, 1)$

*3. Find the volume of the parallelopiped determined by the vectors:
(a) $(1, 1, 1)$, $(1, 0, 2)$, $(3, -1, 2)$ (b) $(2, -1, 2)$, $(1, 1, -1)$, $(1, 5, 5)$

*4. Find the area of the triangle with vertices:
(a) $(0, 0)$, $(4, 3)$, $(-2, 1)$ (b) $(0, 0, 0)$, $(1, 2, 3)$, $(2, -2, 4)$

5. Let e_1, e_2, e_3 be the standard basis of R^3. Calculate $e_1 \times e_2$, $e_2 \times e_3$, and $e_3 \times e_1$.

6. Verify that for all vectors a, b, and c in R^3 and for all scalars α:
(a) $a \times 0 = 0 \times b = 0$ (b) $(\alpha a) \times b = a \times (\alpha b) = \alpha(a \times b)$
(c) $(a + b) \times c = (a \times c) + (b \times c)$ (d) $a \times (b + c) = (a \times b) + (a \times c)$
(e) $a \times b = -(b \times a)$

7. Verify that for all vectors a, b, and c in R^3:
(a) $(a \times b) \times c + (b \times c) \times a + (c \times a) \times b = 0$
(b) $(a \times b) \times c = \langle a, c \rangle b - \langle a, b \rangle c$
Is it always true that $(a \times b) \times c = a \times (b \times c)$? Justify your answer.

8. (Chapter 5 required) Let $T : R^2 \rightarrow R^2$ be a linear transformation and let A be the standard matrix for T. Let Q be the rectangle determined by two nonzero orthogonal vectors a and b. Then T maps Q onto a parallelogram (see Problem 8 of Section 5.2). Show that if A is nonsingular, then the area of this parallelogram is equal to $|\det A| \cdot$ (area of Q). Does this formula hold if A is singular?

7.4   THE ADJOINT MATRIX

In Section 1.7 we discussed a method (based on the Gaussian elimination process) for computing the inverse of a nonsingular square matrix. We now present another method for calculating inverses.

Consider an $n \times n$ matrix

$$A = \begin{bmatrix} a_{11} & a_{12} & \cdots & a_{1n} \\ a_{21} & a_{22} & \cdots & a_{2n} \\ \vdots & \vdots & & \vdots \\ a_{n1} & a_{n2} & \cdots & a_{nn} \end{bmatrix}.$$

For each entry a_{ij} let A_{ij} be the cofactor of a_{ij}. The **adjoint matrix** of A is defined to be the $n \times n$ matrix whose entry in the ith row and jth column is A_{ji}. Thus the adjoint matrix is the transpose of the matrix of cofactors. We will denote this matrix by adj A.

$$\text{adj } A = \begin{bmatrix} A_{11} & A_{21} & \cdots & A_{n1} \\ A_{12} & A_{22} & \cdots & A_{n2} \\ \vdots & \vdots & & \vdots \\ A_{1n} & A_{2n} & \cdots & A_{nn} \end{bmatrix}$$

Note that, for instance, the entry in the first row and second column of adj A is A_{21}, *not* A_{12}.

EXAMPLE 1 The adjoint matrix of

$$A = \begin{bmatrix} 1 & -2 & 4 \\ 2 & -7 & 9 \\ -1 & 8 & -1 \end{bmatrix} \quad \text{is} \quad \text{adj } A = \begin{bmatrix} -65 & 30 & 10 \\ -7 & 3 & -1 \\ 9 & -6 & -3 \end{bmatrix}.$$

(The cofactors of A were calculated in Example 1 of Section 7.2.)

Now consider the product of A with its adjoint,

$$P = \begin{bmatrix} a_{11} & a_{12} & \cdots & a_{1n} \\ a_{21} & a_{22} & \cdots & a_{2n} \\ \vdots & \vdots & & \vdots \\ a_{n1} & a_{n2} & \cdots & a_{nn} \end{bmatrix} \begin{bmatrix} A_{11} & A_{21} & \cdots & A_{n1} \\ A_{12} & A_{22} & \cdots & A_{n2} \\ \vdots & \vdots & & \vdots \\ A_{1n} & A_{2n} & \cdots & A_{nn} \end{bmatrix}.$$

The entry in the ith row and jth column of P is

$$p_{ij} = a_{i1} A_{j1} + a_{i2} A_{j2} + \cdots + a_{in} A_{jn}.$$

If $i = j$, then $p_{ii} = a_{i1} A_{i1} + a_{i2} A_{i2} + \cdots + a_{in} A_{in}$ is the expansion of the determinant of A along the ith row of A (i.e., $p_{ii} = \det A$). Thus all entries on the main diagonal of P are equal to $\det A$. Next suppose that $i \neq j$. We claim that in this case $p_{ij} = 0$. To see this, replace the jth row of A by the ith row and call the resulting matrix B. Now the ith row and the jth row of B are identical, and consequently $\det B = 0$. Note that the changing of the jth row of A has no effect on the cofactors belonging to this row because these cofactors do not involve the entries of the jth row. It follows that $A_{j1} = B_{j1}, A_{j2} = B_{j2}, \ldots, A_{jn} = B_{jn}$. Since the entries in the jth row of B are $a_{i1}, a_{i2}, \ldots, a_{in}$ we conclude that $p_{ij} = a_{i1} A_{j1} + a_{i2} A_{j2} + \cdots + a_{in} A_{jn}$ is the cofactor expansion of $\det B$ along the jth row (i.e., $p_{ij} = \det B = 0$). Hence all entries of P off the main diagonal are zero. It follows that

$$P = \begin{bmatrix} \det A & 0 & \cdots & 0 \\ 0 & \det A & \cdots & 0 \\ \vdots & \vdots & & \vdots \\ 0 & 0 & \cdots & \det A \end{bmatrix} = \det A \begin{bmatrix} 1 & 0 & \cdots & 0 \\ 0 & 1 & \cdots & 0 \\ \vdots & \vdots & & \vdots \\ 0 & 0 & \cdots & 1 \end{bmatrix}.$$

This gives us the equation

$$A \text{ adj } A = (\det A)I, \tag{1}$$

where I is the $n \times n$ identity matrix. Now suppose that A is nonsingular. Then det $A \neq 0$ and dividing the equation (1) by det A yields

$$I = \frac{1}{\det A} \cdot (A \operatorname{adj} A) = A\left(\frac{1}{\det A} \cdot \operatorname{adj} A\right).$$

It follows that the matrix $\dfrac{1}{\det A} \cdot \operatorname{adj} A$ must be the inverse of A.

Result 1   If A is a nonsingular square matrix, then

$$A^{-1} = \frac{1}{\det A} \cdot \operatorname{adj} A.$$

EXAMPLE 2   The determinant of $A = \begin{bmatrix} a & b \\ c & d \end{bmatrix}$ is $ad - bc$ and the adjoint of A is $\begin{bmatrix} d & -b \\ -c & a \end{bmatrix}$. A direct calculation shows that

$$\begin{bmatrix} a & b \\ c & d \end{bmatrix}\begin{bmatrix} d & -b \\ -c & a \end{bmatrix} = \begin{bmatrix} ad - bc & 0 \\ 0 & ad - bc \end{bmatrix} = (ad - bc)\begin{bmatrix} 1 & 0 \\ 0 & 1 \end{bmatrix}.$$

If A is nonsingular, then $ad - bc \neq 0$ and we obtain the familiar formula

$$A^{-1} = \frac{1}{ad - bc}\begin{bmatrix} d & -b \\ -c & a \end{bmatrix}.$$

EXAMPLE 3   The determinant of the matrix

$$A = \begin{bmatrix} 1 & -2 & 4 \\ 2 & -7 & 9 \\ -1 & 8 & -1 \end{bmatrix}$$

is det $A = -15 \neq 0$. Thus A is nonsingular and (see Example 1)

$$A^{-1} = \frac{1}{\det A} \cdot \operatorname{adj} A = \frac{1}{-15}\begin{bmatrix} -65 & 30 & 10 \\ -7 & 3 & -1 \\ 9 & -6 & -3 \end{bmatrix}.$$

EXERCISE 1   Use Result 1 to calculate the inverse of the matrix

$$A = \begin{bmatrix} 2 & -1 & 3 \\ 1 & 5 & -7 \\ 4 & 1 & 1 \end{bmatrix}.$$

Result 1 gives us a precise formula for the inverse of a nonsingular square matrix A in terms of its determinant and cofactors. However, from a computational point of view this result is practically useless. To compute the inverse of an $n \times n$ matrix A we would have to calculate $n^2 + 1$ determinants, the determinant of A, and n^2 cofactors [each cofactor being the determinant of an $(n - 1) \times (n - 1)$ matrix]. For example, if $n = 4$, this would involve the evaluation of the determinant of a 4×4 matrix and of the determinants of sixteen 3×3 matrices, a formidable task. The method for computing inverses by Gaussian elimination is much faster. Only in the case $n = 2$ do we obtain the useful and familiar formula for the inverse of a nonsingular 2×2 matrix (see Example 2).

Using Result 1, we can derive a rule for representing the solution of the linear system $Ax = b$ provided that A is a nonsingular square matrix.

Result 2   (Cramer's Rule) Let A be a nonsingular $n \times n$ matrix and let b be a vector in R^n. Then the components of the unique solution $x = (x_1, x_2, \ldots, x_n)$ of $Ax = b$ are

$$x_i = \frac{\det A_i}{\det A}, \qquad i = 1, 2, \ldots, n,$$

where A_i is the matrix obtained by replacing the ith column of A by b.

Proof   Since $x = A^{-1}b = \dfrac{1}{\det A}(\operatorname{adj} A)b$, it follows that

$$x_i = \frac{1}{\det A}(b_1 A_{1i} + b_2 A_{2i} + \cdots + b_n A_{ni}),$$

where $b = (b_1, b_2, \ldots, b_n)$. But $b_1 A_{1i} + b_2 A_{2i} + \cdots + b_n A_{ni}$ is the expansion of the determinant of A_i down the ith column (which is b).

EXAMPLE 4   Consider the linear system

$$\begin{bmatrix} 1 & 1 & 2 \\ 0 & 2 & -1 \\ 3 & 1 & 2 \end{bmatrix} \begin{bmatrix} x_1 \\ x_2 \\ x_3 \end{bmatrix} = \begin{bmatrix} 1 \\ 1 \\ 2 \end{bmatrix}.$$

The determinant of the coefficient matrix A is equal to -10 and hence A is nonsingular. Now (using the notation of Result 2) we have

$$\det A_1 = \det \begin{bmatrix} 1 & 1 & 2 \\ 1 & 2 & -1 \\ 2 & 1 & 2 \end{bmatrix} = -5,$$

$$\det A_2 = \det \begin{bmatrix} 1 & 1 & 2 \\ 0 & 1 & -1 \\ 3 & 2 & 2 \end{bmatrix} = -5,$$

and

$$\det A_3 = \det \begin{bmatrix} 1 & 1 & 1 \\ 0 & 2 & 1 \\ 3 & 1 & 2 \end{bmatrix} = 0.$$

Hence by Cramer's rule the unique solution is $x_1 = -5/-10 = 1/2$, $x_2 = -5/-10 = 1/2$, $x_3 = 0/-10 = 0$.

Cramer's rule can be applied only to linear systems with nonsingular coefficient matrices. If the system has n equations, we must evaluate the determinants of $n + 1$ matrices of size $n \times n$. One attempt to apply Cramer's rule to a 6×6 system will convince the reader that the Gaussian elimination process is a much more efficient method for solving the system.

PROBLEMS 7.4

*1. Use adjoint matrices to find the inverses of the following nonsingular matrices.

$$\text{(a)}\begin{bmatrix} -1 & 2 & 2 \\ 3 & 0 & 1 \\ 5 & 3 & -2 \end{bmatrix} \quad \text{(b)} \begin{bmatrix} 2 & -1 & 0 \\ 0 & 2 & -1 \\ -5 & -1 & 2 \end{bmatrix} \quad \text{(c)} \begin{bmatrix} \frac{1}{3} & 0 & \frac{1}{3} \\ \frac{1}{2} & 1 & \frac{1}{2} \\ -1 & \frac{1}{3} & \frac{2}{3} \end{bmatrix}$$

*2. Consider the upper triangular matrix

$$A = \begin{bmatrix} a & b & c \\ 0 & d & e \\ 0 & 0 & f \end{bmatrix},$$

where $adf \neq 0$. Find the inverse of A. Can you explain why in general the inverse of a nonsingular upper triangular matrix must also be upper triangular?

3. Let A be a square matrix whose entries are integers. Suppose that $\det A = \pm 1$. Explain why all entries of A^{-1} are also integers (see Problem 6 in Section 7.2).

*4. Use Cramer's rule to solve the following systems.

$$\text{(a)} \begin{bmatrix} 2 & -5 \\ 7 & 1 \end{bmatrix} \begin{bmatrix} x_1 \\ x_2 \end{bmatrix} = \begin{bmatrix} 3 \\ 4 \end{bmatrix} \quad \text{(b)} \begin{bmatrix} 1 & 1 & -3 \\ 2 & 1 & 2 \\ -1 & -2 & 1 \end{bmatrix} \begin{bmatrix} x_1 \\ x_2 \\ x_3 \end{bmatrix} = \begin{bmatrix} 1 \\ 0 \\ 2 \end{bmatrix}$$

Chapter 8

Eigenvalues and Eigenvectors

In Chapter 1 we found that determining the prices in a stable economy is equivalent to solving the equation $Ap = p$. The Markov process example in the same chapter found us interested in a steady-state vector, a vector f such that $Af = f$. Both of these are examples of the fundamental equation

$$Av = \lambda v.$$

We saw another example of this equation in the Leslie population distribution model. There we mentioned that an equilibrium population distribution vector p is a solution of $Ap = \lambda p$ for some λ, and that this λ is the "natural" growth rate. The many-species population prediction model also leads to questions whose answers involved solving this same equation $Ap = \lambda p$.

The equation $Av = \lambda v$ is also closely connected with the problem of raising matrices to powers. Several examples in Chapter 1 involved computing $A^k v$ for large k. If $Av = \lambda v$, then this computation is trivial. It is also easy to compute $A^k v$ when $v = \alpha_1 v_1 + \cdots + \alpha_n v_n$ and $Av_i = \lambda_i v_i$ for $i = 1, \ldots, n$.

Finally, we remark that the solution of systems of differential equations is an important topic in calculus, one with many applications. Certain important and useful systems of differential equations can be changed into sets of very simple differential equations by solving the equation $Av = \lambda v$ for an appropriate matrix A.

The discussion above shows that the equation $Av = \lambda v$ arises in many situations. This chapter is devoted to the study of this important and difficult equation.

237

8.1 EIGENVALUES AND EIGENVECTORS

Definition Let A be an $n \times n$ matrix. A nonzero vector v is called an **eigenvector** of A if there is a number λ such that

$$Av = \lambda v.$$

The number λ is called an **eigenvalue** associated with the eigenvector v.

Since eigenvalues and eigenvectors are defined only for square matrices, we will assume in this chapter that all matrices are square matrices.

EXAMPLE 1 Let $A = \begin{bmatrix} 1 & 2 \\ 0 & 3 \end{bmatrix}$. Since

$$\begin{bmatrix} 1 & 2 \\ 0 & 3 \end{bmatrix}\begin{bmatrix} 1 \\ 1 \end{bmatrix} = \begin{bmatrix} 3 \\ 3 \end{bmatrix} = 3\begin{bmatrix} 1 \\ 1 \end{bmatrix},$$

the vector $(1, 1)$ is an eigenvector of A associated with the eigenvalue 3.

EXERCISE 1 Verify that $(1, 0)$ is an eigenvector of the matrix A in Example 1. What is the associated eigenvalue?

If λ is an eigenvalue of a matrix A and v is an associated eigenvector, then Av is a multiple of v, that is, $Av = \lambda v$. Av and v are collinear vectors and the length of Av is $|\lambda|$ times the length of v. Thus the eigenvectors of A associated with the eigenvalue λ are those vectors in R^n that when multiplied by A remain collinear to themselves but whose length is changed by the eigenvalue λ.

The definition above excludes the zero vector from being an eigenvector. Without this restriction any scalar λ would be an eigenvalue of A having the zero vector as an eigenvector and there would be nothing special about eigenvalues. In addition, by imposing this restriction we have the following result.

Result 1 An eigenvector of a square matrix A is associated with exactly one eigenvalue. In other words, distinct eigenvalues have different eigenvectors.

Proof Suppose that v is a nonzero vector and that $Av = \lambda_1 v$ and $Av = \lambda_2 v$. Then $\lambda_1 v = \lambda_2 v$, and we have

$$0 = \lambda_1 v - \lambda_2 v = (\lambda_1 - \lambda_2)v.$$

Since v is an eigenvector, $v \neq 0$ and it follows that $\lambda_1 - \lambda_2 = 0$. Hence $\lambda_1 = \lambda_2$.

EXAMPLE 2 Although the zero vector is excluded from being an eigenvector, the number zero is *not* excluded from being an eigenvalue. For example, the vector $v = (-2, 1, 0)$ is an eigenvector of the matrix

$$A = \begin{bmatrix} 1 & 2 & 3 \\ 0 & 0 & 1 \\ 0 & 0 & 2 \end{bmatrix}$$

with eigenvalue 0 since $Av = 0 = 0v$.

It is easy to see that the number zero is an eigenvalue of any singular matrix. For if A is a singular matrix, then A has a nontrivial null space. Any nonzero vector v in the null space of A is an eigenvector for the eigenvalue 0 since $Av = 0 = 0v$. In Problem 3 of Section 8.2 we will ask the reader to prove the converse: If 0 is an eigenvalue of a matrix A, then A is singular.

EXERCISE 2 Show that $(-1, 1)$ is an eigenvector of the matrix $A = \begin{bmatrix} 1 & 1 \\ 1 & 1 \end{bmatrix}$ associated with the eigenvalue 0.

Let λ be an eigenvalue of an $n \times n$ matrix A. Then the eigenvectors associated with λ are precisely the nontrivial solutions to the equation $Av = \lambda v$. The set of all solutions to this equation forms a subspace of R^n. To prove this, suppose that v and w are solutions and that α is a scalar. Then

$$A(v + w) = Av + Aw = \lambda v + \lambda w = \lambda(v + w),$$
$$A(\alpha v) = \alpha Av = \alpha \lambda v = \lambda(\alpha v).$$

Thus the sum of two solutions is a solution and a scalar multiple of a solution is a solution. The subspace of R^n consisting of all solutions to $Av = \lambda v$ is called the **eigenspace** of λ and is denoted by $E(\lambda)$. The eigenvectors associated with λ are the nonzero vectors in this space since the definition of an eigenvector specifically excludes the zero vector.

Let us examine some of the consequences that follow from the equation $Av = \lambda v$. Let A be an $n \times n$ matrix and I be the $n \times n$ identity matrix. Since $\lambda Iv = \lambda v$, the equation $Av = \lambda v$ is equivalent to the equation

$$(A - \lambda I)v = 0.$$

Therefore, λ is an eigenvalue of A with an associated eigenvector v if and only if

$$(A - \lambda I)v = 0 \quad \text{and} \quad v \neq 0.$$

This proves the following theorem.

Theorem 8.1 *Let A be a square matrix.*

(a) *λ is an eigenvalue of A if and only if the null space of $A - \lambda I$ is nontrivial.*

(b) *If λ is an eigenvalue, then v is an eigenvector associated with λ if and only if v is a nonzero vector in the null space of $A - \lambda I$.*

It follows from (b) that $E(\lambda) = N(A - \lambda I)$. Since the null space of a matrix is a subspace, we have another proof of the fact that the eigenspace $E(\lambda)$ is a subspace.

Perhaps the most important consequence of (b) is that it provides us with a method for finding the eigenvectors associated with an eigenvalue λ. We simply compute a basis for $N(A - \lambda I)$ using the method presented in Chapter 2. The eigenvectors associated with λ are then the nontrivial linear combinations of these basis vectors.

EXAMPLE 3 $\lambda = 2$ is an eigenvalue of the matrix

$$A = \begin{bmatrix} 2 & -5 & 5 \\ 0 & 3 & -1 \\ 0 & -1 & 3 \end{bmatrix}.$$

To find the associated eigenvectors, we compute the null space of

$$A - 2I = \begin{bmatrix} 0 & -5 & 5 \\ 0 & 1 & -1 \\ 0 & -1 & 1 \end{bmatrix}.$$

The reduced echelon form of this matrix is

$$\begin{bmatrix} 0 & 1 & -1 \\ 0 & 0 & 0 \\ 0 & 0 & 0 \end{bmatrix}.$$

Thus a basis for the eigenspace $E(2)$ is $(1, 0, 0)$ and $(0, 1, 1)$. Any eigenvector associated with the eigenvalue 2 can be expressed as a nontrivial linear combination of these two eigenvectors. Thus the eigenvectors are

$$\alpha(1, 0, 0) + \beta(0, 1, 1) = (\alpha, \beta, \beta),$$

where at least one of α and β is not zero.

EXERCISE 3 $\lambda = 4$ is also an eigenvalue of the matrix A in Example 3. Find a basis for the eigenspace $E(4)$. Describe the eigenvectors associated with the eigenvalue 4.

PROBLEMS 8.1

1. Let $A = \begin{bmatrix} 3 & 1 \\ -3 & 7 \end{bmatrix}$. Show that $\lambda = 4$ and $\lambda = 6$ are eigenvalues of A with eigenvectors $(1, 1)$ and $(1, 3)$.

2. Let $A = \begin{bmatrix} 2 & 0 \\ 0 & 3 \end{bmatrix}$. Show that $\lambda = 2$ and $\lambda = 3$ are eigenvalues of A with eigenvectors $v_1 = (1, 0)$ and $v_2 = (0, 1)$.

*3. Show that $(1, -1)$ and $(-1, 6)$ are eigenvectors of $A = \begin{bmatrix} 3 & 1 \\ -6 & -4 \end{bmatrix}$. What are the associated eigenvalues?

*4. Show that $(1,0,1)$, $(-1,3,0)$, and $(1,0,2)$ are eigenvectors of each of the following matrices.

(a) $\begin{bmatrix} -12 & -5 & 9 \\ 0 & 3 & 0 \\ -18 & -6 & 15 \end{bmatrix}$ (b) $\begin{bmatrix} -4 & -1 & 3 \\ 0 & -1 & 0 \\ -6 & -2 & 5 \end{bmatrix}$

What are the associated eigenvalues?

5. Show that $(1,0,0)$, $(2,1,0)$, and $(3,4,1)$ are eigenvectors of each of the following matrices.

(a) $\begin{bmatrix} 2 & -2 & 2 \\ 0 & 1 & -4 \\ 0 & 0 & 0 \end{bmatrix}$ (b) $\begin{bmatrix} -1 & 0 & 6 \\ 0 & -1 & 8 \\ 0 & 0 & 1 \end{bmatrix}$

What are the associated eigenvalues?

*6. Show that $(1,0,0,0)$, $(1,1,0,0)$, $(1,1,1,0)$, and $(1,1,1,1)$ are eigenvectors of each of the following matrices.

(a) $\begin{bmatrix} -1 & -1 & -1 & -1 \\ 0 & -2 & -1 & -1 \\ 0 & 0 & -3 & -1 \\ 0 & 0 & 0 & -4 \end{bmatrix}$ (b) $\begin{bmatrix} 1 & 0 & 1 & 0 \\ 0 & 1 & 1 & 0 \\ 0 & 0 & 2 & 0 \\ 0 & 0 & 0 & 2 \end{bmatrix}$

What are the associated eigenvalues?

*7. Let λ be an eigenvalue of a matrix A. Show that if v is any eigenvector in $E(\lambda)$, then Av is also in $E(\lambda)$.

8. Let v be an eigenvector of a matrix A. Show that v is an eigenvector of the matrix $B = A - \alpha I$ for any number α. What is the corresponding eigenvalue?

*9. Show that the standard basis vectors $e_1, \ldots, e_n$ for R^n are eigenvectors for any $n \times n$ diagonal matrix. What are the associated eigenvalues?

10. Let A be an orthogonal matrix (i.e., $A^{-1} = A^T$). Show that if λ is an eigenvalue of A, then $|\lambda| = 1$. [*Hint:* Use Problem 17(e) of Section 3.2.]

11. A matrix A is called *nilpotent* if there is a positive integer n such that $A^n = 0$. Show that 0 is the only eigenvalue of a nilpotent matrix.

*12. A matrix A is called *idempotent* is $A^2 = A$. What can you say about the eigenvalues of such a matrix?

8.2   THE CHARACTERISTIC EQUATION

Theorem 8.1(b) provides us with a method for finding the eigenvectors associated with an eigenvalue λ. Once it is known that λ is an eigenvalue, the eigenvectors associated with λ are simply the nonzero vectors in the null space of $A - \lambda I$. We now address the question of how to find the eigenvalues. Theorem 8.1(a) provides the answer. λ is an eigenvalue of A if and only if the null space of $A - \lambda I$ is nontrivial, and this holds if and only if $A - \lambda I$ is singular. Since singular matrices are exactly those matrices whose determinant is zero, we have the following result.

Result 1 λ is an eigenvalue of A if and only if
$$\det(A - \lambda I) = 0.$$

Before we can use this result to find the eigenvalues of a matrix, we must study the equation $\det(A - \lambda I) = 0$. In view of the previous result, the eigenvalues are precisely the zeros of the function $p(\lambda) = \det(A - \lambda I)$. Let us compute $p(\lambda)$ when A is a 2×2 matrix and when A is a 3×3 matrix.

First suppose that A is a 2×2 matrix.

$$A = \begin{bmatrix} a_{11} & a_{12} \\ a_{21} & a_{22} \end{bmatrix}$$

Then

$$A - \lambda I = \begin{bmatrix} a_{11} & a_{12} \\ a_{21} & a_{22} \end{bmatrix} - \lambda \begin{bmatrix} 1 & 0 \\ 0 & 1 \end{bmatrix} = \begin{bmatrix} a_{11} - \lambda & a_{12} \\ a_{21} & a_{22} - \lambda \end{bmatrix}$$

and

$$\det(A - \lambda I) = (a_{11} - \lambda)(a_{22} - \lambda) - a_{12}a_{21}$$
$$= \lambda^2 - (a_{11} + a_{22})\lambda + (a_{11}a_{22} - a_{12}a_{21}).$$

Thus when A is a 2×2 matrix, $p(\lambda) = \det(A - \lambda I)$ is a polynomial of degree 2 in λ.

Now suppose that A is a 3×3 matrix.

$$A = \begin{bmatrix} a_{11} & a_{12} & a_{13} \\ a_{21} & a_{22} & a_{23} \\ a_{31} & a_{32} & a_{33} \end{bmatrix}$$

Then

$$A - \lambda I = \begin{bmatrix} a_{11} & a_{12} & a_{13} \\ a_{21} & a_{22} & a_{23} \\ a_{31} & a_{32} & a_{33} \end{bmatrix} - \lambda \begin{bmatrix} 1 & 0 & 0 \\ 0 & 1 & 0 \\ 0 & 0 & 1 \end{bmatrix}$$
$$= \begin{bmatrix} a_{11} - \lambda & a_{12} & a_{13} \\ a_{21} & a_{22} - \lambda & a_{23} \\ a_{31} & a_{32} & a_{33} - \lambda \end{bmatrix}$$

and

$$\det(A - \lambda I) = (a_{11} - \lambda)\det\begin{bmatrix} a_{22} - \lambda & a_{23} \\ a_{32} & a_{33} - \lambda \end{bmatrix}$$
$$- a_{12}\det\begin{bmatrix} a_{21} & a_{23} \\ a_{31} & a_{33} - \lambda \end{bmatrix} + a_{13}\det\begin{bmatrix} a_{21} & a_{22} - \lambda \\ a_{31} & a_{32} \end{bmatrix}.$$

The first term is a polynomial of degree 3 in λ and the second and third terms are polynomials of degree 1 in λ. Thus when A is a 3×3 matrix, $p(\lambda) = \det(A - \lambda I)$ is a polynomial of degree 3 in λ.

These examples prove the following result when $n = 2$ and $n = 3$. A similar argument works for an $n \times n$ matrix.

Result 2   Let A be an $n \times n$ matrix. The function $p(\lambda) = \det(A - \lambda I)$ is a polynomial of degree n in λ.

The polynomial $p(\lambda) = \det(A - \lambda I)$ is called the **characteristic polynomial** of A and the equation $\det(A - \lambda I) = 0$ is called the **characteristic equation** of A.

Combining our last two results, we obtain the following.

Theorem 8.2 λ *is an eigenvalue of a matrix A if and only if λ is a zero of the characteristic equation* $\det(A - \lambda I) = 0$.

Theorem 8.1(b) showed that the problem of finding the eigenspace $E(\lambda)$ of an eigenvalue λ is equivalent to finding the null space of $A - \lambda I$. Theorem 8.2 shows that the problem of finding eigenvalues of a matrix is equivalent to finding the zeros of the characteristic polynomial.

EXAMPLE 1   Let us find the eigenvalues and the associated eigenvectors of the matrix

$$A = \begin{bmatrix} 1 & 0 & 1 \\ 0 & 1 & 0 \\ 1 & 2 & 1 \end{bmatrix}.$$

The characteristic polynomial of A is

$$\det(A - \lambda I) = \det \begin{bmatrix} 1 - \lambda & 0 & 1 \\ 0 & 1 - \lambda & 0 \\ 1 & 2 & 1 - \lambda \end{bmatrix} = \lambda(1 - \lambda)(\lambda - 2).$$

The zeros of the characteristic equation are 0, 1, and 2 and by the previous result these are the eigenvalues of A. To find the eigenvectors associated with these eigenvalues, we compute the null space of $A - 0I$, $A - 1I$, and $A - 2I$. Using Gaussian elimination, we find that the reduced echelon forms of A, $A - I$, and $A - 2I$ are

$$\begin{bmatrix} 1 & 0 & 1 \\ 0 & 1 & 0 \\ 0 & 0 & 0 \end{bmatrix}, \quad \begin{bmatrix} 1 & 2 & 0 \\ 0 & 0 & 1 \\ 0 & 0 & 0 \end{bmatrix}, \quad \text{and} \quad \begin{bmatrix} 1 & 0 & -1 \\ 0 & 1 & 0 \\ 0 & 0 & 0 \end{bmatrix}.$$

Thus the eigenspace $E(0)$ is spanned by $(-1, 0, 1)$, the eigenspace $E(1)$ is spanned by $(-2, 1, 0)$, and the eigenspace $E(2)$ is spanned by $(1, 0, 1)$.

Eigenvalues	Eigenvectors
0	$\alpha(-1, 0, 1)$, $\alpha \neq 0$
1	$\alpha(-2, 1, 0)$, $\alpha \neq 0$
2	$\alpha(1, 0, 1)$, $\alpha \neq 0$

EXERCISE 1 Show that the 3×3 matrix A in Example 1 has three independent eigenvectors.

EXERCISE 2 Find the eigenvalues and a basis for each of the eigenspaces of the matrix

$$A = \begin{bmatrix} 2 & 3 & -1 \\ -6 & -7 & 5 \\ -6 & -6 & 6 \end{bmatrix}.$$

EXAMPLE 2 Let us verify that $\lambda = 2$ and $\lambda = 4$ are the eigenvalues of the matrix

$$A = \begin{bmatrix} 2 & -5 & 5 \\ 0 & 3 & -1 \\ 0 & -1 & 3 \end{bmatrix}$$

considered in Example 3 of Section 8.1. The characteristic polynomial of A is

$$\det(A - \lambda I) = \det \begin{bmatrix} 2 - \lambda & -5 & 5 \\ 0 & 3 - \lambda & -1 \\ 0 & -1 & 3 - \lambda \end{bmatrix}$$

$$= (2 - \lambda) \det \begin{bmatrix} 3 - \lambda & -1 \\ -1 & 3 - \lambda \end{bmatrix}$$

$$= (2 - \lambda)(\lambda - 4)(\lambda - 2).$$

Thus $\lambda = 2$ and $\lambda = 4$ are the zeros of the characteristic polynomial. (Since $\lambda = 2$ is a double root of the characteristic equation, we say that it has **multiplicity** 2.) In Example 3 of Section 8.1 we also saw that a basis for the eigenspace $E(2)$ is $(0, 1, 1)$ and $(1, 0, 0)$. In Exercise 3 of Section 8.1 the reader was asked to find a basis for the eigenspace $E(4)$. A basis for $E(4)$ is $(5, -1, 1)$.

Eigenvalues	Eigenvectors
2	$\alpha(0, 1, 1) + \beta(1, 0, 0)$, α or $\beta \neq 0$
4	$\alpha(5, -1, 1)$, $\alpha \neq 0$

Note that the eigenvalue $\lambda = 2$ has multiplicity 2 and that the eigenspace $E(2)$ is two-dimensional.

EXERCISE 3 Show that the 3×3 matrix A in Example 2 has three independent eigenvectors.

EXERCISE 4 Find the eigenvalues and a basis for each of the eigenspaces of the matrix

$$A = \begin{bmatrix} 1 & -2 & 2 \\ 0 & -1 & 2 \\ 0 & 0 & 1 \end{bmatrix}.$$

EXAMPLE 3   Let

$$A = \begin{bmatrix} -1 & 1 & 0 \\ 0 & 5 & 0 \\ 4 & -2 & 5 \end{bmatrix}.$$

The characteristic polynomial of A is

$$\det(A - \lambda I) = \det \begin{bmatrix} -1-\lambda & 1 & 0 \\ 0 & 5-\lambda & 0 \\ 4 & -2 & 5-\lambda \end{bmatrix}$$

$$= (-1-\lambda)(5-\lambda)^2,$$

so the eigenvalues of A are $\lambda = -1$ and $\lambda = 5$. The reduced echelon form of

$$A + I = \begin{bmatrix} 0 & 1 & 0 \\ 0 & 6 & 0 \\ 4 & -2 & 6 \end{bmatrix} \quad \text{is} \quad \begin{bmatrix} 1 & 0 & \frac{3}{2} \\ 0 & 1 & 0 \\ 0 & 0 & 0 \end{bmatrix},$$

and of

$$A - 5I = \begin{bmatrix} -6 & 1 & 0 \\ 0 & 0 & 0 \\ 4 & -2 & 0 \end{bmatrix} \quad \text{is} \quad \begin{bmatrix} 1 & 0 & 0 \\ 0 & 1 & 0 \\ 0 & 0 & 0 \end{bmatrix}.$$

Thus the eigenspace $E(-1)$ is spanned by $(-\frac{3}{2}, 0, 1)$ and the eigenspace $E(5)$ is spanned by $(0, 0, 1)$.

Eigenvalues	Eigenvectors
-1	$\alpha(-\frac{3}{2}, 0, 1),\ \alpha \neq 0$
5	$\alpha(0, 0, 1),\ \alpha \neq 0$

EXERCISE 5   Show that the 3×3 matrix A in Example 3 does not have three independent eigenvectors.

EXERCISE 6   Find the eigenvalues and a basis for each of the eigenspaces of

$$A = \begin{bmatrix} -1 & 0 & 4 \\ -1 & 0 & 3 \\ -1 & 1 & 2 \end{bmatrix}.$$

All of the preceding examples were carefully arranged so that it was easy to factor the characteristic polynomial. However, the student is no doubt well aware that it is difficult, if not impossible, to find the zeros of most polynomials. Polynomials of degree 2 are simple; we have the quadratic formula. It is also possible to find the zeros of any polynomial of degree 3 or 4 in the sense that there are formulas, similar to the quadratic formula although much more complicated, which will give us the zeros. But this is as far as it goes. Evariste Galois (1811–1832) showed that there is no analogue of the quadratic formula for polynomials of degree 5 or higher. For such polynomials we must resort to ad hoc methods to find their zeros. In Chapter 9 we discuss some numerical methods for finding the eigenvalues of a matrix that are not based on finding the zeros of a characteristic polynomial.

The discussion thus far has left open one extremely important question: Do polynomials always have zeros? If we insist that all zeros of a polynomial be real numbers, then the answer is definitely no! The polynomial $\lambda^2 + 1$ $\left(\text{which is the characteristic polynomial of the matrix } \begin{bmatrix} 0 & -1 \\ 1 & 0 \end{bmatrix}\right)$ has no real zeros. It does, however, have the two complex zeros $\pm i$, where $i = \sqrt{-1}$. In fact, if we allow complex zeros, then the following theorem answers our question. *Every polynomial of degree $n \geq 1$ has n zeros.* [Note that these zeros need not be distinct. If the polynomial has a repeated zero (i.e., a zero with multiplicity greater than 1), then this zero must be counted the number of times it is repeated. For example, the three zeros of $(\lambda - 1)^2(\lambda + 1)$ are $\lambda = 1$, $\lambda = 1$, and $\lambda = -1$. The multiplicity of a zero is the number of times the zero is repeated.] This result, called the Fundamental Theorem of Algebra, was first proved by Gauss in 1799. Its proof is very difficult and will not be given here.

Thus if we allow complex numbers to be eigenvalues, any $n \times n$ matrix A has exactly n eigenvalues. They are the n zeros of the characteristic equation $\det(A - \lambda I) = 0$. As we have just pointed out, this equation is in general very difficult to solve, and it may have complex roots. When an eigenvalue λ is a complex number, the eigenvectors associated with λ will have complex entries. Since the eigenspace $E(\lambda)$ is the null space of $A - \lambda I$, the procedure for finding the eigenspace associated with a complex eigenvalue λ is simply that of applying Gaussian elimination to a matrix with complex entries. We refer the reader to the appendix where we develop the necessary arithmetic of complex numbers and give several examples. *In the remainder of this chapter we deal only with matrices all of whose eigenvalues are real.*

One situation where it is exceptionally easy to find the eigenvalues is described in the following result.

Result 3 If A is a triangular matrix, then the eigenvalues of A are its diagonal entries.

This result applies to either upper or lower triangular matrices. In particular, it applies to diagonal matrices.

Proof If A is upper triangular, say

$$A = \begin{bmatrix} a_{11} & a_{12} & \cdots & a_{1n} \\ 0 & a_{21} & \cdots & a_{2n} \\ \vdots & \vdots & & \vdots \\ 0 & 0 & \cdots & a_{nn} \end{bmatrix},$$

then so is $A - \lambda I$.

$$A - \lambda I = \begin{bmatrix} a_{11} - \lambda & a_{12} & \cdots & a_{1n} \\ 0 & a_{22} - \lambda & \cdots & a_{2n} \\ \vdots & \vdots & & \vdots \\ 0 & 0 & \cdots & a_{nn} - \lambda \end{bmatrix}.$$

Consequently, $\det(A - \lambda I) = (a_{11} - \lambda)(a_{22} - \lambda) \cdots (a_{nn} - \lambda)$ and the zeros of this polynomial are $a_{11}, a_{22}, \ldots, a_{nn}$. The essence of this proof is the following. If A is triangular, then $A - \lambda I$ is triangular and the determinant of a triangular matrix is the product of its diagonal entries.

We conclude this section with two useful facts that relate the eigenvalues of a matrix to its diagonal entries and its determinant.

Result 4   Let A be an $n \times n$ matrix with the eigenvalues $\lambda_1, \ldots, \lambda_n$. Then the sum of the eigenvalues is the sum of the diagonal entries of A,

$$\lambda_1 + \lambda_2 + \cdots + \lambda_n = a_{11} + a_{22} + \cdots + a_{nn},$$

and the product of the eigenvalues is the determinant of A,

$$\lambda_1 \lambda_2 \cdots \lambda_n = \det A.$$

The proof of this result will be omitted. It is most useful as a numerical check on the eigenvalues. If one of these equations fails, there must be a mistake. However, these equations are not sufficient to guarantee that the eigenvalues are correct. That is, these equations may be satisfied by incorrect eigenvalues.

EXAMPLE 4   In Example 2 we showed that the eigenvalues of the matrix

$$A = \begin{bmatrix} 2 & -5 & 5 \\ 0 & 3 & -1 \\ 0 & -1 & 3 \end{bmatrix}$$

are $\lambda_1 = 2$, $\lambda_2 = 2$, and $\lambda_3 = 4$. Thus $\lambda_1 + \lambda_2 + \lambda_3 = 8$ and the sum of the diagonal entries of A is $2 + 3 + 3 = 8$. Similarly, $\lambda_1 \lambda_2 \lambda_3 = 16 = \det A$. However, suppose that we had erroneously obtained $1 + \sqrt{5}$, $1 + \sqrt{5}$, and $6 - 2\sqrt{5}$ for the eigenvalues. Since $(1 + \sqrt{5}) + (1 + \sqrt{5}) + (6 - 2\sqrt{5}) = 8$ and $(1 + \sqrt{5})(1 + \sqrt{5})(6 - 2\sqrt{5}) = 16$, these equations would not have detected the error.

PROBLEMS 8.2

*1. Find the eigenvalues and a basis for each of the corresponding eigenspaces.

(a) $\begin{bmatrix} 2 & 0 \\ -1 & 3 \end{bmatrix}$ (b) $\begin{bmatrix} 1 & -2 \\ -1 & 2 \end{bmatrix}$ (c) $\begin{bmatrix} 0 & -1 \\ -3 & 2 \end{bmatrix}$

(d) $\begin{bmatrix} 1 & 0 & 1 \\ 1 & 1 & 1 \\ 0 & 0 & -3 \end{bmatrix}$ (e) $\begin{bmatrix} 1 & 1 & 0 \\ 1 & 5 & -2 \\ 1 & 3 & -1 \end{bmatrix}$ (f) $\begin{bmatrix} 1 & 1 & 1 \\ 1 & -1 & 3 \\ 0 & 1 & 0 \end{bmatrix}$

(g) $\begin{bmatrix} 0 & 0 & 0 \\ 0 & 1 & 0 \\ 1 & 0 & 1 \end{bmatrix}$ (h) $\begin{bmatrix} 2 & 0 & 1 \\ 0 & 1 & 2 \\ 0 & 0 & 1 \end{bmatrix}$ (i) $\begin{bmatrix} 0 & 1 & 1 \\ 0 & 0 & 2 \\ 0 & 0 & 1 \end{bmatrix}$

(j) $\begin{bmatrix} 0 & 0 & 2 \\ 0 & 0 & 0 \\ 2 & 0 & 0 \end{bmatrix}$ (k) $\begin{bmatrix} 1 & -1 & 0 \\ -1 & 2 & -1 \\ 0 & -1 & 1 \end{bmatrix}$ (l) $\begin{bmatrix} -2 & -2 & -4 \\ 2 & 3 & 2 \\ 3 & 2 & 5 \end{bmatrix}$

(m) $\begin{bmatrix} 3 & 1 & 0 \\ 0 & 2 & 1 \\ 0 & 1 & 2 \end{bmatrix}$ (n) $\begin{bmatrix} 2 & 0 & 0 \\ 0 & 2 & 7 \\ 0 & 0 & -5 \end{bmatrix}$ (o) $\begin{bmatrix} 1 & -1 & 4 \\ 3 & 2 & -1 \\ 2 & 1 & -1 \end{bmatrix}$

(p) $\begin{bmatrix} 1 & 1 & 0 & 0 \\ 0 & 1 & 0 & 0 \\ 0 & 0 & 1 & 1 \\ 0 & 0 & 0 & 1 \end{bmatrix}$ (q) $\begin{bmatrix} 1 & 1 & 0 & 0 \\ 0 & 1 & 0 & 0 \\ 0 & 0 & -2 & 0 \\ 0 & 0 & 0 & -2 \end{bmatrix}$

(r) $\begin{bmatrix} 2 & 1 & 0 & 0 \\ 0 & 2 & 1 & 0 \\ 0 & 0 & 2 & 0 \\ 0 & 0 & 0 & -1 \end{bmatrix}$

2. Verify that Result 4 holds for each of the matrices in Problem 1.

3. Show that 0 is an eigenvalue of a matrix A if and only if A is singular.

*4. Let A be an invertible matrix. Show that λ is an eigenvalue of A if and only if λ^{-1} is an eigenvalue of A^{-1}. How are the eigenvectors related?

*5. Show that A and A^T have the same eigenvalues. (*Hint:* Show that A and A^T have the same characteristic equation.) Do they have the same eigenvectors? If so, give a proof. If not, give a counterexample.

6. Let λ be an eigenvalue of A.
   (a) Show that λ^2 is an eigenvalue of A^2, λ^3 is an eigenvalue of A^3, and so on.
   (b) If v is an eigenvector of A associated with λ, show that v is an eigenvector of $A^2, A^3, \ldots$ associated with $\lambda^2, \lambda^3, \ldots$.

7. Let A be an $n \times n$ matrix whose rows all add up to 2. Show that $\lambda = 2$ is an eigenvalue of A. What is an associated eigenvector? Is there anything special about the number 2?

8.3 DIAGONALIZATION

In Chapter 1 we saw several examples of situations where difference equations arise naturally. Recall that a difference equation is an equation of the form

$$x_{k+1} = Ax_k, \qquad k = 0, 1, 2, \ldots, \tag{1}$$

where A is an $n \times n$ matrix and $x_0, x_1, x_2, \ldots$ are n-vectors. Since $x_1 = Ax_0$,

$x_2 = Ax_1 = A^2 x_0, \ldots,$ it is clear that the solution to the difference equation (1) is

$$x_k = A^k x_0.$$

Yet this solution leaves a lot to be desired. If we are required to find x_{25}, we must raise A to the 25th power. That is a lot of work (even when A is only a 2×2 matrix).

There is, however, one situation where this computation is relatively simple. Suppose that the matrix A in equation (1) is diagonal, say

$$A = \begin{bmatrix} d_1 & & \\ & \ddots & \\ & & d_n \end{bmatrix}.$$

Then it is easy to compute the powers of A. We have

$$A^2 = \begin{bmatrix} d_1^2 & & \\ & \ddots & \\ & & d_n^2 \end{bmatrix}, \qquad A^3 = \begin{bmatrix} d_1^3 & & \\ & \ddots & \\ & & d_n^3 \end{bmatrix},$$

and in general

$$A^k = \begin{bmatrix} d_1^k & & \\ & \ddots & \\ & & d_n^k \end{bmatrix}.$$

Thus the kth power of a diagonal matrix A is simply the diagonal matrix having as its diagonal entries the kth powers of the diagonal entries of A.

This discussion suggests that when the matrix A in equation (1) is not a diagonal matrix, we might try to make a substitution (i.e., a change of variables) that will yield a new difference equation, one with a diagonal matrix. As we have seen above, we can easily find the solution to this new equation. Then by making the inverse substitution, we can use this solution to find the solution to the original difference equation.

Suppose that we are given the difference equation

$$x_{k+1} = Ax_k, \tag{2}$$

where A is not a diagonal matrix. If we substitute

$$x_k = Py_k \tag{3}$$

into equation (2), we obtain

$$Py_{k+1} = APy_k.$$

Assuming that P is nonsingular, this last equation becomes

$$y_{k+1} = P^{-1}APy_k.$$

From this equation it is clear that we want to choose P to be a nonsingular matrix with the property that $\Lambda = P^{-1}AP$ is a diagonal matrix. Assuming for the moment that we can find such a matrix P, we can then solve the difference equation

$$y_{k+1} = \Lambda y_k$$

to obtain

$$y_k = \Lambda^k y_0.$$

Now make the inverse substitution

$$y_k = P^{-1}x_k$$

of (3) in this equation to obtain

$$P^{-1}x_k = \Lambda^k P^{-1}x_0 \quad \text{or} \quad x_k = P\Lambda^k P^{-1}x_0. \tag{4}$$

This is the solution to our original difference equation (2).

Our whole discussion thus far rests on the following question: Given an $n \times n$ matrix A, does there exist a nonsingular matrix P such that $\Lambda = P^{-1}AP$ is a diagonal matrix? Unfortunately, there are matrices A for which no such P exists. In this case we might look for a matrix P such that $P^{-1}AP$ is as diagonal as possible. We return to this question at the end of the next section.

Definition   An $n \times n$ matrix A is said to be **diagonalizable** if there is a nonsingular matrix P such that

$$\Lambda = P^{-1}AP$$

is a diagonal matrix. The matrix P is called a **diagonalizing matrix** for A.

Suppose that A is an $n \times n$ diagonalizable matrix. Then there is a nonsingular matrix P such that $\Lambda = P^{-1}AP$ is a diagonal matrix. Thus

$$AP = P\Lambda$$

where P is a nonsingular matrix and Λ is a diagonal matrix. Let

$$P = \begin{bmatrix} | & & | \\ w_1 & \cdots & w_n \\ | & & | \end{bmatrix} \quad \text{and} \quad \Lambda = \begin{bmatrix} \mu_1 & & \\ & \ddots & \\ & & \mu_n \end{bmatrix}.$$

Then

$$AP = \begin{bmatrix} | & & | \\ Aw_1 & \cdots & Aw_n \\ | & & | \end{bmatrix}$$

and

$$
P\Lambda = \begin{bmatrix} | & & | \\ w_1 & \cdots & w_n \\ | & & | \end{bmatrix} \begin{bmatrix} \mu_1 & & \\ & \ddots & \\ & & \mu_n \end{bmatrix} = \begin{bmatrix} | & & | \\ \mu_1 w_1 & \cdots & \mu_n w_n \\ | & & | \end{bmatrix}.
$$

Since $AP = P\Lambda$, we obtain

$$
Aw_1 = \mu_1 w_1, \ldots, Aw_n = \mu_n w_n.
$$

Because P is nonsingular, its columns $w_1, \ldots, w_n$ are independent and nonzero. Thus the vectors $w_1, \ldots, w_n$ are independent eigenvectors of A and $\mu_1, \ldots, \mu_n$ are the associated eigenvalues.

We have proved that if A is an $n \times n$ diagonalizable matrix, then A has n independent eigenvectors. In fact, we have proved much more. If P is a nonsingular matrix and Λ is a diagonal matrix such that

$$
AP = P\Lambda,
$$

then the columns of P are independent eigenvectors of A and the ith diagonal entry of the diagonal matrix Λ is the eigenvalue for the eigenvector in the ith column of P.

Is the converse of this result true? That is, if A is the $n \times n$ matrix that has n independent eigenvectors, is A diagonalizable? In view of our previous discussion, it is obvious how we should investigate this question. We simply let P and Λ be as above and then ask whether P is a nonsingular matrix with $AP = P\Lambda$. Let $v_1, \ldots, v_n$ be independent eigenvectors of A, $\lambda_1, \ldots, \lambda_n$ be associated eigenvalues, and

$$
P = \begin{bmatrix} | & & | \\ v_1 & \cdots & v_n \\ | & & | \end{bmatrix} \quad \text{and} \quad \Lambda = \begin{bmatrix} \lambda_1 & & \\ & \ddots & \\ & & \lambda_n \end{bmatrix}.
$$

Since $Av_1 = \lambda_1 v_1, \ldots, Av_n = \lambda_n v_n$, it follows that

$$
AP = \begin{bmatrix} | & & | \\ Av_1 & \cdots & Av_n \\ | & & | \end{bmatrix} = \begin{bmatrix} | & & | \\ \lambda_1 v_1 & \cdots & \lambda_n v_n \\ | & & | \end{bmatrix} = P\Lambda.
$$

P is nonsingular because its columns are independent. thus

$$
\Lambda = P^{-1}AP
$$

and A is diagonalizable.

Theorem 8.3 *An $n \times n$ matrix A is diagonalizable if and only if it has n independent eigenvectors. Any diagonalizing matrix P has independent eigenvectors of A as its columns and the diagonal matrix Λ has the eigenvalues of A as its diagonal entries, the i th diagonal entry of Λ being the eigenvalue for the eigenvector in the i th column of P.*

EXAMPLE 1 Let A be the matrix

$$A = \begin{bmatrix} 1 & 0 & 1 \\ 0 & 1 & 0 \\ 1 & 2 & 1 \end{bmatrix}.$$

In Example 1 in Section 8.2 we found that 0, 1, and 2 are the eigenvalues of A and that the eigenspaces $E(0)$, $E(1)$, and $E(2)$ are spanned by the vectors $(-1, 0, 1)$, $(-2, 1, 0)$, and $(1, 0, 1)$, respectively. In Exercise 1 of Section 8.2 we asked the student to show that these eigenvectors are independent. Thus A is a diagonalizable matrix and

$$P = \begin{bmatrix} -1 & -2 & 1 \\ 0 & 1 & 0 \\ 1 & 0 & 1 \end{bmatrix} \quad \text{and} \quad \Lambda = \begin{bmatrix} 0 & & \\ & 1 & \\ & & 2 \end{bmatrix}.$$

Let us verify for this example that $A = P\Lambda P^{-1}$. Since

$$P^{-1} = \frac{1}{2} \begin{bmatrix} -1 & -2 & 1 \\ 0 & 2 & 0 \\ 1 & 2 & 1 \end{bmatrix},$$

$$P\Lambda P^{-1} = \frac{1}{2} \begin{bmatrix} -1 & -2 & 1 \\ 0 & 1 & 0 \\ 1 & 0 & 1 \end{bmatrix} \begin{bmatrix} 0 & & \\ & 1 & \\ & & 2 \end{bmatrix} \begin{bmatrix} -1 & -2 & 1 \\ 0 & 2 & 0 \\ 1 & 2 & 1 \end{bmatrix}$$

$$= \frac{1}{2} \begin{bmatrix} 0 & -2 & 2 \\ 0 & 1 & 0 \\ 0 & 0 & 2 \end{bmatrix} \begin{bmatrix} -1 & -2 & 1 \\ 0 & 2 & 0 \\ 1 & 2 & 1 \end{bmatrix} = \begin{bmatrix} 1 & 0 & 1 \\ 0 & 1 & 0 \\ 1 & 2 & 1 \end{bmatrix} = A.$$

EXERCISE 1 Verify that the matrix given in Example 2 in Section 8.2 is diagonalizable and that the matrix given in Example 3 in Section 8.2 is not diagonalizable.

A natural question to ask at this point is whether there are any "nice" conditions which guarantee that A has n independent eigenvectors. We begin to answer this question with the following result.

Result 1 Let A be an $n \times n$ matrix. If A has n distinct eigenvalues, then A has n independent eigenvectors. Hence A is diagonalizable.

This result follows immediately from the following more general theorem.

Theorem 8.4 *Let A be an $n \times n$ matrix and let $\lambda_1, \ldots, \lambda_r$ be distinct eigenvalues of A with associated eigenvectors $v_1, \ldots, v_r$. Then $v_1, \ldots, v_r$ are independent eigenvectors.*

In other words, eigenvectors associated with distinct eigenvalues are independent. If the zero vector were allowed to be an eigenvector, this would not be true. Why?

Proof First, we prove that v_1 and v_2 are independent. Suppose that there are scalars α_1 and α_2 such that

$$\alpha_1 v_1 + \alpha_2 v_2 = 0. \tag{5}$$

Multiplying by A, we obtain (since $Av_i = \lambda_i v_i$, $i = 1, 2$)

$$\alpha_1 \lambda_1 v_1 + \alpha_2 \lambda_2 v_2 = 0.$$

Now subtract λ_2 times equation (5) from this equation to obtain

$$\alpha_1 (\lambda_1 - \lambda_2) v_1 = 0.$$

Since v_1 is an eigenvector, $v_1 \neq 0$. Hence $\alpha_1(\lambda_1 - \lambda_2) = 0$ and because $\lambda_1 \neq \lambda_2$ we must have $\alpha_1 = 0$. From (5) we now find that $\alpha_2 = 0$ since $v_2 \neq 0$. Therefore, $\alpha_1 = \alpha_2 = 0$ and v_1 and v_2 are independent.

Next we show that v_1, v_2, and v_3 are independent. Suppose that there are scalars α_1, α_2, and α_3 such that

$$\alpha_1 v_1 + \alpha_2 v_2 + \alpha_3 v_3 = 0. \tag{6}$$

As above, we multiply this equation by A to obtain

$$\alpha_1 \lambda_1 v_1 + \alpha_2 \lambda_2 v_2 + \alpha_3 \lambda_3 v_3 = 0.$$

Now subtract λ_3 times equation (6) from this equation and obtain

$$\alpha_1 (\lambda_1 - \lambda_3) v_1 + \alpha_2 (\lambda_2 - \lambda_3) v_2 = 0.$$

Because v_1 and v_2 are independent it follows that

$$\alpha_1 (\lambda_1 - \lambda_3) = \alpha_2 (\lambda_2 - \lambda_3) = 0.$$

Since the λ's are distinct, $\alpha_1 = \alpha_2 = 0$. From (6) we now find that $\alpha_3 = 0$. Thus v_1, v_2, and v_3 are independent. Repeating this argument proves the result.

We saw at the beginning of this section that the solution to the difference equation $x_{k+1} = Ax_k$ is

$$x_k = A^k x_0.$$

We also saw that when there is a nonsingular matrix P such that $\Lambda = P^{-1}AP$ is a diagonal matrix, then we could express this solution in the form [see equation (4)]

$$x_k = P\Lambda^k P^{-1} x_0.$$

This suggests (but does not prove—why?) that

$$A^k = P\Lambda^k P^{-1}.$$

To prove this result, we proceed as follows. Since $\Lambda = P^{-1}AP$,

$$A = P\Lambda P^{-1},$$
$$A^2 = (P\Lambda P^{-1})(P\Lambda P^{-1}) = P\Lambda^2 P^{-1},$$
$$A^3 = AA^2 = (P\Lambda P^{-1})(P\Lambda^2 P^{-1}) = P\Lambda^3 P^{-1}.$$

It is clear that by repeating this process, we obtain

$$A^k = P\Lambda^k P^{-1}.$$

EXAMPLE 2   In Example 1 we found that the matrix

$$A = \begin{bmatrix} 1 & 0 & 1 \\ 0 & 1 & 0 \\ 1 & 2 & 1 \end{bmatrix}$$

can be written as

$$A = \begin{bmatrix} -1 & -2 & 1 \\ 0 & 1 & 0 \\ 1 & 0 & 1 \end{bmatrix} \begin{bmatrix} 0 & & \\ & 1 & \\ & & 2 \end{bmatrix} \begin{bmatrix} -\frac{1}{2} & -1 & \frac{1}{2} \\ 0 & 1 & 0 \\ \frac{1}{2} & 1 & \frac{1}{2} \end{bmatrix}$$
$$= P\Lambda P^{-1}.$$

Thus for $m \geq 1$

$$A^m = \begin{bmatrix} -1 & -2 & 1 \\ 0 & 1 & 0 \\ 1 & 0 & 1 \end{bmatrix} \begin{bmatrix} 0 & & \\ & 1 & \\ & & 2^m \end{bmatrix} \begin{bmatrix} -\frac{1}{2} & -1 & \frac{1}{2} \\ 0 & 1 & 0 \\ \frac{1}{2} & 1 & \frac{1}{2} \end{bmatrix}$$
$$= \begin{bmatrix} 2^{m-1} & 2^m - 2 & 2^{m-1} \\ 0 & 1 & 0 \\ 2^{m-1} & 2^m & 2^{m-1} \end{bmatrix}.$$

EXAMPLE 3   In this example we apply the process of diagonalization to the problem of predicting the age distribution of a certain fish population. This population has been divided into three age groups, the group of fish in their first year of life, their second year of life, and their third year of life, respectively. Each year $\frac{1}{32}$ of the fish in the first group survive to become members of the second group. The fish in the second group have an average of 26 offspring per individual and $\frac{1}{5}$ of the fish in the second group survive to become members of the third group. The fish in the third group have an average of 30 offspring per individual and then die. Let us denote the number of fish in the first, second, and third groups after k years by a_k, b_k, and c_k, respectively. Let $N_k = a_k + b_k + c_k$ be the total number of fish after k years. Then $a_{k+1} = 26b_k + 30c_k$, $b_{k+1} = a_k/32$, and $c_{k+1} = b_k/5$. Setting $x_k = (a_k, b_k, c_k)$ and

$$A = \begin{bmatrix} 0 & 26 & 30 \\ \frac{1}{32} & 0 & 0 \\ 0 & \frac{1}{5} & 0 \end{bmatrix},$$

we obtain the difference equation

$$x_{k+1} = Ax_k, \qquad k = 0, 1, 2, \dots, \tag{7}$$

where $x_0 = (a_0, b_0, c_0)$ is the initial age distribution of the population. The matrix A is called the Leslie matrix for the population (see Section 1.8). The solution to (7) is

$$x_k = A^k x_0, \qquad k = 0, 1, 2, \ldots .$$

We are interested in the long-range behavior of the population (i.e., in x_k as k becomes very large). The characteristic equation of A is

$$0 = \det(A - \lambda I) = -\lambda^3 + \tfrac{13}{16}\lambda^2 + \tfrac{3}{16}.$$

This equation has three distinct solutions: $\lambda_1 = 1$, $\lambda_2 = -\tfrac{1}{4}$, and $\lambda_3 = -\tfrac{3}{4}$. Hence A is diagonalizable. The vectors $v_1 = (160, 5, 1)$, $v_2 = (40, -5, 4)$, and $v_3 = (360, -15, 4)$ are eigenvectors associated with the eigenvalues 1, $-\tfrac{1}{4}$, and $-\tfrac{3}{4}$, respectively. Thus $A = P\Lambda P^{-1}$, where

$$P = \begin{bmatrix} 160 & 40 & 360 \\ 5 & -5 & -15 \\ 1 & 4 & 4 \end{bmatrix} \quad \text{and} \quad \Lambda = \begin{bmatrix} 1 & 0 & 0 \\ 0 & -\tfrac{1}{4} & 0 \\ 0 & 0 & -\tfrac{3}{4} \end{bmatrix}.$$

This yields $x_k = A^k x_0 = P\Lambda^k P^{-1} x_0$.
Setting

$$\begin{bmatrix} \alpha \\ \beta \\ \gamma \end{bmatrix} = P^{-1} x_0 = P^{-1} \begin{bmatrix} a_0 \\ b_0 \\ c_0 \end{bmatrix},$$

we obtain

$$x_k = P\Lambda^k \begin{bmatrix} \alpha \\ \beta \\ \gamma \end{bmatrix} = P \begin{bmatrix} \alpha \\ \left(-\tfrac{1}{4}\right)^k \beta \\ \left(-\tfrac{3}{4}\right)^k \gamma \end{bmatrix} = \alpha v_1 + \left(-\tfrac{1}{4}\right)^k \beta v_2 + \left(-\tfrac{3}{4}\right)^k \gamma v_3.$$

For large k, $\left(-\tfrac{1}{4}\right)^k$ and $\left(-\tfrac{3}{4}\right)^k$ are very close to zero. Therefore, x_k approaches

$$x_\infty = \alpha v_1 = \alpha(160, 5, 1)$$

and the number of fish N_k approaches

$$N_\infty = \alpha(160 + 5 + 1) = 166\alpha.$$

If we solve

$$P \begin{bmatrix} \alpha \\ \beta \\ \gamma \end{bmatrix} = \begin{bmatrix} a_0 \\ b_0 \\ c_0 \end{bmatrix}$$

we obtain $\alpha = (a_0 + 32b_0 + 30c_0)/350$. We conclude that in the long run the population approaches a stable state. In this state the total number of fish is 166α and of these fish about 96.4% belong to the first group, 3.0% belong to the second group, and 0.6% belong to the third group. Notice that in contrast to the number $N_\infty = 166\alpha$, these percentages do not depend upon the initial values a_0, b_0, and c_0.

EXERCISE 2 Referring to Example 3, express N_∞ in terms of N_0 when

(a) $a_0 = b_0 = c_0 = N_0/3$
(b) $a_0 = N_0,\ b_0 = c_0 = 0$
(c) $a_0 = c_0 = 0,\ b_0 = N_0$

EXERCISE 3 Referring to Example 3, show that when x_0 is a scalar multiple of $(160, 5, 1)$, then $x_k = x_0$ for all k (i.e., the population is completely stable).

PROBLEMS 8.3

*1. Determine which of the matrices in Problem 1 of Section 8.2 are diagonalizable. For each matrix A that is diagonalizable, find a nonsingular matrix P and a diagonal matrix Λ such that $\Lambda = P^{-1}AP$.

*2. Find a matrix with eigenvalues 0, 1, and 3 and corresponding eigenvectors $(1,0,0)$, $(1,1,0)$, and $(1,1,1)$.

*3. Is the converse of Theorem 8.4 true?

4. Prove Result 1 using Theorem 8.4.

*5. If A is diagonalizable, is the diagonalizing matrix unique? If so, why? If not, how are the diagonalizing matrices related (if at all)?

*6. If A is diagonalizable, is the diagonal matrix unique? If so, why? If not, how are the diagonal matrices related (if at all)?

*7. Suppose that A is an $n \times n$ matrix with exactly one eigenvalue with multiplicity n. Show that if A is diagonalizable, then A is a diagonal matrix. Find A.

8. Let A be a diagonalizable matrix. Prove that a diagonalizable matrix B commutes with A (i.e., $AB = BA$) if A and B have the same eigenvectors. (*Hint:* If P is the matrix of eigenvectors of A and B, then there are diagonal matrices Λ_1 and Λ_2 such that

$$A = P\Lambda_1 P^{-1} \quad \text{and} \quad B = P\Lambda_2 P^{-1}.$$

The condition that $AB = BA$ implies

$$AP\Lambda_2 P^{-1} = P\Lambda_2 P^{-1}A.$$

Now multiply this equation on the left by P^{-1} and on the right by P.)

9. Show that if A has distinct eigenvalues, and B is a matrix that commutes with A, then A and B have the same eigenvectors. (*Hint:* Let v be an eigenvector of A and λ be an associated eigenvalue. Then $ABv = BAv = B\lambda v = \lambda Bv$. Now use the fact that the eigenspaces are one-dimensional.)

10. Show that if A has distinct eigenvalues, and B is a matrix that commutes with A, then B is diagonalizable.

11. Recall that two matrices A and B are called *similar* if there exists a nonsingular matrix S such that $A = S^{-1}BS$.
*(a) Show that similar matrices have the same eigenvalues.

*(b) Is the converse of part (a) true? That is, if two matrices have the same eigenvalues, each with the same multiplicity, are they necessarily similar?

*(c) If A is $n \times n$ and if A is similar to a diagonal matrix, must it follow that A has n independent eigenvectors?

*(d) Are the matrices

$$\begin{bmatrix} 1 & & \\ & 5 & \\ & & -2 \end{bmatrix} \quad \text{and} \quad \begin{bmatrix} 5 & & \\ & 1 & \\ & & -2 \end{bmatrix}$$

similar?

(e) Show that any two diagonal matrices that have the same diagonal entries (counting multiplicity) are similar.

(f) If A and B are diagonalizable and have the same eigenvalues, show that A and B are similar matrices.

*(g) What matrices are similar to the identity matrix?

*(h) How are the eigenvectors of similar matrices related?

(i) Show that if A and B are similar, then $\operatorname{tr} A = \operatorname{tr} B$. [*Hint:* Use Problem 17(c) of Section 1.7.]

12. If $A \neq 0$ and A is nilpotent, is A similar to a diagonal matrix? In other words, can A be diagonalized? (*Hint:* Use Problem 11 of Section 8.1.)

13. Can an idempotent matrix be diagonalized? (*Hint:* Use Problem 12 of Section 8.1.)

14. Show that if A is diagonalizable, the eigenvalues of A^2 are exactly the squares of the eigenvalues of A and that A and A^2 have the same eigenvectors.

15. Using the notation of Example 3, suppose that the Leslie matrix for a certain fish population is

$$A = \begin{bmatrix} 0 & 19 & 15 \\ \frac{1}{16} & 0 & 0 \\ 0 & \frac{1}{2} & 0 \end{bmatrix}.$$

(a) Show that the eigenvalues of A are $\frac{5}{4}$, $-\frac{1}{2}$, and $-\frac{3}{4}$.

(b) Show that for large k, x_k is approximately $\alpha(\frac{5}{4})^k(100, 5, 2)$, where α is a constant.

(c) Show that in the long run about 93% of the fish belong to the first group, about 5% belong to the second group, and about 2% belong to the third group.

(d) Describe how this population grows.

8.4   MATRICES WITH REPEATED EIGENVALUES

In Section 8.3 we proved that if the eigenvalues of an $n \times n$ matrix A are all distinct, then A is diagonalizable. This is the best result we can obtain by just looking at the eigenvalues of A. Matrices with repeated eigenvalues may or may not be diagonalizable. In Exercise 1 of Section 8.3 we saw an example of a 3×3 matrix with a repeated eigenvalue which is diagonalizable and an example of a matrix with a repeated eigenvalue which is not diagonalizable. Is there a necessary and sufficient condition that a matrix with multiple eigenvalues be diagonalizable?

EXERCISE 1   Show that any diagonal matrix is diagonalizable. (*Hint:* Show that the standard basis vectors $e_1, \ldots, e_n$ are n independent eigenvectors of any diagonal matrix.)

The heart of the matter lies in the relationship between the multiplicity of λ as a zero of the characteristic polynomial and the number of independent eigenvectors associated with λ, that is, the dimension of the eigenspace $E(\lambda)$. The multiplicity of λ as a zero of the characteristic polynomial is called the **algebraic multiplicity** of λ. The dimension of the eigenspace $E(\lambda)$ is called the **geometric multiplicity** of λ.

Result 1   The geometric multiplicity of λ is always less than or equal to the algebraic multiplicity of λ.

The proof of this result is difficult and will be omitted. With this result we can prove the following theorem, which gives the best result obtainable in the case of repeated roots.

Theorem 8.5   *Let A be an $n \times n$ matrix and let $\lambda_1, \ldots, \lambda_r$ be the distinct eigenvalues of A. Then A is diagonalizable if and only if the geometric multiplicity of each eigenvalue is equal to its algebraic multiplicity.*

Proof   Let $d_1, \ldots, d_r$ be the algebraic multiplicities of $\lambda_1, \ldots, \lambda_r$ respectively. Then $d_1 + \cdots + d_r = n$ since the degree of the characteristic polynomial is n. If A is diagonalizable, then it has n independent eigenvectors. Each of these eigenvectors is associated with a unique eigenvalue and by the previous result, at most d_1 are associated with λ_1, d_2 with λ_2, and so on. Therefore, exactly d_i of the eigenvectors are associated with the eigenvalue λ_i. Hence the geometric multiplicity of each eigenvalue is equal to its algebraic multiplicity. Conversely, if this condition holds, then reversing the steps of this argument shows that A has n independent eigenvectors and hence is diagonalizable.

The proof of Theorem 8.5 amounts to showing that the matrix A has n independent eigenvectors if and only if the dimension of each eigenspace $E(\lambda)$ is equal to the algebraic multiplicity of λ. But to find the dimension of the eigenspace $E(\lambda)$, we must find the number of independent eigenvectors associated with the eigenvalue λ. Thus to establish the condition mentioned in the theorem we must in effect find n independent eigenvectors, and we already know (Theorem 8.3) that this is a necessary and sufficient condition for A to be diagonalizable. However, Theorem 8.5 is useful for showing that a matrix is not diagonalizable. For if we are able to find an eigenvalue λ of a matrix A whose geometric multiplicity is less than its algebraic multiplicity, then the matrix A is not diagonalizable.

EXAMPLE 1   The eigenvalue 5 of matrix A in Example 3 of Section 8.2 has algebraic multiplicity 2 and geometric multiplicity 1. Therefore, A is not diagonalizable.

If a matrix is not diagonalizable, the best we can hope to accomplish is to find a nonsingular matrix P such that $J = P^{-1}AP$ is as diagonal as possible. The best possible matrix J is called the *Jordan normal form* of A. It is a matrix that has the following properties:

1. The eigenvalues of A are on the main diagonal.
2. The diagonal above the main diagonal contains only zeros and ones.
3. All other entries are zero.

The proof of this theorem and its applications are beyond the scope of this book.

PROBLEMS 8.4

*1. Let

$$A = \begin{bmatrix} 2 & 1 & 0 & 0 \\ 0 & 2 & 0 & 0 \\ 0 & 0 & 2 & 0 \\ 0 & 0 & 0 & 1 \end{bmatrix}.$$

(a) Show that $\lambda = 2$ is an eigenvalue of A with algebraic multiplicity 3 and geometric multiplicity 2.
(b) What is the other eigenvalue of A?

2. Let

$$A = \begin{bmatrix} 5 & 1 & 0 & 0 \\ 0 & 5 & 1 & 0 \\ 0 & 0 & 5 & 0 \\ 0 & 0 & 0 & 5 \end{bmatrix} \quad \text{and} \quad B = \begin{bmatrix} 5 & 1 & 0 & 0 \\ 0 & 5 & 0 & 0 \\ 0 & 0 & 5 & 1 \\ 0 & 0 & 0 & 5 \end{bmatrix}.$$

Show that $\lambda = 5$ is an eigenvalue of A and B with algebraic multiplicity 4 and geometric multiplicity 2.

*3. Let

$$A = \begin{bmatrix} 4 & 1 & 0 & 0 & 0 \\ 0 & 4 & 0 & 0 & 0 \\ 0 & 0 & -1 & 1 & 0 \\ 0 & 0 & 0 & -1 & 1 \\ 0 & 0 & 0 & 0 & -1 \end{bmatrix}.$$

(a) Show that $\lambda = 4$ is an eigenvalue of A with algebraic multiplicity 2 and geometric multiplicity 1.

(b) Show that $\lambda = -1$ is an eigenvalue of A with algebraic multiplicity 3 and geometric multiplicity 1.

4. Let

$$D = \begin{bmatrix} 1 & & \\ & 1 & \\ & & 2 \end{bmatrix}.$$

Show that $\lambda = 1$ is an eigenvalue of D with algebraic and geometric multiplicity 2.

5. Let D be a diagonal matrix. Show that any eigenvalue of D with algebraic multiplicity r also has geometric multiplicity r.

6. Show that if λ has algebraic multiplicity 1, then it also has geometric multiplicity 1. Use this result and Theorem 8.5 to prove Result 1 of Section 8.3.

8.5   SYMMETRIC MATRICES

In Section 8.4 we investigated the relationship between the eigenvalues of a matrix and the diagonalizability of the matrix. We turn now to the question of whether there are any reasonable conditions on the matrix itself which will guarantee that it is diagonalizable.

An $n \times n$ matrix A is called **symmetric** if $A = A^T$, that is, if $a_{ij} = a_{ji}$ for all i and j (see Problem 19 in Section 1.7). There are two surprising and important facts about symmetric matrices: All of their eigenvalues are real, and they are diagonalizable. The proofs of these statements are difficult and will be omitted.

Theorem 8.6   *Let A be a symmetric matrix.*

(a) All of the eigenvalues of A are real.
(b) A is diagonalizable.

Still more is true about symmetric matrices. Suppose that A is a symmetric matrix and that v_1 and v_2 are eigenvectors of A associated with different eigenvalues λ_1 and λ_2. We have

$$Av_1 = \lambda_1 v_1, \qquad Av_2 = \lambda_2 v_2, \qquad \text{and} \qquad \lambda_1 \neq \lambda_2.$$

Multiply the second equation on the left by v_1^T to obtain

$$v_1^T A v_2 = \lambda_2 v_1^T v_2 = \lambda_2 \langle v_1, v_2 \rangle. \tag{1}$$

Transpose the first equation and use the fact that A is symmetric to obtain

$$v_1^T A = \lambda_1 v_1^T.$$

Now multiply this equation on the right by v_2 to obtain

$$v_1^T A v_2 = \lambda_1 v_1^T v_2 = \lambda_1 \langle v_1, v_2 \rangle. \tag{2}$$

From (1) and (2) we see that

$$\lambda_1 \langle v_1, v_2 \rangle = \lambda_2 \langle v_1, v_2 \rangle.$$

Since $\lambda_1 \neq \lambda_2$, it follows that $\langle v_1, v_2 \rangle = 0$. This proves the following result.

Result 1 Let A be a symmetric matrix. Eigenvectors of A associated with different eigenvalues are orthogonal.

The student should compare this result with Result 1 of Section 8.3. We know that for any matrix the eigenvectors associated with different eigenvalues are independent. For symmetric matrices, they are not only independent, they are also orthogonal.

EXAMPLE 1 Let us find the eigenvalues and eigenvectors of the matrix

$$A = \begin{bmatrix} 1 & -2 & 2 \\ -2 & 1 & 2 \\ 2 & 2 & 1 \end{bmatrix}$$

and verify that the eigenvectors associated with different eigenvalues are orthogonal. The characteristic equation of A is

$$\det(A - \lambda I) = \det \begin{bmatrix} 1-\lambda & -2 & 2 \\ -2 & 1-\lambda & 2 \\ 2 & 2 & 1-\lambda \end{bmatrix}$$

$$= -(\lambda - 3)^2(\lambda + 3) = 0.$$

Thus the eigenvalues of A are 3, 3, and -3. The reduced echelon form of $A - 3I$ and $A + 3I$ is, respectively,

$$\begin{bmatrix} 1 & 1 & -1 \\ 0 & 0 & 0 \\ 0 & 0 & 0 \end{bmatrix} \quad \text{and} \quad \begin{bmatrix} 1 & -1 & 0 \\ 0 & 1 & 1 \\ 0 & 0 & 0 \end{bmatrix}.$$

Thus a basis for the eigenspace $E(3)$ is $(1, 0, 1)$ and $(-1, 1, 0)$ and a basis for the eigenspace $E(-3)$ is $(-1, -1, 1)$. Therefore, the eigenvectors associated with the eigenvalue 3 are the vectors of the form

$$\alpha(1, 0, 1) + \beta(-1, 1, 0) = (\alpha - \beta, \beta, \alpha)$$

with not both α and β equal to zero. Similarly, the eigenvectors associated with the eigenvalue -3 are the vectors of the form $(-\gamma, -\gamma, \gamma)$, $\gamma \neq 0$. Since

$$\langle (\alpha - \beta, \beta, \alpha), (-\gamma, -\gamma, \gamma) \rangle = -\gamma(\alpha - \beta) - \beta\gamma + \alpha\gamma = 0,$$

these vectors are indeed orthogonal.

EXERCISE 1   Find the eigenvalues and eigenvectors of the matrix

$$A = \begin{bmatrix} 3 & 0 & 1 \\ 0 & 2 & 0 \\ 1 & 0 & 3 \end{bmatrix}$$

and verify that the eigenvectors associated with distinct eigenvalues are orthogonal.

There is one more important result about symmetric matrices that we will prove.

Theorem 8.7 *Let A be an n × n symmetric matrix. Then A has n orthonormal eigenvectors.*

Proof Let $\lambda_1, \ldots, \lambda_r$ be the distinct eigenvalues of A, and let $d_1, \ldots, d_r$ be the algebraic multiplicities of $\lambda_1, \ldots, \lambda_r$, respectively. Since A is diagonalizable, the dimension of each of the eigenspaces $E(\lambda_i)$ is d_i, $i = 1, \ldots, r$. Using the Gram–Schmidt process we can find an orthogonal basis for each of the eigenspaces. But since eigenvectors associated with different eigenvalues are orthogonal, the collection of all of these eigenvectors is a collection of n orthogonal vectors. Thus A has n orthogonal eigenvectors. Dividing each of these vectors by its length, we obtain n orthonormal eigenvectors.

ORTHONORMAL EIGENVECTORS OF A SYMMETRIC MATRIX

Given
An $n \times n$ symmetric matrix A.

Goal
To find n orthonormal eigenvectors of A.

Procedure
1. Find the eigenvalues of A. (They are all real.)
2. Find a basis for each of the eigenspaces in the usual way.
3. Apply the Gram–Schmidt process to obtain an orthonormal basis for each of the eigenspaces. The collection of all eigenvectors obtained in this way forms a collection of n orthonormal eigenvectors.

EXAMPLE 2   In Example 1 we found that $(1, 0, 1)$ and $(-1, 1, 0)$ formed a basis for the eigenspace $E(3)$ and $(-1, -1, 1)$ formed a basis for the eigenspace $E(-3)$. Applying the Gram–Schmidt process to the vectors $(1, 0, 1)$ and $(-1, 1, 0)$, we obtain the orthonormal vectors

$$\frac{1}{\sqrt{2}}(1, 0, 1) \quad \text{and} \quad \frac{1}{\sqrt{6}}(-1, 2, 1).$$

Thus

$$\frac{1}{\sqrt{2}}(1,0,1), \qquad \frac{1}{\sqrt{6}}(-1,2,1), \qquad \text{and} \qquad \frac{1}{\sqrt{3}}(-1,-1,1)$$

are three orthonormal eigenvectors of A.

EXERCISE 2   Find three orthonormal eigenvectors of the symmetric matrix in Exercise 1.

   Given an $n \times n$ symmetric matrix A, we know that it has n orthonormal eigenvectors. Let P be the matrix whose columns are these orthonormal eigenvectors. Then $\Lambda = P^{-1}AP$ is a diagonal matrix whose diagonal entries are the eigenvalues of A. But because P has orthonormal columns, it has the additional property that [see Problem 17(a) of Section 3.2]

$$P^TP = I = PP^T. \tag{3}$$

Thus $P^{-1} = P^T$ and hence $\Lambda = P^TAP$. Matrices that satisfy (3) are called **orthogonal** matrices. Matrices that can be diagonalized by an orthogonal matrix are called **orthogonally diagonalizable**. Using this terminology, we can restate Theorem 8.7 as follows.

Theorem 8.7'   *Symmetric matrices are orthogonally diagonalizable.*

EXAMPLE 3   In Example 2 we found that

$$\frac{1}{\sqrt{2}}(1,0,1), \qquad \frac{1}{\sqrt{6}}(-1,2,1), \qquad \text{and} \qquad \frac{1}{\sqrt{3}}(-1,-1,1)$$

are orthonormal eigenvectors of the matrix A given in Example 1. Letting

$$P = \begin{bmatrix} \dfrac{1}{\sqrt{2}} & \dfrac{-1}{\sqrt{6}} & \dfrac{-1}{\sqrt{3}} \\[2mm] 0 & \dfrac{2}{\sqrt{6}} & \dfrac{-1}{\sqrt{3}} \\[2mm] \dfrac{1}{\sqrt{2}} & \dfrac{1}{\sqrt{6}} & \dfrac{1}{\sqrt{3}} \end{bmatrix},$$

it is easy to verify that $P^TP = I$ and that

$$\begin{bmatrix} 3 & & \\ & 3 & \\ & & -3 \end{bmatrix} = P^TAP.$$

EXERCISE 3   Find an orthogonal matrix P which orthogonally diagonalizes the matrix in Exercise 1.

If A is an $n \times n$ symmetric matrix and $v_1, \ldots, v_n$ are n orthogonal eigenvectors of A with corresponding eigenvalues $\lambda_1, \ldots, \lambda_n$, then

$$Ax = P \Lambda P^{-1} x$$

$$= \begin{bmatrix} | & & | \\ v_1 & \cdots & v_n \\ | & & | \end{bmatrix} \begin{bmatrix} \lambda_1 & & \\ & \ddots & \\ & & \lambda_n \end{bmatrix} \begin{bmatrix} -v_1- \\ \vdots \\ -v_n- \end{bmatrix} \begin{bmatrix} | \\ x \\ | \end{bmatrix}$$

$$= \begin{bmatrix} | & & | \\ \lambda_1 v_1 & \cdots & \lambda_n v_n \\ | & & | \end{bmatrix} \begin{bmatrix} \langle v_1, x \rangle \\ \vdots \\ \langle v_n, x \rangle \end{bmatrix}$$

$$= \lambda_1 v_1 \langle v_1, x \rangle + \cdots + \lambda_n v_n \langle v_n, x \rangle$$

$$= \lambda_1 v_1 v_1^T x + \cdots + \lambda_n v_n v_n^T x$$

$$= \left(\lambda_1 v_1 v_1^T + \cdots + \lambda_n v_n v_n^T \right) x.$$

From our work in Chapter 3 we recognize that $v_i v_i^T$ is a projection matrix that projects a vector x onto the vector v_i. Thus Ax is a linear combination of the projections of x onto the orthogonal eigenvectors $v_1, \ldots, v_n$. This decomposition is called the spectral decomposition for symmetric matrices.

Theorem 8.8 (*Spectral Theorem for Symmetric Matrices*) *Let A be an $n \times n$ symmetric matrix. Then A can be decomposed into*

$$A = \lambda_1 v_1 v_1^T + \cdots + \lambda_n v_n v_n^T, \tag{4}$$

where $v_1, \ldots, v_n$ are orthogonal eigenvectors and $\lambda_1, \ldots, \lambda_n$ are associated eigenvalues.

EXERCISE 4   Show that if A is an $n \times n$ symmetric matrix with spectral decomposition (4), then

$$A^k = \lambda_1^k v_1 v_1^T + \cdots + \lambda_n^k v_n v_n^T.$$

PROBLEMS 8.5

1. Find an orthogonal matrix that orthogonally diagonalizes each of the following symmetric matrices.

*(a) $\begin{bmatrix} 1 & 2 \\ 2 & 1 \end{bmatrix}$ (b) $\begin{bmatrix} 1 & 0 \\ 0 & -1 \end{bmatrix}$ (c) $\begin{bmatrix} 3 & 1 \\ 1 & 2 \end{bmatrix}$

*(d) $\begin{bmatrix} -1 & 1 & 1 \\ 1 & -1 & 1 \\ 1 & 1 & -1 \end{bmatrix}$ (e) $\begin{bmatrix} 5 & 2 & 2 \\ 2 & 5 & 2 \\ 2 & 2 & 5 \end{bmatrix}$ *(f) $\begin{bmatrix} 1 & 2 & 0 \\ 2 & 1 & 0 \\ 0 & 0 & -1 \end{bmatrix}$

$$
\text{(g)} \begin{bmatrix} 3 & 0 & -1 \\ 0 & 3 & 0 \\ -1 & 0 & 3 \end{bmatrix} \quad \text{(h)} \begin{bmatrix} 3 & 0 & -1 \\ 0 & 2 & 0 \\ -1 & 0 & 3 \end{bmatrix} \quad \text{(i)} \begin{bmatrix} 1 & 1 & 0 \\ 1 & 0 & -1 \\ 0 & -1 & 1 \end{bmatrix}
$$

$$
\text{(j)} \begin{bmatrix} 1 & 1 & 1 & 1 \\ 1 & 1 & 1 & 1 \\ 1 & 1 & 1 & 1 \\ 1 & 1 & 1 & 1 \end{bmatrix} \quad \text{*(k)} \begin{bmatrix} 1 & 0 & 1 & 0 \\ 0 & 1 & 0 & 1 \\ 1 & 0 & 1 & 0 \\ 0 & 1 & 0 & 1 \end{bmatrix}
$$

2. Find a symmetric matrix A with:
   (a) Eigenvalues 1, 3 and corresponding eigenvectors $(1, 1)$ and $(-2, 2)$
   *(b) Eigenvalues 1, 1, 2 and corresponding eigenvectors $(-1, 2, 2)$, $(2, -1, 2)$, $(2, 2, -1)$
   (c) Eigenvalues $-1, 0, 1$ and corresponding eigenvectors $(1, 1, 1)$, $(1, -1, 0)$, $(1, 1, -2)$

*3. Prove the converse of Theorem 8.7′ by showing that if A is orthogonally diagonalizable, then A is a symmetric matrix.

4. Show that if A is a symmetric matrix, then $A^2, A^3, \dots$ are also symmetric matrices.

*5. Let A be a symmetric matrix with one eigenvalue λ. Show that $A = \lambda I$.

6. Suppose that the eigenspaces of a diagonalizable matrix A are orthogonal. (See Problem 10 of Section 3.4 for the definition of orthogonal subspaces.) Prove that the matrix A is symmetric.

7. Let A be a symmetric matrix with the property that $A^2 = A$ (see Problem 16 of Section 3.5).
   (a) Show that the eigenvalues of A are 0 and 1.
   (b) Show that the eigenspace $E(1)$ is the column space of A.
   (c) Show that the eigenspace $E(0)$ is the orthogonal complement of the eigenspace $E(1)$.

*8. Two matrices A and B are called *orthogonally similar* if $B = P^T A P$, where P is an orthogonal matrix. Show that if A and B are orthogonally similar, then A is symmetric if and only if B is symmetric.

8.6   DIAGONAL REPRESENTATION OF A LINEAR TRANSFORMATION

In Chapter 5 we saw that there is a remarkable connection between $n \times n$ matrices and the linear transformations from R^n into itself. Every $n \times n$ matrix A defines a linear transformation $T : R^n \to R^n$ via the equation $T(x) = Ax$. Conversely, given a linear transformation $T : R^n \to R^n$, and given a basis for R^n, T is represented by a matrix with respect to this basis. If the basis is the standard basis for R^n, then the resulting matrix A which represents T is called the standard matrix.

Since it is possible to represent a given transformation T with respect to different bases, it is natural to ask whether there is some basis for R^n that will give rise to a particularly simple matrix representation for T. For example, is there a basis for R^n such that the matrix for T with respect to this basis is a diagonal matrix? And if there is such a basis, how do we find it? It should come as no surprise that these questions are intimately related to the question of whether the standard matrix for T is diagonalizable.

Theorem 8.9 *A linear transformation $T : R^n \to R^n$ can be represented by a diagonal matrix if and only if the standard matrix of T is diagonalizable.*

Proof Let $T : R^n \to R^n$ be a linear transformation and let A be its standard matrix. Suppose that A is diagonalizable. Then A has n independent eigenvectors $v_1, \ldots, v_n$. If P is the matrix whose columns are these eigenvectors, then

$$\Lambda = P^{-1}AP$$

is a diagonal matrix whose diagonal entries are the eigenvalues $\lambda_1, \ldots, \lambda_n$ of A, the ith diagonal entry λ_i being the eigenvalue for the eigenvector v_i in the ith column of P. Since $T(v_i) = Av_i = \lambda_i v_i$ and since the coordinate vector of $\lambda_i v_i$ with respect to the basis $v_1, \ldots, v_n$ is $\lambda_i e_i$, it follows that the matrix of T with respect to the basis $v_1, \ldots, v_n$ is Λ.

Conversely, suppose that $T : R^n \to R^n$ is a linear transformation and that there is a basis $v_1, \ldots, v_n$ of R^n such that the matrix of T with respect to this basis is a diagonal matrix, say

$$\Lambda = \begin{bmatrix} \lambda_1 & & \\ & \ddots & \\ & & \lambda_n \end{bmatrix}.$$

Since the coordinate vector of v_i with respect to the basis $v_1, \ldots, v_n$ is e_i, it follows that the coordinate vector of $T(v_i)$ with respect to this basis is $\lambda_i e_i$. Therefore, $T(v_i) = \lambda_i v_i$. Now if A is the standard matrix of T, then $T(v_i) = Av_i$, and hence $Av_i = \lambda_i v_i$. This proves that $v_1, \ldots, v_n$ are eigenvectors of A. Since they are independent, A is diagonalizable.

EXAMPLE 1 Consider the linear transformation $T : R^3 \to R^3$ defined by

$$T(x_1, x_2, x_3) = (-2x_1 + 9x_3, -2x_2 + 12x_3, x_3).$$

The standard matrix of T is

$$A = \begin{bmatrix} | & | & | \\ T(e_1) & T(e_2) & T(e_3) \\ | & | & | \end{bmatrix} = \begin{bmatrix} -2 & 0 & 9 \\ 0 & -2 & 12 \\ 0 & 0 & 1 \end{bmatrix}.$$

Since A is a triangular matrix, its eigenvalues are its diagonal entries, namely -2, -2, and 1. The reduced echelon form of $A + 2I$, $A - I$ is, respectively,

$$\begin{bmatrix} 0 & 0 & 1 \\ 0 & 0 & 0 \\ 0 & 0 & 0 \end{bmatrix}, \qquad \begin{bmatrix} 1 & 0 & -3 \\ 0 & 1 & -4 \\ 0 & 0 & 0 \end{bmatrix}.$$

Thus a basis for the eigenspace $E(-2)$ is $(1,0,0)$, $(0,1,0)$, and a basis for the eigenspace $E(1)$ is $(3,4,1)$. Therefore, A has three independent eigenvectors and is diagonalizable. If these eigenvectors are used as a basis for R^3, then T is represented by the matrix

$$\Lambda = \begin{bmatrix} -2 & & \\ & -2 & \\ & & 1 \end{bmatrix}.$$

EXERCISE 1   Let $T : R^3 \to R^3$ be defined by

$$T(x_1, x_2, x_3) = (3x_1 + x_3, 2x_3, x_1 + 3x_3).$$

Find a basis for R^3 such that the matrix of T with respect to this basis is a diagonal matrix.

By Theorem 8.9, a linear transformation $T : R^n \to R^n$ can be represented by a diagonal matrix Λ if and only if there is a nonsingular matrix P such that Λ and the standard matrix A of T are related by the equation

$$\Lambda = P^{-1}AP.$$

What is the role of the matrix P in this equation? Given any vector x in R^n, Λ represents T with respect to the basis $v_1, \dots, v_n$ by giving the coordinates of $T(x)$ with respect to this basis. On the other hand, A represents T by giving the coordinates of $T(x)$ with respect to the standard basis. This suggests that the matrix P is a change of coordinate matrix. That is, it suggests that the matrix P changes the coordinates of a vector with respect to the basis $v_1, \dots, v_n$ to the coordinates of the vector with respect to the standard matrix, and that the matrix P^{-1} changes the coordinates of a vector with respect to the standard basis to the coordinates of the vector with respect to the basis $v_1, \dots, v_n$. In fact, this is exactly what the matrix P does. For if $x = (x_1, \dots, x_n)$ is a vector in R^n, and $(x'_1, \dots, x'_n)$ is the coordinate vector of x with respect to the basis $v_1, \dots, v_n$ then

$$x = x'_1 v_1 + \cdots + x'_n v_n.$$

Since the matrix form of this equation is

$$\begin{bmatrix} x_1 \\ \vdots \\ x_n \end{bmatrix} = P \begin{bmatrix} x'_1 \\ \vdots \\ x'_n \end{bmatrix},$$

we see that P does indeed change the coordinates of the vector x with respect to the basis $v_1, \ldots, v_n$ to the coordinates of x with respect to the standard basis.

EXAMPLE 2   In Example 1 we found that the linear transformation $T : R^3 \rightarrow R^3$ defined by

$$T(x_1, x_2, x_3) = (-2x_1 + 9x_3, -2x_2 + 12x_3, x_3)$$

can be represented by the diagonal matrix

$$\Lambda = \begin{bmatrix} -2 & & \\ & -2 & \\ & & 1 \end{bmatrix}$$

with respect to the basis $(1, 0, 0)$, $(0, 1, 0)$, and $(3, 4, 1)$. Thus Λ and the standard matrix of T

$$A = \begin{bmatrix} -2 & 0 & 9 \\ 0 & -2 & 12 \\ 0 & 0 & 1 \end{bmatrix}$$

are related by the equation $\Lambda = P^{-1}AP$, where

$$P = \begin{bmatrix} 1 & 0 & 3 \\ 0 & 1 & 4 \\ 0 & 0 & 1 \end{bmatrix}.$$

If $x = (1, 2, -1)$, then the coordinate vector of x with respect to the basis $(1, 0, 0)$, $(0, 1, 0)$, and $(3, 4, 1)$ is

$$P^{-1}x = \begin{bmatrix} 1 & 0 & -3 \\ 0 & 1 & -4 \\ 0 & 0 & 1 \end{bmatrix} \begin{bmatrix} 1 \\ 2 \\ -1 \end{bmatrix} = \begin{bmatrix} 4 \\ 6 \\ -1 \end{bmatrix}.$$

Therefore, the coordinate vector of $T(x)$ with respect to this basis is

$$\begin{bmatrix} -2 & & \\ & -2 & \\ & & 1 \end{bmatrix} \begin{bmatrix} 4 \\ 6 \\ -1 \end{bmatrix} = \begin{bmatrix} -8 \\ -12 \\ -1 \end{bmatrix}.$$

We conclude this section by indicating how these results can be generalized. In Theorem 8.9 we stated that a linear transformation from R^n into itself can be represented by a diagonal matrix if and only if its standard matrix can be diagonalized. This result holds if any matrix that represents the transformation can be diagonalized. This is because any two matrices A and B which represent the same linear transformation with respect to two different bases are related by the equation

$$B = P^{-1}AP,$$

where P is the change of coordinates matrix from one basis to the other (see Problem 3). Finally, all of these remarks hold for a transformation of any finite-dimensional vector space V into itself.

PROBLEMS 8.6

1. Determine which of the following linear transformations T can be represented by a diagonal matrix. If the linear transformation can be represented by a diagonal matrix, find a diagonal matrix that represents T and find the basis that gives rise to this matrix.
   *(a) $T(x_1, x_2) = (2x_1 + 2x_2, 2x_1 + 5x_2)$
   (b) $T(x_1, x_2) = (x_1 + x_2, x_2)$
   *(c) $T(x_1, x_2, x_3) = (2x_1 + x_2 + x_3, 2x_1 + 3x_2 + 2x_3, 3x_1 + 3x_2 + 4x_3)$
   (d) $T(x_1, x_2, x_3) = (2x_1, 3x_1 + 2x_2, 5x_1 + 2x_2 - x_3)$

2. Let V be a subspace of R^n.
   (a) Show that the projection of R^n onto V can be represented by a diagonal matrix.
   (b) Show that the reflection across V can be represented by a diagonal matrix.

*3. Let $v_1, \ldots, v_n$ and $w_1, \ldots, w_n$ be two bases for R^n. Find a change of coordinate matrix P. That is, find a matrix P with the following property. If v is a vector in R^n and the coordinate vectors of v with respect to the bases $v_1, \ldots, v_n$ and $w_1, \ldots, w_n$ are v' and v'', respectively, then $v' = Pv''$.

8.7 RICHARDSON'S MODEL FOR AN ARMS RACE

In this section we study Richardson's model for an arms race between two nations. The formulation of this model leads naturally to a system of differential equations that will be studied in detail in the next four sections.

Suppose that there are two nations and that each nation is worried about an attempted takeover by the other. In such a situation each nation keeps a watchful eye on the armament expenditures of the other nation. The reason, of course, is obvious. If one nation increases its armament expenditures, then it increases its war potential and hence it becomes a greater threat to the security of the other nation. In an attempt to defend itself, the other nation may think it necessary to increase its own war potential and hence it may increase its armament expenditures. What effect will this have on the first nation? The increase in the expenditures of the second nation may lead the first nation to increase its expenditures even more. The process is self-perpetuating. However, since there are many other demands on the nations' resources, the cost of armaments will clearly have a restraining effect on this process.

Let $y = y(t)$ and $z = z(t)$ denote the armament expenditures of these two nations, and let t stand for time in years. Our assumption about the foreign policies of these two nations is that the more one nation spends on armaments, the more the other nation is inclined to spend on armaments. There are many possible ways to interpret this assumption. We will interpret it to mean that each nation increases its armaments expenditures at a rate that is proportional to the existing armaments expenditures of the other nation. Similarly, we will assume that the cost of armaments creates a restraining effect that is proportional to the current armament expenditures. These assumptions lead to the following differential equations:

$$\frac{dy}{dt} = -ay + bz, \qquad \frac{dz}{dt} = cy - dz \qquad (a, b, c, d > 0).$$

This is a *system* of two differential equations. Our objective is to find the functions $y = y(t)$ and $z = z(t)$ that solve this system and hence describe the arms race between the two nations. We now turn our attention to this problem.

First, we rewrite the system in matrix notation. If we let $x = (y, z)$, then $\frac{dx}{dt} = \left(\frac{dy}{dt}, \frac{dz}{dt} \right)$ and the system becomes

$$\frac{dx}{dt} = Ax, \qquad \text{where} \qquad A = \begin{bmatrix} -a & b \\ c & -d \end{bmatrix}. \qquad (1)$$

If we ignore momentarily the fact that this is a matrix equation, it looks exactly like the differential equation in Example 5 of Section 4.1. We found there that any solution to $\frac{dx}{dt} = \alpha x$ has the form $x(t) = ce^{\alpha t}$, where c is a constant. This suggests in the present situation that any solution to $\frac{dx}{dt} = Ax$ should be of the form $x(t) = ce^{At}$, where c is a vector. The only problem with this is that we do not know how to raise e to a matrix power.

We must try something else. Perhaps our next best guess is a solution of the form $x(t) = e^{\lambda t}v$, where λ is a scalar and v is a constant vector. If v is the zero vector, then $x(t) = e^{\lambda t}v = 0$ is a solution to (1). The interpretation of this solution is that there will be no arms race provided that the initial armament expenditure level of each nation was zero [for $x(0) = (y(0), z(0))$ gives the initial armament levels of each nation]. If either nation has armaments at the beginning of our study, we must find a solution $x(t)$ such that $x(0) \neq 0$ and hence we must suppose that $v \neq 0$. Let y_0 and z_0 denote the initial armament expenditures of each nation. The formulation of our model leads to the *initial value problem*

$$\frac{dx}{dt} = Ax, \qquad x(0) = \begin{bmatrix} y_0 \\ z_0 \end{bmatrix}. \qquad (2)$$

If $x(t) = e^{\lambda t}v$, then

$$\frac{dx}{dt} = \frac{d}{dt} \begin{bmatrix} e^{\lambda t}v_1 \\ e^{\lambda t}v_2 \end{bmatrix} = \begin{bmatrix} \lambda e^{\lambda t}v_1 \\ \lambda e^{\lambda t}v_2 \end{bmatrix} = \lambda e^{\lambda t}v,$$

$$Ax = Ae^{\lambda t}v = e^{\lambda t}Av,$$

and our differential equation $\frac{dx}{dt} = Ax$ becomes $\lambda e^{\lambda t}v = e^{\lambda t}Av$. Since $e^{\lambda t}$ is never zero, we may divide both sides of this equation by $e^{\lambda t}$ and obtain

$$Av = \lambda v.$$

We have obtained partial success; $x(t) = e^{\lambda t}v$ is a solution to the differential equation (1) if and only if λ is an eigenvalue of A with eigenvector v.

Let us suppose we have determined that the matrix A in (2) is

$$A = \begin{bmatrix} -2 & 4 \\ 1 & -2 \end{bmatrix}.$$

Since $\det(A - \lambda I) = \det \begin{bmatrix} -2-\lambda & 4 \\ 1 & -2-\lambda \end{bmatrix} = \lambda(\lambda + 4)$, the eigenvalues of A are $\lambda = 0$ and $\lambda = -4$. The eigenspaces $E(0)$ and $E(-4)$ are spanned by the eigenvectors $(2, 1)$ and $(-2, 1)$, respectively.

Thus we have arrived at two solutions to our differential equation: namely,

$$x(t) = e^{0t}(2, 1) = (2, 1) \quad \text{and} \quad x(t) = e^{-4t}(-2, 1).$$

Furthermore, it is clear that any linear combination

$$x(t) = c_1(2, 1) + c_2 e^{-4t}(-2, 1)$$

of these two solutions is also a solution. (Verify this.)

This last observation is important for the following reason. If we suppose that $y_0 = 2$ and $z_0 = 2$, we find that it is impossible to choose a constant c such that

$$x(t) = c(2, 1) \quad \text{or} \quad x(t) = ce^{-4t}(-2, 1)$$

will satisfy $x(0) = (2, 2)$. However, it is possible to choose two constants c_1 and c_2 such that

$$x(t) = c_1(2, 1) + c_2 e^{-4t}(-2, 1)$$

satisfies $x(0) = (2, 2)$. For

$$c_1(2, 1) + c_2(-2, 1) = (2, 2)$$

yields the following system of equations:

$$\begin{bmatrix} 2 & -2 \\ 1 & 1 \end{bmatrix} \begin{bmatrix} c_1 \\ c_2 \end{bmatrix} = \begin{bmatrix} 2 \\ 2 \end{bmatrix}.$$

The solution of this system is $c_1 = \frac{3}{2}$ and $c_2 = \frac{1}{2}$, and hence the solution to the initial value problem (2) is

$$x(t) = \frac{3}{2}(2, 1) + \frac{1}{2}e^{-4t}(-2, 1)$$

or

$$y(t) = 3 - e^{-4t}$$
$$z(t) = \frac{3}{2} + \frac{1}{2}e^{-4t}.$$

This is the mathematical solution to our model. Since

$$\lim_{t \to \infty} y(t) = 3 \quad \text{and} \quad \lim_{t \to \infty} z(t) = \frac{3}{2},$$

it predicts that the amount that each nation will spend on armaments will approach a limit. The first nation will tend to increase its expenditures by 50% and the second nation will decrease its expenditures by 50%.

PROBLEMS 8.7

*1. Formulate a model for an arms race between three nations.

2. What changes in the assumptions of our model would have to be made in order to replace (1) by

$$\frac{dx}{dt} = Ax, \quad \text{where } A = \begin{bmatrix} a & b \\ c & d \end{bmatrix}, \quad a, b, c, d > 0.$$

3. Let λ_1 and λ_2 be distinct eigenvalues of the matrix

$$A = \begin{bmatrix} -a & b \\ c & -d \end{bmatrix},$$

where $a, b, c, d > 0$. Show that Richardson's model makes the following predictions.
(a) If $\lambda_1 > 0$ and λ_2 is arbitrary, then there will be a runaway arms race.
(b) If $\lambda_1 < 0$ and $\lambda_2 \leq 0$, then stability occurs.
(c) If $\lambda_1 < 0$ and $\lambda_2 < 0$, then each nation disarms.

4. Let $x_1(t)$ and $x_2(t)$ be the populations of two species at time t. Discuss the assumptions that would lead to the equation $\frac{dx}{dt} = Ax$ with

(a) $A = \begin{bmatrix} a & b \\ c & d \end{bmatrix}$ (b) $A = \begin{bmatrix} a & -b \\ -c & d \end{bmatrix}$ (c) $A = \begin{bmatrix} a & -b \\ c & -d \end{bmatrix}$

Here $a, b, c, d > 0$.

8.8 SYSTEMS OF DIFFERENTIAL EQUATIONS

Richardson's model led to a system of differential equations of the form

$$\frac{dx_1}{dt} = a_{11}x_1 + a_{12}x_2$$

$$\frac{dx_2}{dt} = a_{21}x_1 + a_{22}x_2.$$

We saw that this system is represented in matrix notation by

$$\frac{dx}{dt} = Ax,$$

where

$$A = \begin{bmatrix} a_{11} & a_{12} \\ a_{21} & a_{22} \end{bmatrix}, \quad x = \begin{bmatrix} x_1 \\ x_2 \end{bmatrix}, \quad \text{and} \quad \frac{dx}{dt} = \begin{bmatrix} dx_1/dt \\ dx_2/dt \end{bmatrix}.$$

In general, *a system of n differential equations* is a system of the form

$$\frac{dx_1}{dt} = a_{11}x_1 + \cdots + a_{1n}x_n$$

$$\vdots \tag{1}$$

$$\frac{dx_n}{dt} = a_{n1}x_1 + \cdots + a_{nn}x_n.$$

Using matrix notation, this system can be expressed by the single matrix equation

$$\frac{dx}{dt} = Ax, \tag{1'}$$

where

$$A = \begin{bmatrix} a_{11} & \cdots & a_{1n} \\ \vdots & & \vdots \\ a_{n1} & \cdots & a_{nn} \end{bmatrix}, \quad x = \begin{bmatrix} x_1 \\ \vdots \\ x_n \end{bmatrix}, \quad \text{and} \quad \frac{dx}{dt} = \begin{bmatrix} dx_1/dt \\ \vdots \\ dx_n/dt \end{bmatrix}.$$

A *solution* of (1) is a set of n functions $x_1 = x_1(t), \ldots, x_n = x_n(t)$ which satisfy the system (1). A *solution* of (1') is a vector-valued function $x(t) = (x_1(t), \ldots, x_n(t))$ which satisfies (1').

EXERCISE 1   Consider the system of two differential equations

$$\frac{dx_1}{dt} = 2x_1 + 2x_2$$

$$\frac{dx_2}{dt} = x_1 + 3x_2.$$

Show that a solution of this system is given by

$$x_1(t) = -2c_1e^t + c_2e^{4t}$$
$$x_2(t) = c_1e^t + c_2e^{4t}$$

where c_1 and c_2 are arbitrary constants.

EXAMPLE 1   In matrix notation the system considered in Exercise 1 becomes

$$\frac{dx}{dt} = Ax, \quad \text{where} \quad A = \begin{bmatrix} 2 & 2 \\ 1 & 3 \end{bmatrix} \quad \text{and} \quad x = \begin{bmatrix} x_1 \\ x_2 \end{bmatrix}.$$

If $x(t) = c_1e^t \begin{bmatrix} -2 \\ 1 \end{bmatrix} + c_2e^{4t} \begin{bmatrix} 1 \\ 1 \end{bmatrix}$, then

$$\frac{dx}{dt} = c_1e^t \begin{bmatrix} -2 \\ 1 \end{bmatrix} + 4c_2e^{4t} \begin{bmatrix} 1 \\ 1 \end{bmatrix}$$

and

$$Ax = c_1 e^t \begin{bmatrix} 2 & 2 \\ 1 & 3 \end{bmatrix} \begin{bmatrix} -2 \\ 1 \end{bmatrix} + c_2 e^{4t} \begin{bmatrix} 2 & 2 \\ 1 & 3 \end{bmatrix} \begin{bmatrix} 1 \\ 1 \end{bmatrix}$$

$$= c_1 e^t \begin{bmatrix} -2 \\ 1 \end{bmatrix} + c_2 e^{4t} \begin{bmatrix} 4 \\ 4 \end{bmatrix}$$

$$= \frac{dx}{dt}.$$

Thus $x(t)$ is a solution.

EXERCISE 2 Verify that $x(t) = (c_1 e^{2t} + c_2 e^{4t}, - c_1 e^{2t} + c_2 e^{4t})$ is a solution to the system of differential equations

$$\frac{dx_1}{dt} = 3x_1 + x_2$$

$$\frac{dx_2}{dt} = x_1 + 3x_2.$$

EXERCISE 3 Write the system of differential equations in Exercise 2 in matrix notation and verify that $x(t) = c_1 e^{2t} \begin{bmatrix} 1 \\ -1 \end{bmatrix} + c_2 e^{4t} \begin{bmatrix} 1 \\ 1 \end{bmatrix}$ is a solution.

As we found in the study of Richardson's model, it is sometimes necessary to impose **initial conditions** on the system (1) of the form

$$x_1(t_0) = x_1^{(0)}, \ldots, x_n(t_0) = x_n^{(0)}, \tag{2}$$

where t_0 is a specified value and $x_1^{(0)}, \ldots, x_n^{(0)}$ are specified numbers. The system (1) together with the initial conditions (2) is called an **initial value problem**. A solution to an initial value problem is a solution of (1) that satisfies the initial conditions (2).

EXAMPLE 2 The function $x(t) = (-2e^t + e^{4t}, e^t + e^{4t})$ is a solution of the initial value problem

$$\frac{dx}{dt} = \begin{bmatrix} 2 & 2 \\ 1 & 3 \end{bmatrix} x, \qquad x(0) = \begin{bmatrix} -1 \\ 2 \end{bmatrix}.$$

In Exercise 1 we asked the reader to verify that it is a solution to the differential equation. Since $x(0) = (-1, 2)$, it also satisfies the initial condition.

EXERCISE 4 Show that $x(t) = (e^{2t} + 2e^{4t}, - e^{2t} + 2e^{4t})$ is a solution to the initial value problem $\frac{dx}{dt} = Ax$, $x(0) = (3, 1)$, where $\frac{dx}{dt} = Ax$ is the system in Exercise 2.

Given a system of differential equations, the goal is of course to find a solution. This immediately raises two questions: (1) Does the system of differential equations always have a solution? (2) If so, does there exist a general method for finding all the solutions?

The answer to the first question as posed is obvious. The zero function $[x(t) = 0$ for all $t]$ is clearly a solution; it is called the **trivial solution**. Hence the real question that we should ask is whether the system of differential equations has a **nontrivial** solution (i.e., a solution that is different from the trivial solution).

The following important result settles this question; nontrivial solutions always exist. We answer the second question in Section 8.10.

Theorem 8.10   (*The Existence–Uniqueness Theorem*)

(*a*) *The differential equation* $\dfrac{dx}{dt} = Ax$ *always has a nontrivial solution.*

(*b*) *The initial value problem*

$$\frac{dx}{dt} = Ax, \qquad x(t_0) = x^{(0)} = \left(x_1^{(0)}, \ldots, x_n^{(0)} \right),$$

has one and only one solution.

The proof is difficult and will be omitted. However, we ask the reader to think about why the first part of the theorem follows immediately from the second part.

We conclude this section with a simple but subtle application of the existence–uniqueness theorem. The theorem states that the initial value problem

$$\frac{dx}{dt} = Ax, \qquad x(t_0) = \left(x_1^{(0)}, \ldots, x_n^{(0)} \right)$$

has one and only one solution. Let $y(t)$ be this unique solution and suppose that for some t_1, $y(t_1) = 0$. Now consider the initial value problem

$$\frac{dx}{dt} = Ax, \qquad x(t_1) = 0.$$

The solution $y(t)$ to the first initial value problem is also a solution to this initial value problem. But so is the trivial solution $x(t) = 0$ for all t. Since there can be only one solution to the second initial value problem, $y(t)$ must be the trivial solution $y(t) = 0$ for all t. This proves the following result.

Result 1   Let $y(t)$ be a solution of $\dfrac{dx}{dt} = Ax$. Then either $y(t)$ is the trivial solution or $y(t) \neq 0$ for all t.

PROBLEMS 8.8

1. In each of the following problems, express the system of differential equations in matrix form and verify that the given functions are solutions.

(a) $\dfrac{dx_1}{dt} = x_2,$ $x(t) = e^t \begin{bmatrix} 1 \\ 1 \end{bmatrix}$

$\dfrac{dx_2}{dt} = 2x_1 - x_2,$ $x(t) = e^{-2t} \begin{bmatrix} 1 \\ -2 \end{bmatrix}$

(b) $\dfrac{dx_1}{dt} = x_1 + x_3,$ $x(t) = 4 \begin{bmatrix} -1 \\ 0 \\ 1 \end{bmatrix}$

$\dfrac{dx_2}{dt} = x_2,$ $x(t) = 2e^t \begin{bmatrix} -2 \\ 1 \\ 0 \end{bmatrix}$

$\dfrac{dx_3}{dt} = x_1 + 2x_2 + x_3,$ $x(t) = e^{2t} \begin{bmatrix} 1 \\ 0 \\ 1 \end{bmatrix}$

(c) $\dfrac{dx_1}{dt} = 2x_1 - 5x_2 + 5x_3,$ $x(t) = e^{2t} \begin{bmatrix} 0 \\ 1 \\ 1 \end{bmatrix}$

$\dfrac{dx_2}{dt} = -3x_2 - x_3,$ $x(t) = e^{2t} \begin{bmatrix} 1 \\ 0 \\ 0 \end{bmatrix}$

$\dfrac{dx_3}{dt} = -x_2 + 3x_3,$ $x(t) = e^{4t} \begin{bmatrix} 5 \\ -1 \\ 1 \end{bmatrix}$

2. Consider the nth-order homogeneous differential equation

$$\frac{d^n y}{dt^n} + a_{n-1}\frac{d^{n-1}y}{dt^{n-1}} + \cdots + a_1\frac{dy}{dt} + a_0 y = 0.$$

(a) Show that the substitution

$$x_1 = y, \; x_2 = \frac{dy}{dt}, \ldots, \; x_n = \frac{d^{n-1}y}{dt^{n-1}}$$

converts this nth-order differential equation to the following system of n first-order equations

$$\frac{dx_1}{dt} = x_2$$

$$\frac{dx_2}{dt} = x_3$$

$$\vdots$$

$$\frac{dx_{n-1}}{dt} = x_n$$

$$\frac{dx_n}{dt} = -(a_0 x_1 + \cdots + a_{n-1}x_n)$$

(b) Show that $x(t) = (x_1(t), \ldots, x_n(t))$ is a solution of this system if and only if $y(t) = x_1(t)$ is a solution of the nth-order equation.

3. Convert each of the following equations to a system of first-order equations.
   *(a) $y'' - y = 0$ (b) $y'' + 2y' - 3y = 0$
   *(c) $y''' + y'' - 2y' = 0$ (d) $y''' + 4y'' - 7y' - 10y = 0$
   *(e) $y'''' - 2y''' - y'' + 2y = 0$

8.9 APPLICATIONS OF LINEAR ALGEBRA TO SYSTEMS OF DIFFERENTIAL EQUATIONS

This section is devoted to the study of the algebraic properties of the set of solutions to a system of differential equations. We begin by introducing some notation. When dealing with a single solution $x(t)$ to a system, we denote the components of $x(t)$ by $x_1(t), \ldots, x_n(t)$. That is,

$$x(t) = (x_1(t), \ldots, x_n(t)).$$

Thus when dealing with several solutions to a system, it is impossible to use subscripts to label them. For if we did, we would be using $x_j(t)$ to denote both the jth component of a single solution and the jth solution. To remedy this situation we will use superscripts in parentheses to denote different solutions and subscripts to denote the components of these solutions. Thus our notation for the jth solution will be

$$x^{(j)}(t) = \left(x_1^{(j)}(t), \ldots, x_n^{(j)}(t)\right).$$

Theorem 8.11 *The set of solutions of* $\dfrac{dx}{dt} = Ax$ *is a function space.*

Proof Let $x^{(1)}(t)$ and $x^{(2)}(t)$ be two solutions and let $x(t) = c_1 x^{(1)}(t) + c_2 x^{(2)}(t)$, where c_1 and c_2 are scalars. Then

$$\frac{dx}{dt} = c_1 \frac{dx^{(1)}}{dt} + c_2 \frac{dx^{(2)}}{dt} = c_1 A x^{(1)} + c_2 A x^{(2)}$$
$$= A\left(c_1 x^{(1)} + c_2 x^{(2)}\right) = Ax.$$

Thus $x(t)$ is a solution and the set of solutions is closed under addition and scalar multiplication.

In view of this result, we call the set of solutions to $\dfrac{dx}{dt} = Ax$ the **solution space**. The fact that the solution space of $\dfrac{dx}{dt} = Ax$ is a function space is important. For if we are given a basis for the solution space, we can express any solution to the differential equation as a linear combination of the solutions in the basis. Thus the problem of finding all solutions is reduced to the problem of finding a basis for the solution space. If $x^{(1)}(t), \ldots, x^{(n)}(t)$ form a basis for the solution space of $\dfrac{dx}{dt} = Ax$, then any solution of this equation can be written in the form

$$x(t) = c_1 x^{(1)}(t) + \cdots + c_n x^{(n)}(t).$$

This last expression is called the **general solution**.

EXAMPLE 1 We will show in the next section that $x^{(1)}(t) = (-2e^t, e^t)$ and $x^{(2)}(t) = (e^{4t}, e^{4t})$ form a basis for the solution space of $\dfrac{dx}{dt} = Ax$, where

$$A = \begin{bmatrix} 2 & 2 \\ 1 & 3 \end{bmatrix}.$$

Hence the general solution to the equation is

$$\begin{aligned} x(t) &= c_1(-2e^t, e^t) + c_2(e^{4t}, e^{4t}) \\ &= (-2c_1 e^t + c_2 e^{4t}, c_1 e^t + c_2 e^{4t}). \end{aligned}$$

EXERCISE 1 It can be shown that the solutions $x^{(1)}(t) = (e^{2t}, -e^{2t})$ and $x^{(2)}(t) = (e^{4t}, e^{4t})$ form a basis for the solution space of the system of differential equations given in Exercise 2 of Section 8.8. Find the general solution to this system.

A basis for a function space is a set of independent functions that span the space. It is in general difficult to recognize when a set of functions in a function space is independent. However, when the function space is the solution space of a system of differential equations, there is an elegant and remarkably simple method for determining whether or not a set of solutions is independent.

Result 1 (Test for Independence of Solutions) Let $x^{(1)}(t), \ldots, x^{(m)}(t)$ be solutions of $\dfrac{dx}{dt} = Ax$, and let t_0 be any number. Then $x^{(1)}(t), \ldots, x^{(m)}(t)$ are linearly independent solutions if and only if $x^{(1)}(t_0), \ldots, x^{(m)}(t_0)$ are linearly independent vectors.

Proof Let $x^{(1)}(t), \ldots, x^{(m)}(t)$ be linearly independent solutions and let t_0 be a number. Suppose that $c_1, \ldots, c_m$ are scalars such that

$$c_1 x^{(1)}(t_0) + \cdots + c_m x^{(m)}(t_0) = 0.$$

Let

$$x(t) = c_1 x^{(1)}(t) + \cdots + c_m x^{(m)}(t).$$

Since $x(t)$ is a linear combination of solutions, it is a solution. But $x(t_0) = 0$, so Result 1 of Section 8.9 implies that $x(t)$ is the trivial solution;

that is, $x(t) = 0$ for all t. Thus

$$c_1 x^{(1)}(t) + \cdots + c_m x^{(m)}(t) = 0 \qquad \text{for all } t.$$

Since $x^{(1)}(t), \ldots, x^{(m)}(t)$ are independent solutions, this equation implies that $c_1 = \cdots = c_m = 0$. Thus $x^{(1)}(t_0), \ldots, x^{(m)}(t_0)$ are independent.

Conversely suppose that for some number t_0 the vectors $x^{(1)}(t_0)$, $\ldots, x^{(m)}(t_0)$ are independent. If $c_1, \ldots, c_m$ are scalars such that

$$c_1 x^{(1)}(t) + \cdots + c_m x^{(m)}(t) = 0 \qquad \text{for all } t,$$

then letting $t = t_0$, we obtain

$$c_1 x^{(1)}(t_0) + \cdots + c_m x^{(m)}(t_0) = 0.$$

Since $x^{(1)}(t_0), \ldots, x^{(m)}(t_0)$ are independent, it follows that $c_1 = \cdots = c_m = 0$. Thus $x^{(1)}(t), \ldots, x^{(m)}(t)$ are independent.

EXAMPLE 2   Consider the differential equation $\dfrac{dx}{dt} = Ax$, where

$$A = \begin{bmatrix} 0 & 1 \\ 2 & -1 \end{bmatrix}.$$

The student can easily verify that $x^{(1)}(t) = (e^t, e^t)$ and $x^{(2)}(t) = (e^{-2t}, -2e^{-2t})$ are solutions of this equation. Since $x^{(1)}(0) = (1, 1)$ and $x^{(2)}(0) = (1, -2)$ are independent vectors in R^2, it follows from Result 1 that $x^{(1)}(t)$ and $x^{(2)}(t)$ are independent solutions.

EXERCISE 2   Use Result 1 to verify that the solutions given in Example 1 and Exercise 1 are linearly independent.

Our next result gives the dimension of the solution space in terms of the size of the square matrix A.

Theorem 8.12   *Let A be an $n \times n$ matrix. The dimension of the solution space of $\dfrac{dx}{dt} = Ax$ is n.*

Proof   We prove the result in two steps. First we show that there exist n independent solutions. Then we show that any set consisting of more than n solutions is dependent.

1. *There exist n independent solutions.* Consider the n different initial value problems

$$\frac{dx}{dt} = Ax, \qquad x(0) = e_j, \qquad j = 1, \ldots, n.$$

By the existence–uniqueness theorem, each of these initial value problems has a unique solution which we denote by $x^{(j)}(t)$, $j = 1, \ldots, n$. By the previous result these n solutions are linearly independent, since $x^{(j)}(0) = e_j$, $j = 1, \ldots, n$, and the vectors $e_1, \ldots, e_n$ are independent in R^n.

2. *Any set with more than n solutions is dependent.* If $x^{(1)}(t), \ldots, x^{(m)}(t)$, $m > n$, are solutions of $\dfrac{dx}{dt} = Ax$, then $x^{(1)}(0), \ldots, x^{(m)}(0)$ are m vectors in R^n. Since $m > n$, they are dependent. Thus, by the previous result, $x^{(1)}(t), \ldots, x^{(m)}(t)$ are dependent.

EXAMPLE 3   In Example 2 we found two independent solutions to the system $\dfrac{dx}{dt} = Ax$, where A was a 2×2 matrix. In view of this theorem, these solutions form a basis for the solution space.

PROBLEMS 8.9

1. In Problem 1 of Section 8.8 we gave solutions to certain differential equations. For each of these differential equations, show that the given solutions form a basis for the solution space.

*2. In each of the following, determine whether the given solutions of $\dfrac{dx}{dt} = Ax$ form a basis for the solution space.

(a) $A = \begin{bmatrix} 1 & 6 \\ 0 & -1 \end{bmatrix}$, $x^{(1)}(t) = e^t \begin{bmatrix} 1 \\ 0 \end{bmatrix}$, $x^{(2)}(t) = e^{-t} \begin{bmatrix} -3 \\ 1 \end{bmatrix}$

(b) $A = \begin{bmatrix} 1 & 1 & 1 \\ 0 & 0 & 1 \\ 0 & 0 & -1 \end{bmatrix}$, $x^{(1)}(t) = \begin{bmatrix} 2 \\ -2 \\ 0 \end{bmatrix}$, $x^{(2)}(t) = e^t \begin{bmatrix} 4 \\ 0 \\ 0 \end{bmatrix}$,

$x^{(3)}(t) = e^{-t} \begin{bmatrix} 0 \\ -3 \\ 3 \end{bmatrix}$

(c) $A = \begin{bmatrix} 2 & 0 & 0 \\ 0 & 3 & -1 \\ 0 & -1 & 3 \end{bmatrix}$, $x^{(1)}(t) = e^{2t} \begin{bmatrix} -1 \\ 0 \\ 0 \end{bmatrix}$, $x^{(2)}(t) = e^{2t} \begin{bmatrix} 0 \\ 1 \\ 1 \end{bmatrix}$,

$x^{(3)}(t) = e^{4t} \begin{bmatrix} 0 \\ 1 \\ -1 \end{bmatrix}$

3. Show that the conclusion of Result 1 fails for the functions
$$f(t) = (t, t) \quad \text{and} \quad g(t) = (t^2, t^2).$$
What can you conclude from this?

4. Consider the nth-order homogeneous differential equation
$$\frac{d^n y}{dt^n} + a_{n-1} \frac{d^{n-1} y}{dt^{n-1}} + \cdots + a_0 y = 0.$$

(a) Show that the set of solutions to this equation forms a subspace of $C^n(R)$.
(b) Show that the dimension of this subspace is n.

5. Let A be an $n \times n$ matrix, let b be an n-vector, and let $x^{(1)}(t), \ldots, x^{(n)}(t)$ be n linearly independent solutions of the differential equation $\dfrac{dx}{dt} = Ax$. If $x^{(p)}(t)$

is a solution to the nonhomogeneous equation $\dfrac{dx}{dt} = Ax + b$, show that any solution to this differential equation can be written as

$$x(t) = x^{(p)}(t) + c_1 x^{(1)}(t) + \cdots + c_n x^{(n)}(t)$$

where $c_1, \ldots, c_n$ are scalars.

8.10   EIGENVALUES AND DIFFERENTIAL EQUATIONS

Let us now turn our attention to the problem of finding a basis for the solution space. It follows from Theorem 8.12 that a basis for the solution space of $\dfrac{dy}{dx} = Ax$, A an $n \times n$ matrix, consists of n independent solutions. Once we have found solutions, we can use Result 1 of Section 7 to determine whether or not they are independent. What remains to be done, therefore, is to develop a method for finding solutions.

In Section 8.7 we explained that it is reasonable to try to find a solution of the form $x(t) = e^{\lambda t}v$, where λ is a scalar and v is a nonzero constant n-vector. Since

$$\frac{d}{dt}(e^{\lambda t}v) = \lambda e^{\lambda t}v \qquad \text{and} \qquad A(e^{\lambda t}v) = e^{\lambda t}Av,$$

it follows that $x(t) = e^{\lambda t}v$ is a solution of $\dfrac{dx}{dt} = Ax$ if and only if $\lambda e^{\lambda t}v = e^{\lambda t}Av$. Since $e^{\lambda t}$ is never zero, we may divide both sides of this equation by $e^{\lambda t}$ to obtain $Av = \lambda v$. This proves the following result.

Result 1   Let A be an $n \times n$ matrix, v a nonzero n-vector, and λ a scalar. The function $x(t) = e^{\lambda t}v$ is a solution of $\dfrac{dx}{dt} = Ax$ if and only if λ is an eigenvalue of A and v is an eigenvector associated with λ.

Thus each eigenvalue λ of A determines an infinite number of solutions to $\dfrac{dx}{dt} = Ax$, namely, $x(t) = e^{\lambda t}v$, where v is any eigenvector associated with λ. The number of independent solutions of this form depends on the number of independent eigenvectors associated with λ. This follows immediately from Result 1 of Section 8.9 since the solutions

$$x^{(1)}(t) = e^{\lambda t}v_1, \ldots, x^{(k)}(t) = e^{\lambda t}v_k$$

are independent if and only if

$$x^{(1)}(0) = v_1, \ldots, x^{(k)}(0) = v_k$$

are independent.

Result 2   Let A be an $n \times n$ matrix that has n independent eigenvectors $v_1, \ldots, v_n$. Let $\lambda_1, \ldots, \lambda_n$ be the corresponding eigenvalues. Then the solutions

$$x^{(1)}(t) = e^{\lambda_1 t} v_1, \ldots, x^{(n)}(t) = e^{\lambda_n t} v_n$$

form a basis for the solution space of $\dfrac{dx}{dt} = Ax$. Therefore, the general form of the solution is

$$x(t) = c_1 e^{\lambda_1 t} v_1 + \cdots + c_n e^{\lambda_n t} v_n.$$

Proof   Since $x^{(1)}(0) = v_1, \ldots, x^{(n)}(0) = v_n$ and $v_1, \ldots, v_n$ are independent vectors, the result follows from Result 1 of Section 8.9 and Theorem 8.12.

EXAMPLE 1   Consider the differential equation

$$\frac{dx}{dt} = Ax \qquad \text{where} \qquad A = \begin{bmatrix} 1 & 0 & 1 \\ 0 & 1 & 0 \\ 1 & 2 & 1 \end{bmatrix}.$$

In Example 1 of Section 8.2 we showed that the eigenvalues of A are 0, 1, and 2 and that the eigenspaces $E(0)$, $E(1)$, and $E(2)$ are spanned by $(-1, 0, 1)$, $(-2, 1, 0)$, and $(1, 0, 1)$, respectively. These eigenvectors are independent. Since A is a 3×3 matrix with three independent eigenvectors, the general solution of $\dfrac{dx}{dt} = Ax$ is

$$x(t) = c_1 e^{0t}(-1, 0, 1) + c_2 e^t(-2, 1, 0) + c_3 e^{2t}(1, 0, 1)$$
$$= \left(-c_1 - 2c_2 e^t + c_3 e^{2t}, c_2 e^t, c_1 + c_3 e^{2t} \right).$$

EXAMPLE 2   Consider the differential equation

$$\frac{dx}{dt} = Ax \qquad \text{where} \qquad A = \begin{bmatrix} 2 & -5 & 5 \\ 0 & 3 & -1 \\ 0 & -1 & 3 \end{bmatrix}.$$

From Example 2 and Exercise 3 of Section 8.2 we know that A has three independent eigenvectors $(0, 1, 1)$, $(1, 0, 0)$, and $(5, -1, 1)$ with corresponding eigenvalues 2, 2, and 4, respectively. By the previous result, the general solution of $\dfrac{dx}{dt} = Ax$ is

$$x(t) = c_1 e^{2t}(0, 1, 1) + c_2 e^{2t}(1, 0, 0) + c_3 e^{4t}(5, -1, 1)$$
$$= \left(c_2 e^{2t} + 5c_3 e^{4t}, c_1 e^{2t} - c_3 e^{4t}, c_1 e^{2t} + c_3 e^{4t} \right).$$

PROBLEMS 8.10

1. Find the general solution to the differential equation $\dfrac{dx}{dt} = Ax$ for each diagonalizable matrix in Problem 1 of Section 8.2.

2. Suppose that A is a diagonalizable matrix, $\Lambda = P^{-1}AP$, where Λ is a diagonal matrix.

(a) Show that the substitution $x = Py$ reduces the equation $\dfrac{dx}{dt} = Ax$ to $\dfrac{dy}{dt} = \Lambda y$.

(b) Find the general solution to $\dfrac{dy}{dt} = \Lambda y$.

(c) Show that if y is the general solution to $\dfrac{dy}{dt} = \Lambda y$, then $x = Py$ is the general solution to $\dfrac{dx}{dt} = Ax$.

(d) Show that the above gives a different proof of Result 2.

3. Apply the results of Problem 2 to each of the diagonalizable matrices in Problem 1 of Section 8.2.

4. Suppose that A is a diagonalizable matrix, $A = P\Lambda P^{-1}$, with $\lambda_1, \ldots, \lambda_n$ being the diagonal entries of Λ. Show that the solution of the initial value problem

$$\frac{dx}{dt} = Ax, x(0) = x_0$$

is

$$x(t) = P \begin{bmatrix} e^{\lambda_1 t} & & \\ & \ddots & \\ & & e^{\lambda_n t} \end{bmatrix} P^{-1}x_0.$$

5. We define $e^{At} = I + At + \dfrac{1}{2!}(At)^2 + \dfrac{1}{3!}(At)^3 + \cdots$. It can be shown that this series converges for any matrix A. Note that e^{At} is an $n \times n$ matrix.

(a) Show that

$$e^{\Lambda t} = \begin{bmatrix} e^{\lambda_1 t} & & \\ & \ddots & \\ & & e^{\lambda_n t} \end{bmatrix} \quad \text{if} \quad \Lambda = \begin{bmatrix} \lambda_1 & & \\ & \ddots & \\ & & \lambda_n \end{bmatrix}.$$

(b) Show that if A is a diagonalizable matrix, $A = P\Lambda P^{-1}$, where Λ is a diagonal matrix, then

$$e^{At} = Pe^{\Lambda t}P^{-1}.$$

From this and Problem 4 it follows that the solution to the initial value problem $\dfrac{dx}{dt} = Ax, x(0) = x_0$, is

$$x(t) = e^{At}x_0.$$

(c) Find e^{At} for each of the diagonalizable matrices in Problem 1 of Section 8.2.

(d) Show that:
  (i) e^{-At} is the inverse of e^{At}.
  (ii) If λ is an eigenvalue of A with associated eigenvector v, then $e^{\lambda t}$ is an eigenvalue of e^{At} with associated eigenvector v. Hence zero is not an eigenvalue of e^{At}.

It follows from either (i) or (ii) that e^{At} is nonsingular for any value of t.

Chapter 9

Numerical Methods

In previous chapters we have ignored virtually all questions concerning the speed and accuracy of computation. In addition, the matrices in our examples, exercises, and problems have been somewhat unrealistic. Both the size of the matrices and the numbers used in them have been chosen so that solutions could be reasonably calculated by hand. These restrictions are both unnecessary and unwise in the age of computers. Computers enable us to solve problems involving large matrices very quickly.

In this chapter we discuss how to maximize the speed and increase the accuracy with which we solve systems, perform matrix operations, and compute eigenvalues and eigenvectors. We also discuss some new methods for solving these problems, methods that are particularly appropriate for use on computers.

9.1 OPERATION COUNTS

This section is concerned with estimating the time required to complete certain computations. Such estimates have two important uses. First, we can determine how a change in the size of the problem affects the time required for its solution. This knowledge helps us determine what size problems can be solved in a reasonable amount of time. Second, we can compare the times required by two different procedures. This helps us decide which of the procedures is more efficient.

The procedures we have developed in this book involve three types of operations: multiplication-type operations (multiplications and divisions), addition-type operations (additions and subtractions), and movement-type operations (interchanging numbers). Since multiplications and divisions are much more time consuming than the other operations, it is appropriate to use the number of multiplication-type operations as a rough measure of the time necessary to complete a procedure. Of course, if the other operations are many times more numerous than the multiplication-type operations, this measure may be in error. However, in the cases that we consider, no error of this kind will arise.

Let us begin our analysis by examining the Gaussian elimination procedure. We must count the total number of operations required to solve an $n \times n$ system $Ax = b$ which has a unique solution. The system is solved by performing row operations on the augmented matrix $[A : b]$, a matrix with n rows and $n + 1$ columns. In the interest of simplicity we will assume that the system can be solved without any pivoting. This assumption has no effect on our results.

The first step in forward elimination is to eliminate the first entry from each of the last $n - 1$ rows. Each of these eliminations involves computing a multiplier for the initial first row (one division), multiplying every entry in the initial row (except the first) by this multiplier (n multiplications), and subtracting each of these n products from the appropriate entry in some row (n subtractions). Of course, we do not count or perform operations which are not needed. We already know that the new first entries in each row will be zero, so it is pointless to compute them. Each elimination therefore involves $n + 1$ multiplication type operations and the $n - 1$ different eliminations involve $(n - 1)(n + 1) = n^2 - 1$ multiplication-type operations.

The next step in the forward elimination process is to eliminate the second entry from each of the last $n - 2$ rows.

EXERCISE 1 Show that each of these eliminations requires one division, $n - 1$ multiplications, and $n - 1$ subtractions.

Exercise 1 shows that the second step of the forward elimination process takes

$$n(n - 2) = [(n - 1) + 1][(n - 1) - 1] = (n - 1)^2 - 1$$

multiplication-type operations. The following table presents a summary of the operations required for the complete forward elimination process.

The total number of operations required for forward elimination can be computed from the table using the following formulas.

$$1 + 2 + 3 + \cdots + (k - 1) + k = \frac{k(k + 1)}{2}$$

$$1^2 + 2^2 + 3^2 + \cdots + (k - 1)^2 + k^2 = \frac{k(k + 1)(2k + 1)}{6} \tag{1}$$

Gaussian Elimination
Forward Elimination Operation Count

	Each elimination				Total operation count	
		requires		Elimina-	Addition	Multiplication
Step	$\div s$	$\times s$	$-s$	tions	type	type
1	1	n	n	$n-1$	$n(n-1)$	$(n+1)(n-1) = \qquad n^2 - 1$
2	1	$n-1$	$n-1$	$n-2$	$(n-1)(n-2)$	$n(n-2) = (n-1)^2 - 1$
3	1	$n-2$	$n-2$	$n-3$	$(n-2)(n-3)$	$(n-1)(n-3) = (n-2)^2 - 1$
⋮						
$n-2$	1	3	3	2	$3 \cdot 2$	$4 \cdot 2 = 3^2 - 1$
$n-1$	1	2	2	1	$2 \cdot 1$	$3 \cdot 1 = 2^2 - 1$
n	—	—	—	0	0	$0 = 1^2 - 1$

First we compute the number of multiplication-type operations. The last column of the table gives this as

$$\left(n^2 - 1\right) + \left(\left[n - 1\right]^2 - 1\right) + \cdots + \left(2^2 - 1\right) + \left(1^2 - 1\right)$$

which equals

$$1^2 + 2^2 + 3^2 + \cdots + (n-1)^2 + n^2 - n.$$

Using the second formula in (1) this expression reduces to

$$\frac{2n^3 + 3n^2 - 5n}{6}.$$

To count the number of addition-type operations, we first note that $k(k-1) = k^2 - k$. Since the terms we must add up are all of this form, we see that the total number of addition-type operations is

$$\left(n^2 - n\right) + \left(\left[n-1\right]^2 - \left[n-1\right]\right) + \cdots + \left(2^2 - 2\right) + \left(1^2 - 1\right).$$

Using the formulas in (1) this expression reduces to

$$\frac{n(n+1)(2n+1)}{6} - \frac{n(n+1)}{2} = \frac{n^3 - n}{3}.$$

The complete solution of the system is now obtained by back substitution. Solving for x_n requires one division. Solving for x_{n-1} requires one multiplication, one subtraction, and one division. Solving for x_{n-2} requires two multiplications, two subtractions, and one division. The last step is to solve for x_1, which takes $n - 1$ multiplications, $n - 1$ subtractions, and one division. (The reader should, of course, verify these statements by examining the equations that are being solved.) The total number of multiplication-type operations needed for back substitution is thus

$$1 + 2 + 3 + \cdots + (n-1) \text{ multiplications} + n \text{ divisions}$$

which equals $(n^2 + n)/2$.

EXERCISE 2   Show that the number of addition-type operations needed for back substitution is

$$1 + 2 + 3 + \cdots + (n - 1) = \frac{n^2 - n}{2}$$

Since the total number of operations required by Gaussian elimination is the number of operations required for forward elimination plus the number of operations required for back substitution, we have the following result.

Result 1   Gaussian elimination requires $(n^3 + 3n^2 - n)/3$ multiplication-type operations and $(2n^3 + 3n^2 - 5n)/6$ addition-type operations to solve an $n \times n$ system which has a nonsingular coefficient matrix.

We can use this result to estimate the time required to solve a system of a given size. The following table gives the number of operations required for systems of various sizes.

Size of system (number of equations and variables)	Operation count (multiplication-type operations)
5	65
10	430
20	3,060
30	9,890
40	22,920
50	44,150
100	343,300
200	2,706,600
1000	334,333,000

If each multiplication requires 10 seconds to do by hand, we can deduce from this table that a 20×20 system will require more than 8 hours to solve by hand. (This estimate is high if the numbers are all one-digit integers but low if we have to perform multiplications such as 2.435 times 8.732.) Even a 10×10 problem takes at least 1 hour to complete. This explains why square systems with 10 or more variables are conspicuously absent from the problems in this book. Problems with 10 or more variables are not reasonable problems to solve by hand calculation.

If we perform our calculations using a computer, the picture changes considerably. A conservative estimate of the speed of a computer is 100,000 multiplications per second. Under this assumption a 20×20 system can be solved in slightly more than 0.03 second and a 100×100 system takes less than 4 seconds. Faster machines (millions of multiplications per second!) will solve these systems in correspondingly less time.

In the discussion above we have ignored the time taken by additions, subtractions, and data movement. Addition-type operations can be done 10 to 100 times faster than multiplications. Since the number of addition-type operations is about the same as the number of multiplication-type operations, we would have to increase our time estimates by no more than 10%. Thus the time taken by the addition-type operations can be safely ignored. Pivoting takes hardly any time and can also be ignored.

Let us now compute the operation count for Gauss–Jordan elimination. This procedure also requires us to perform row operations on the augmented matrix. At each step of the Gauss–Jordan process we must first make the pivot equal to one and then eliminate $n - 1$ entries since we must eliminate every entry in the pivot column except the pivot itself. The first step involves dividing n entries by the pivot and then performing $n - 1$ eliminations each requiring n multiplications. This gives a total of n^2 multiplication-type operations. The next step involves dividing $n - 1$ entries by the pivot and then performing $n - 1$ eliminations each requiring $n - 1$ multiplications. This gives $n(n - 1)$ multiplication-type operations. The total number of multiplication-type operations for the whole process is

$$n^2 + n(n - 1) + n(n - 2) + \dots + n \cdot 1.$$

Using equations (1) to simplify this sum, we obtain the following result.

Result 2 Gauss–Jordan elimination requires $(n^3 + n^2)/2$ multiplication-type operations to solve an $n \times n$ system with a nonsingular coefficient matrix.

We now use Results 1 and 2 to compare Gaussian elimination with Gauss–Jordan elimination so that we may determine the more efficient procedure. These two results give precise operation counts for the two procedures. But for the comparison we have in mind it is easier to use approximate values for the operation counts. When n is moderately large (say 10), n^3 is quite a bit larger than both n^2 and n. This says that the n^3 term in the results is the important (or dominant) term. This means that Gaussian elimination requires about $n^3/3$ multiplication-type operations. Gauss–Jordan elimination, on the other hand, requires about $n^3/2$ multiplication-type operations. Thus Gauss–Jordan elimination requires 50% more operations than Gaussian elimination. This justifies our remark in Chapter 1 that Gaussian elimination is the more efficient procedure.

We conclude this section with a discussion of the inversion process. One way to measure the work required to invert an $n \times n$ matrix by Gaussian elimination is to count the operations required to solve n different systems with identical coefficient matrices (see Problem 4). But when we invert a matrix, the n columns on the right side of the augmented matrix are very special. If we take advantage of this and omit multiplications by 0 and 1, we can easily compute the number of operations required. The following two tables give the operation counts.

Forward Elimination

Step	Each elimination Division	Each elimination Multiplications	Number of eliminations	Total
1	1	$n - 1$	$n - 1$	$(n - 1)n$
2	1	$n - 1$	$n - 2$	$(n - 2)n$
$\vdots$				
$n - 1$	1	$n - 1$	1	$1n$
n	—	—	0	0

Grand total $(n^3 - n^2)/2$

Back Substitution

Step	To make pivot 1 divisions	Each elimination multiplications	Number of eliminations	Total
1	n	n	$n - 1$	$n(n - 1) + n$
2	n	n	$n - 2$	$n(n - 2) + n$
3	n	n	$n - 3$	$n(n - 3) + n$
$\vdots$				
$n - 1$	n	n	1	$n(1) + n$
n	n	—	0	$0 + n$

Grand total $(n^3 + n^2)/2$

Result 3   Gaussian elimination requires n^3 multiplication-type operations to invert an $n \times n$ nonsingular matrix.

It is important to compare Results 1 and 3. Solving $Ax = b$ by Gaussian elimination requires approximately $n^3/3$ multiplication-type operations. Just computing A^{-1} requires three times as many operations. Thus solving the system $Ax = b$ by computing A^{-1} and then finding $A^{-1}b$ is much less efficient than solving by the Gaussian elimination procedure. A related question is discussed in Problem 6.

PROBLEMS 9.1

*1. Let v be an n-vector, A and B be $n \times n$ matrices. Determine the number of multiplications and the number of additions required to compute the following.
(a) $\langle v, v \rangle$ (b) Av
(c) AB (d) A^2

(e) A^3 (f) $(A^2)^2$

(g) $A(A(AA))$

2. Show that the most efficient way to compute A^7 is to compute $(A^2)^2 A^2 A$.

3. Show that Gauss–Jordan elimination requires $(n^3 - n)/2$ addition-type operations to solve an $n \times n$ system with a nonsingular coefficient matrix.

*4. The solution of k different systems $Ax = b_1$, $Ax = b_2, \ldots$, $Ax = b_k$ with identical $n \times n$ nonsingular coefficient matrices can be accomplished by using row operations on the augmented matrix $[A : b_1 \quad b_2 \quad \cdots \quad b_k]$.
   (a) Determine how many multiplication-type operations are required when using Gaussian elimination.
   (b) Determine how many multiplication-type operations are required when using Gauss–Jordan elimination.

5. Determine the number of multiplications and divisions required to invert an $n \times n$ nonsingular matrix using Gauss–Jordan elimination.

*6. (a) One method for solving the k different systems, $Ax = b_1$, $Ax = b_2, \ldots$, $Ax = b_k$, is to determine A^{-1} and then compute the k products $A^{-1}b_1$, $A^{-1}b_2, \ldots$, $A^{-1}b_k$. Combine Result 3 and the answer to Problem 1(b) to find the number of multiplication-type operations required by this method.
   (b) Compare the answer to part (a) with the number given in Problem 4(a). Under what circumstances will Gaussian elimination (applied to the augmented matrix with k different right-hand columns) be a more efficient procedure than the invert-and-multiply method of part (a)?

9.2 PIVOTING STRATEGIES

In this section we discuss how numbers are represented in a computer, the nature of the errors involved in computer computations, and some strategies that are used to improve the accuracy of computer solutions.

Round-off Error

Given an integer $b > 1$ called the **base**, any nonzero number can be uniquely represented in the form

$$s \cdot b^e \cdot m,$$

where $s = \pm 1$ is the **sign**, e is the **exponent**, and m $(b^{-1} \leq m < +1)$ is the **mantissa**. In this representation the first digit of m after the decimal point is never equal to zero. Representing a number by storing its sign, its exponent, and its mantissa is called **floating-point representation**. This is the usual way numbers are represented and stored in computers.

EXAMPLE 1 The following table gives some examples of floating-point representation.

Number	Sign	Base	Exponent	Mantissa
35.7	+1	10	2	0.357
−0.005	−1	10	−2	0.5
15	+1	10	2	0.15
4	+1	2	3	0.1
15	+1	2	4	0.1111

Notice that a different choice of base leads to a different representation for the number.

There is an unavoidable error involved in the floating-point representation of numbers. It arises from the finite number of digits (in whatever base is being used) available to represent the mantissa. The fixed length of the mantissa means that many numbers, indeed all numbers in an actual interval, will all have the same representation. For example, if the base is 10 and four digits are used to represent the mantissa, then 9.87649 and 9.8762 will both have 0.9876 as their mantissa. If mantissas are *rounded* to four digits, then 9.8756 also has the same mantissa. *Truncation* of all digits after the fourth means that 9.8769 would also have 0.9876 as its mantissa. All of the errors that result from the floating-point representation of numbers are called **round-off errors**.

Round-off errors are not necessarily small. Representing 278,342,000,000 in base 10 with a four-digit mantissa involves an error of 42 million. This example shows that the **absolute error** (the actual value of the error) is not a good measure of the significance of the error. An error of 42 million in 10^{100} is trivial; an error of 42 million in 100 million is disastrous. We use **relative error** (the ratio of the error to the number itself) to measure the error because it gives a clearer indication of the actual importance of the error.

One subtle form of round-off error results from the use of different bases to represent numbers. Computers usually use 2 for a base instead of 10. A number that can be represented exactly in base 10 may not be represented exactly in base 2. For example, 0.1 in base 10 has a repeating representation in base 2 (0.0001100110011...). Thus it can never be represented exactly in base 2, no matter how many digits are available for the mantissa.

Round-off error can also occur when two numbers are added. (In the remainder of this chapter we will assume that the base is 10 and that four digits are available to represent the mantissa; that is, we will work to four significant decimal digits.) For example,

$$0.1234 \cdot 10^1 + 0.5123 \cdot 10^{-2} = 1.234 + 0.005123 = 1.239123,$$

which will be represented as $0.1239 \cdot 10^1$. Notice that all the information in the last three digits of the second number $0.5123 \cdot 10^{-2}$ has been lost.

A much more serious form of round-off error occurs when two nearly equal number are subtracted.

$$0.9876 - 0.9875 = 0.0001 = 0.1000 \cdot 10^{-4}.$$

Because of the inaccuracy of the representation of the original number, the true answer is somewhere between 0 and 0.0002. When we subtracted we lost more than three significant digits. The relative error is 100%.

Ill-conditioned Systems

Certain systems are very sensitive to small changes in the constant vector. This makes them very difficult to solve. As an example, let us consider the matrix $A = \begin{bmatrix} 1.000 & 1.001 \\ 1.000 & 1.000 \end{bmatrix}$. We wish to study how the solution of $Ax = b$ is affected by small changes in b. The following table gives the solution for several different choices of the vector b.

b	Solution of $Ax = b$
(1.002, 1.001)	(0.001, 1.000)
(1.001, 1.001)	(1.001, 1.000)
(1.001, 1.000)	(0.000, 1.000)
(1.000, 1.000)	(1.000, 0.000)
(1.000, 1.001)	(2.001, -1.000)

This table gives striking evidence that the system $Ax = b$ is extremely sensitive to small changes in b. Small changes in b, changes that are barely detectable when our numbers are represented with four-digit mantissas, cause changes in the first significant digits of the answer. This behavior is caused by the coefficient matrix of the system. A matrix that gives rise to such an unstable system is called **ill-conditioned**.

There is no cure for the numerical instability caused by ill-conditioned matrices. Whenever the coefficient matrix is ill-conditioned the number of significant digits in the answer will be much less than the number of significant digits in the given constant vector. But even though we cannot cure this problem, we should at least be able to recognize it.

At first glance one might suppose that the instability in this problem has to do with the determinant being close to 0 (det $A = -0.001$). But the conditioning of a matrix cannot be measured by the determinant because the matrix $1000A$ has a determinant that is quite far from 0 (det $1000A = -1000$) and this matrix is just as ill-conditioned as A is.

EXERCISE 1   Verify that $1000A$ is just as ill-conditioned as A is by solving $(1000A)x = b$ for the following vectors.
(a) (1002, 1001)   (b) (1001, 1001)   (c) (1001, 1000)
Compare with the table of solutions of $Ax = b$ given above.

The proper way to measure the conditioning of a matrix A is to look at what A (considered as a linear transformation) does to the set of all unit vectors (the unit sphere). Every linear transformation takes the unit sphere into an "ellipsoid" and if A produces a distorted ellipsoid, then it is ill-conditioned. A useful measure of the distortion of the ellipsoid is the ratio of the longest axis to the shortest axis. This number is called the **condition number** of A, denoted cond A. If λ_{max} is the eigenvalue with the largest absolute value and λ_{min} is the eigenvalue with the smallest absolute value, then cond $A = |(\lambda_{max})/(\lambda_{min})|$. When cond A is large, A is ill-conditioned. Indeed, the size of the condition number of A gives a very good indication of how much accuracy is lost when solving the system $Ax = b$.

EXAMPLE 2 The eigenvalues of the matrix A are the roots of $x^2 - 2x - 0.001 = 0$. Correct to four significant figures, cond $A = 4002$.

Pivoting Strategies

There is always the possibility that the round-off errors that occur while solving a system may "build up" to such magnitudes as to produce significant errors in the answer. We must find some way to minimize the total effect of the round-off errors.

Let us illustrate this buildup of round-off errors in the following example. Let

$$Ax = \begin{bmatrix} 0.1000 \cdot 10^{-5} & 0.1000 \cdot 10^1 \\ 0.1000 \cdot 10^1 & 0.1000 \cdot 10^1 \end{bmatrix} \begin{bmatrix} x_1 \\ x_2 \end{bmatrix} = \begin{bmatrix} 0.1000 \cdot 10^1 \\ 0.2000 \cdot 10^1 \end{bmatrix} = b. \quad (1)$$

We apply Gaussian elimination to solve the system, using four-digit mantissas as we proceed. Forward elimination gives

$$\begin{bmatrix} 0.1000 \cdot 10^{-5} & 0.1000 \cdot 10^1 & \vdots & 0.1000 \cdot 10^1 \\ 0 & -0.1000 \cdot 10^7 & \vdots & -0.1000 \cdot 10^7 \end{bmatrix}$$

where the nonzero entries in the bottom row have been rounded to four significant figures. Back substitution yields $x_2 = 0.1000 \cdot 10^1$, $x_1 = 0$ as the solution.

The best that can be said for this answer is that it is wrong. $A\begin{bmatrix} 0 \\ 1 \end{bmatrix} = \begin{bmatrix} 1 \\ 1 \end{bmatrix} \neq \begin{bmatrix} 1 \\ 2 \end{bmatrix}$. The problem is that the small error in the computation of x_2 has led to a large error in the computation of x_1. The (small) initial error has been magnified by a subsequent computation.

If we interchange the equations, this problem does not arise. We start with the equivalent system whose augmented matrix is

$$\begin{bmatrix} 0.1000 \cdot 10^1 & 0.1000 \cdot 10^1 & \vdots & 0.2000 \cdot 10^1 \\ 0.1000 \cdot 10^{-5} & 0.1000 \cdot 10^1 & \vdots & 0.1000 \cdot 10^1 \end{bmatrix}. \quad (1^*)$$

Forward elimination (using four-significant-digit arithmetic) gives us

$$\begin{bmatrix} 0.1000 \cdot 10^1 & 0.1000 \cdot 10^1 & \vdots & 0.2000 \cdot 10^1 \\ 0 & 0.1000 \cdot 10^1 & \vdots & 0.1000 \cdot 10^1 \end{bmatrix}.$$

Back substitution then yields the answer $x_1 = x_2 = 0.1000 \cdot 10^1$. This answer *is* correct to four significant digits. (The answer correct to seven significant digits is $x_1 = 0.1000001 \cdot 10^1$, $x_2 = 0.9999990 \cdot 10^0$.)

Clearly, if changing the order of the equations can change the effects of round-off error, then some pivoting strategy must be adopted. A common strategy is to choose the pivot equation to be that equation which has the largest (in absolute value) coefficient of the pivot variable. This process is called **partial pivoting**.

PARTIAL PIVOTING

Given

A system that is to be solved using Gaussian elimination.

Goal

To minimize the effects of round-off error.

Procedure

At each step choose as the pivot equation that equation which has the largest (in absolute value) coefficient of the pivot variable. In other words, at each step maximize the absolute value of the pivot.

For example, partial pivoting applied to system (1) leads to system (1*). This system was solved above and gave us an answer correct to four significant digits.

Partial pivoting uses the largest coefficient (in absolute value) of the "next" variable to determine the pivot equation. Another pivoting strategy called *total pivoting* involves choosing the pivot variable as well as the pivot equation. This strategy requires a search for the largest coefficient (in absolute value) in the remaining equations. The location of this coefficient determines both the pivot variable and the pivot equation. When using total pivoting it is essential that we remember which variables have been used as pivot variables as well as the order in which these variables have been used.

Scaling

Partial pivoting does not resolve all of the difficulties that can arise. The system (1) is equivalent to

$$\begin{bmatrix} 0.1000 \cdot 10^1 & 0.1000 \cdot 10^7 \\ 0.5000 \cdot 10^0 & 0.5000 \cdot 10^0 \end{bmatrix} \begin{bmatrix} x_1 \\ x_2 \end{bmatrix} = \begin{bmatrix} 0.1000 \cdot 10^7 \\ 0.1000 \cdot 10^1 \end{bmatrix}. \tag{2}$$

Applying partial pivoting to this system gets us into the same problems as before. The problem here is that the number we pivot on, the one in the first column, is small in comparison to the other entries in that row.

EXERCISE 2 Solve system (2) using partial pivoting and four-significant-digit arithmetic.

To avoid this problem we must make sure that the numbers in our problem are in some sense about the same size. Procedures that accomplish this are called **scaling** techniques. One simple (although not inevitably successful) method to scale a system properly is to first multiply each equation by a constant so that the largest coefficient (in absolute value) in each equation is 1. If this is done to (2), then partial pivoting will succeed in finding a good solution.

EXERCISE 3 Show that the scaling procedure described above transforms system (2) into system (1).

PROBLEMS 9.2

*1. Determine the mantissa and the exponent if the base is 10 and four digits are available for the mantissa.
(a) 0.78764　(b) 787.64　(c) $\frac{5}{8}$
(d) $\frac{1}{3}$　(e) $\frac{1}{7}$　(f) $\frac{1}{70}$
(g) 7000　(h) $\frac{1}{7000}$

*2. What is the largest relative error that can result from representation of numbers in base 10 with four-digit mantissas?

*3. Let $A = \begin{bmatrix} 4,999 & 15,000 \\ 1,664 & 4,993 \end{bmatrix}$.

(a) Compute Ax for each of the following vectors using four-significant-digit arithmetic.

$$x = (300,400), \quad x = (303,399), \quad x = (330,390), \quad \text{and} \quad x = (600.1,300).$$

(b) Estimate cond A. How does this explain the results in part (a)?

4. For each of the following systems, describe the scaling and pivoting that should be done before eliminating the first variable. Do not solve the systems.
(a) $0.01x_1 + 10x_2 = 55$
　$0.1x_1 - x_2 = 26$
(b) $10x_1 + 3x_2 + 10x_3 = 3$
　$10x_1 + 7x_2 - 11x_3 = -4$
　$7x_1 - 2x_2 + 15x_3 = 21$

9.3　SOLVING SYSTEMS ITERATIVELY

There are times when Gaussian elimination is an inappropriate method to solve a linear system. Situations arise in some applications where the number of variables may be too large for the system to be solved by elimination. There are also other systems that arise in applications where most of the entries in the coefficient matrix are small or zero. In this section we study two methods for solving systems which are at times more efficient or more practical than Gaussian elimination.

Jacobi Iteration

Given an $n \times n$ system

$$
\begin{aligned}
a_{11}x_1 + a_{12}x_2 + \cdots + a_{1n}x_n &= b_1 \\
a_{21}x_1 + a_{22}x_2 + \cdots + a_{2n}x_n &= b_2 \\
\vdots \qquad\qquad \vdots \qquad\qquad \vdots \qquad \vdots \\
a_{n1}x_1 + a_{n2}x_2 + \cdots + a_{nn}x_n &= b_n
\end{aligned}
\tag{1}
$$

with all diagonal entries nonzero, then for each i we can solve the ith equation for x_i. We obtain the following equations.

$$
\begin{aligned}
x_1 &= \frac{b_1 - a_{12}x_2 - a_{23}x_3 - \cdots - a_{1n}x_n}{a_{11}} \\[2mm]
x_2 &= \frac{b_2 - a_{21}x_1 - a_{23}x_3 - \cdots - a_{2n}x_n}{a_{22}} \\[2mm]
\vdots \qquad \vdots \\[2mm]
x_n &= \frac{b_n - a_{n1}x_1 - a_{n2}x_2 - \cdots - a_{n,n-1}x_{n-1}}{a_{nn}}
\end{aligned}
\tag{2}
$$

If we have some way to make a guess at an approximate solution to (1) we can use equations (2) to refine our solution. We substitute the values of $x_1, x_2, \ldots, x_n$ (the guesses) in the right-hand sides of (2) and get new values for $x_1, x_2, \ldots, x_n$. In certain circumstances, the answers we obtain will be closer to the true solution than the original guesses. The process is then repeated (*iterated*) to get another approximation to the true solution. When these successive approximations get closer to the true answer, they are said to *converge*. Of course, there is no guarantee that this will happen. The successive approximations could *diverge*; that is, they could fail to get closer to the true answer. We will shortly give a sufficient condition for convergence.

EXAMPLE 1   Consider the system

$$
\begin{aligned}
21x_1 + \quad x_2 - \quad x_3 &= 16 \\
x_1 - 12x_2 + 2x_3 &= 31 \\
-2x_1 + \quad x_2 + 20x_3 &= 56.
\end{aligned}
\tag{3}
$$

Rewriting this system in the form of (2), we obtain

$$x_1 = \frac{16 - x_2 + x_3}{21}$$

$$x_2 = \frac{-31 + x_1 + 2x_3}{12} \tag{4}$$

$$x_3 = \frac{56 + 2x_1 - x_2}{20}.$$

Let us take $x_1 = x_2 = x_3 = 0$ as our initial guess. We now use equations (4) to obtain new values for x_1, x_2, and x_3: $x_1 = 0.7619$, $x_2 = -2.583$, and $x_3 = 2.800$.

We now iterate. We take these new values for the x_i's and substitute them in the right-hand side of equations (4) and determine new values for x_1, x_2 and x_3.

$$x_1 = \frac{16 + 2.583 + 2.800}{21} = 1.018$$

$$x_2 = \frac{-31 + 0.7619 + 2 \cdot 2.800}{12} = -2.053$$

$$x_3 = \frac{56 + 2 \cdot 0.7619 + 2.583}{20} = 3.005$$

The following table gives the results of several iterations. Four significant digits were used throughout the computation.

Iteration	x_1	x_2	x_3
0	0	0	0
1	0.7619	-2.583	2.800
2	1.018	-2.053	3.005
3	1.003	-1.998	3.004
4	1.000	-1.999	3.000
5	1.000	-2.000	3.000

After only five iterations we have found the answer correct to four significant digits. (Actually, this answer is exact.) This iterative method to solve systems is called **Jacobi iteration.**

In this example the approximations converge to the true answer because the diagonal terms are so much larger than the other terms.

Definition   A matrix is called **diagonally dominant** if for each row the absolute value of the diagonal entry is greater than the sum of the absolute values of all other entries in that row.

The coefficient matrix for system (3) is diagonally dominant because $|21| > |1| + |-1|$, $|-12| > |1| + |2|$, and $|20| > |-2| + |1|$. The following result, which we present without proof, explains why Jacobi iteration works for system (3).

Result 1   If a system has a diagonally dominant coefficient matrix, then the approximations obtained by Jacobi iteration will converge to the solution of the system.

It should be noted that although this condition ensures convergence, it is not a necessary condition. Convergence can occur when the coefficient matrix is not diagonally dominant.

Gauss–Seidel Iteration

There is a simple modification of the Jacobi iteration process that sometimes converges more rapidly to the solution. The Jacobi process uses equations (2) to find new values for the x_i's simultaneously. All the old x's are used at one time to find all the new x's. But if the new value of x_1 is "better" than the old one, why not use it when we compute x_2? There is no reason why we cannot use a new value for x_i which has already been computed whenever it is needed in a subsequent equation. This method of solving systems is called **Gauss–Seidel iteration**.

EXAMPLE 2   Let us solve system (3) by Gauss–Seidel iteration. We take $x_1 = x_2 = x_3 = 0$ as our initial approximation. The first iteration gives

$$x_1 = \frac{16 - 0 + 0}{21} = 0.7619,$$

$$x_2 = \frac{-31 + 0.7619 + 2 \cdot 0}{12} = -2.520,$$

$$x_3 = \frac{56 + 2 \cdot 0.7619 + 2.520}{20} = 3.002.$$

The second iteration gives

$$x_1 = \frac{16 + 2.520 + 3.002}{21} = 1.025,$$

$$x_2 = \frac{-31 + 1.025 + 2 \cdot 3.002}{12} = -1.998,$$

$$x_3 = \frac{56 + 2 \cdot 1.025 + 1.998}{20} = 3.002.$$

The following table summarizes the first three iterations.

Iteration	x_1	x_2	x_3
0	0	0	0
1	0.7619	-2.520	3.002
2	1.025	-1.998	3.002
3	1.000	-2.000	3.000

This example shows that Gauss–Seidel iteration can be more rapid than Jacobi iteration. However, the reader should be aware that Jacobi iteration sometimes converges more rapidly than Gauss–Seidel iteration. Even more amazing is the existence of systems where one method converges and the other one does not. There are systems where Jacobi iteration converges while Gauss–Seidel diverges and other systems where the exact opposite occurs. A more thorough analysis of these matters will be found in books on numerical analysis. Here we only mention that Result 1 holds for Gauss–Seidel iteration as well. If the coefficient matrix is diagonally dominant, then Gauss–Seidel iteration will converge.

Before ending this section it is appropriate to say a few words about operation counts when using iterative methods. Each iteration (Jacobi or Gauss–Seidel) requires n^2 multiplication-type operations. However, computing an operation count is impossible because we have no idea of the number of iterations required. Even though this problem cannot be overcome, we should note that the number of iterations does not depend on the number of variables in the problem but on the nature of the equations (2). This means that k iterations require kn^2 operations, whereas Gaussian elimination requires $n^3/3$. If convergence occurs before $n/3$ iterations have been made (i.e., if $k < n/3$), then iteration is a more efficient method. When n is 100 and convergence occurs within 33 iterations, then iteration is more efficient than elimination.

PROBLEMS 9.3

*1. Solve each system by Jacobi iteration.

(a)
$$100x_1 + x_2 + x_3 = 72$$
$$x_1 + 200x_2 - x_3 = 301$$
$$x_1 - x_2 + 10x_3 = 42$$

(b)
$$1000x_1 + 900x_2 + x_3 = 10$$
$$x_1 + 10x_2 + x_3 = 43$$
$$-10x_1 + 45x_2 - 370x_3 = 27$$

(c)
$$370x_1 - 45x_2 + 10x_3 = 35$$
$$x_1 + 10x_2 + x_3 = 47$$
$$10x_1 + 900x_2 - 1000x_3 = 75$$

2. Solve the systems in Problem 1 by Gauss–Seidel iteration.

3. Transform each of the following systems into a diagonally dominant system.

(a) $\begin{aligned} 2x_1 + 50x_2 - \quad x_3 &= 5 \\ 50x_1 - 10x_2 + \quad x_3 &= 4 \\ x_1 + \quad x_2 + 10x_3 &= 5 \end{aligned}$

(b) $\begin{aligned} x_1 + \quad 2x_2 + 30x_3 &= 2 \\ 150x_1 + 100x_2 + \quad x_3 &= 40 \\ 10x_1 + \quad 80x_2 - \quad 3x_3 &= 15 \end{aligned}$

4. Modify the definition of a diagonally dominant system so that the systems in Problem 3 are included in the definition.

5. In Problem 7 of Section 3.8 the QR decomposition of a square matrix A was used to reduce the normal equations $A^T A x = A^T b$ to the equivalent system $Rx = Q^T b$.

   (a) Show that the system $Rx = Q^T b$ is a triangular system and hence can be solved by back substitution.

   (b) Compare the two methods for solving the normal equations (Gaussian elimination versus QR decomposition) for numerical stability.

9.4   EIGENVALUES AND EIGENVECTORS

In this section we present a method for approximating the largest (in absolute value) eigenvalue of a matrix. We could, of course, compute the eigenvalues by determining the characteristic polynomial and then finding its roots. Unfortunately, this method is so lengthy and so numerically unstable that it is virtually useless for large matrices. The method we present here (like most numerical methods for computing eigenvalues) takes advantage of special characteristics of the matrix and will not work in all cases.

Let A be diagonalizable with eigenvalues $\lambda_1, \lambda_2, \ldots, \lambda_n$. Let $v_1, v_2, \ldots, v_n$ be eigenvectors associated with each of these eigenvalues. Suppose in addition that

$$|\lambda_1| > |\lambda_2| \geq |\lambda_3| \geq \cdots \geq |\lambda_n|.$$

Since $v_1, v_2, \ldots, v_n$ form a basis for R^n, we can express any vector u in R^n in the form

$$u = \alpha_1 v_1 + \alpha_1 v_2 + \cdots + \alpha_n v_n.$$

Since the v's are eigenvectors, we have

$$\begin{aligned} Au &= \alpha_1 A v_1 + \alpha_2 A v_2 + \cdots + \alpha_n A v_n \\ &= \alpha_1 \lambda_1 v_1 + \alpha_2 \lambda_2 v_2 + \cdots + \alpha_n \lambda_n v_n. \end{aligned}$$

It is not hard to verify that for any positive integer k

$$A^k u = \alpha_1 \lambda_1^k v_1 + \alpha_2 \lambda_2^k v_2 + \cdots + \alpha_n \lambda_n^k v_n.$$

Now $|\lambda_1|$ is larger than $|\lambda_i|$, $i = 2, \ldots, n$, so the first term of this expression for $A^k u$ will eventually dominate the others. This means that when k is large enough, $A^k u$ is approximately $\alpha_1 \lambda_1^k v_1$, that is, $A^k u$ is approximately an eigenvector associated with the eigenvalue λ_1. (We are of course assuming here that $\alpha_1 \neq 0$.)

If the vector u is close to an eigenvector v associated with the eigenvalue λ_1, then

$$\frac{\langle u, Au \rangle}{\langle u, u \rangle} \quad \text{is approximately} \quad \frac{\langle v, Av \rangle}{\langle v, v \rangle} = \lambda_1.$$

This quotient

$$\frac{\langle u, Au \rangle}{\langle u, u \rangle}$$

is called the **Rayleigh quotient**.

We can use the discussion above to approximate λ_1. The **power method** for approximating the dominant eigenvalue consists of computing the Rayleigh quotient for the vectors

$$u_0 = u, \quad u_1 = Au_0, \quad u_2 = A^2u = Au_1, \quad u_3 = A^3u = Au_2, \ldots, u_n = Au_{n-1}$$

until the Rayleigh quotients get close enough to λ_1.

EXAMPLE 1 Let $A = \begin{bmatrix} 9 & -4 \\ 4 & -1 \end{bmatrix}$ and $u_0 = (1, 1)$. The following table gives the results of the power method.

Iteration (n)	u_n	$\langle u_{n-1}, Au_n \rangle / \langle u_{n-1}, u_{n-1} \rangle$
0	(1, 1)	—
1	(5, 3)	4
2	(33, 17)	6.353
3	(229, 115)	6.903
4	(1,601, 801)	6.986
5	(11,210, 5,603)	6.998
6	(78,480, 39,240)	7.001
7	(549,400, 274,700)	7.000
8	(3,846,000, 1,923,000)	7.000

It is clear that the eigenvalue is approximately 7 and that any of the last three u's would be a good approximation to an eigenvector associated with this eigenvalue. [The eigenvectors associated with 7 are the nonzero multiples of (2, 1).]

Example 1 illustrates the unpleasantly large numbers that are frequently produced by the power method. These numbers can be avoided. We merely multiply each u by a scalar after computing the Rayleigh quotient and use this new vector in the iteration. A common procedure is to choose the scalar so that the absolute value of every coordinate in u is ≤ 1.

EXAMPLE 2 Let us repeat Example 1 using the idea above. The following table summarizes the procedure.

u	Au	Rayleigh quotient	New u
$(1,1)$	$(5.000, 3.000)$	4.000	$(1, 0.6000)$
$(1, 0.6000)$	$(6.600, 3.400)$	6.353	$(1, 0.5152)$
$(1, 0.5152)$	$(6.939, 3.485)$	6.902	$(1, 0.5022)$
$(1, 0.5022)$	$(6.991, 3.498)$	6.986	$(1, 0.5003)$
$(1, 0.5003)$	$(6.999, 3.500)$	6.998	$(1, 0.5000)$
$(1, 0.5000)$	$(7.000, 3.500)$	7.000	$(1, 0.5000)$

The power method finds a single eigenvalue. Fortunately, there are many problems where we are interested only in determining this one eigenvalue. Procedures to find the remaining eigenvalues are treated in books on numerical analysis.

PROBLEMS 9.4

*1. Approximate the dominant eigenvalue and an associated eigenvector by the power method.

$$(a) \begin{bmatrix} 1 & 0 & 1 \\ 0 & 1 & 0 \\ 1 & 2 & 1 \end{bmatrix} \qquad (b) \begin{bmatrix} 1 & 2 & 0 \\ 2 & 2 & 1 \\ 0 & 1 & 3 \end{bmatrix} \qquad (c) \begin{bmatrix} 1 & 1 & 0 \\ 1 & 5 & -2 \\ 1 & 3 & -1 \end{bmatrix}$$

2. Let

$$A = \begin{bmatrix} 1 & -1 & 0 \\ -1 & 2 & -1 \\ 0 & -1 & 1 \end{bmatrix}.$$

(a) Approximate the dominant eigenvalue of A by the power method using $(1, 1, 1)$ as u_0.

(b) Approximate the dominant eigenvalue of A by the power method using $(1, 2, 3)$ as u_0.

(c) Explain the differences in the results in the preceding parts.

3. Let $A = \begin{bmatrix} 344 & 1029 \\ -98 & -293 \end{bmatrix}$ and $B = \begin{bmatrix} 9 & 21 \\ -2 & -4 \end{bmatrix}$.

*(a) Find the eigenvalues and eigenvectors of both A and B by solving the appropriate equations given in Chapter 8.

(b) Use the power method to find the dominant eigenvalue of A and an associated eigenvector. How many iterations are necessary?

(c) Use the power method to find the dominant eigenvalue of B and an associated eigenvector. How many iterations are necessary?

*(d) Can you explain the difference in the rates of convergence in parts (b) and (c)?

Appendix

Complex Numbers

There is no single system of numbers that is appropriate for all problems. The counting numbers, $1, 2, 3, \ldots$, are useful for counting distinct objects but cannot be used to measure ratios. Even though the rational numbers can be used to measure ratios, they are inadequate for the study of continuous phenomena. As we require more properties of our numbers, we are forced to enlarge the number system we use. In this appendix we describe an extension of the real numbers called the complex number system. We also indicate how complex numbers can be used in linear algebra.

A.1 COMPLEX NUMBERS

Two counting numbers can always be added or multiplied. In standard terminology the counting numbers are *closed* under the operations of addition and multiplication. As long as we restrict ourselves to these two operations, the counting numbers are sufficient.

If we are required to divide numbers, we must use the rational number system. The rational numbers are closed under four operations; addition, subtraction, multiplication, and division by a nonzero number. The rational numbers are an example of a very important mathematical structure; they form a *field*. In a field the four operations of addition, subtraction, multiplication, and division are defined and the usual rules of arithmetic hold. But these operations are not always sufficient.

Let us consider the sequence of rational numbers 1.4, 1.41, 1.414, 1.4142,..., where the nth number in the sequence, a_n, satisfies $a_n < 2$ and $[a_n - \sqrt{2}] < 10^{-n}$. The limit of this sequence is $\sqrt{2}$. However, $\sqrt{2}$ is not a rational number even though every number in the sequence is rational. If we want the limit of a sequence of numbers to be a number of the same kind, we must expand the rational number system. This expansion results in the real number system. The limit of a sequence of real numbers is always a real number.

The real number system has been used throughout this book. It is the number system that is used in Euclidean geometry. Since the real number system is closed under the limit operation, it can be used to study continuous phenomena. However, there is another useful operation which cannot always be performed within the real number system, the determination of roots of polynomials.

The problem of finding roots (or zeros) of polynomials arises frequently in mathematics. The solution of many mathematical problems, both pure and applied, can be reduced to finding the solutions of an equation of the form $P(x) = 0$, where $P(x) = a_n x^n + a_{n-1} x^{n-1} + \cdots + a_1 x + a_0$ $(a_n \neq 0)$ is a polynomial of degree n. There are polynomials with real coefficients that do not have roots within the field of real numbers. For example, $x^2 + 1 = 0$ and $x^2 + 2x + 2 = 0$ do not have any real roots because we cannot find a real number x such that $x^2 = -1$ or $(x + 1)^2 = -1$. If we want every polynomial of degree n to have exactly n roots, we must work in an extension of the real number system called the complex number system.

In order to describe the complex numbers we need to create a symbol for a special number—the square root of minus one. We define i to be the square root of minus one. Since $i = \sqrt{-1}$, we have $i^2 = -1$. Using this symbol, the complex numbers are all the numbers of the form $a + bi$, where a and b are real numbers. The number a is called the **real part** of $a + bi$ and b is called the **imaginary part**.

$$\text{Re}\,(a + bi) = a.$$
$$\text{Im}\,(a + bi) = b.$$

Two complex numbers are equal if and only if their real parts are equal and their imaginary parts are equal. In symbols, $a + bi = c + di$ if and only if $a = c$ and $b = d$. The real number a is the same as the complex number $a + 0i$.

Since numbers without arithmetic operations are useless, we need to define the rules for the normal arithmetic operations on the complex numbers. Addition, subtraction, and multiplication are performed by treating each complex number as a "polynomial in the symbol i" and doing the natural thing to these "polynomials." Using this simple procedure and replacing i^2 by -1 whenever it occurs, we have the following rules:

$$(a + bi) + (c + di) = (a + c) + (b + d)i$$
$$(a + bi) - (c + di) = (a - c) + (b - d)i$$
$$(a + bi) \cdot (c + di) = ac + (ad + bc)i + bdi^2$$
$$= (ac - bd) + (ad + bc)i.$$

Dividing $a + bi$ by $c + di$ involves finding a complex number $x + yi$ that satisfies the equation

$$(x + yi) \cdot (c + di) = a + bi.$$

This is the same as

$$(cx - dy) + (dx + cy)i = a + bi.$$

Since two complex numbers are equal if and only if their real and imaginary parts are equal, we have

$$cx - dy = a \qquad \text{(the real parts are equal)}$$
$$dx + cy = b \qquad \text{(the imaginary parts are equal)}.$$

This system of equations will have a unique solution unless c and d are both 0 (see Problem 9). Thus we can divide $a + bi$ by any nonzero complex number.

In practice there is a much easier method for computing the quotient of two complex numbers which does not require solving a system of equations. The procedure used is the following:

$$\frac{a + bi}{c + di} = \frac{(a + bi)(c - di)}{(c + di)(c - di)}$$

$$= \frac{(ac + bd) + (bc - ad)i}{c^2 + d^2}$$

$$= \frac{ac + bd}{c^2 + d^2} + \frac{bc - ad}{c^2 + d^2}i.$$

The complex number, $c - di$, which we used above to make the denominator a real number is called the **complex conjugate** of $c + di$. If $z = a + bi$ is a complex number, its complex conjugate $a - bi$ is denoted by $\bar{z}$. Since $z\bar{z} = a^2 + b^2$, the product $z\bar{z}$ of a complex number and its conjugate is always a nonnegative real number and is zero if and only if $z = 0$.

There is a nice geometric way to view the complex numbers. Every ordered pair (a, b) corresponds to a single complex number $a + bi$ and every complex number $c + di$ corresponds to a single ordered pair (c, d). We thus have a one-to-one correspondence between the ordered pairs of real numbers and the complex numbers. Using this correspondence, we can associate a point in the plane with each complex number (see Figure A.1). In terms of this geometric description, $\bar{z}$ is the reflection of z through the

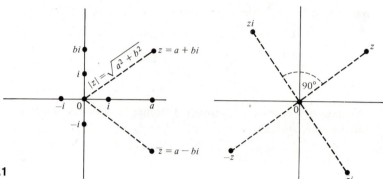

Figure A.1

x-axis and $\sqrt{z\bar{z}}$ is the distance of z from the origin (which we denote by $|z|$). Multiplying by i moves a complex number 90° counterclockwise around 0. Multiplying by -1 rotates it by 180°.

The complex numbers do not have the order properties which the real numbers have. If the complex numbers were ordered in the manner that the reals are ordered, then either $i > 0$ or $i < 0$. But in either case $i^2 (= -1) > 0$, which is impossible. This shows that the complex numbers cannot be linearly ordered.

PROBLEMS A.1

*1. Perform the indicated operations.

(a) $(3 + 4i) \cdot (2 - 4i)$ (b) $(1 + 3i) + (2 - 6i)$
(c) $(1 - 4i)/(3 - 4i)$ (d) $(1 + i) \cdot (1 - i)$
(e) $(1 + i) \cdot (1 + i)$ (f) $(1 + i)^n/(1 - i)^n$

*2. Find the indicated quantities.

(a) $\operatorname{Re}(-3 + 4i)$ (b) $\operatorname{Im}(-3 + 4i)$
(c) $|-3 + 4i|$ (d) $|i|$
(e) $\operatorname{Re}(-3 - 4i)$ (f) $\operatorname{Im}(-3 - 4i)$
(g) $|-3 - 4i|$ (h) $|-i|$

3. *(a) Find two square roots of i by solving $(x + yi)^2 = i$ for x and y.
　　*(b) Find two square roots of $-i$ by solving $(x + yi)^2 = -i$ for x and y.
　　(c) The answers in parts (a) and (b) are the four fourth roots of -1. Plot these four roots in the complex plane.

4. The quadratic formula gives the roots of $x^2 + x + 1$ as

$$w_1 = -\frac{1}{2} + \frac{\sqrt{3}}{2}i \quad \text{and} \quad w_2 = -\frac{1}{2} - \frac{\sqrt{3}}{2}i.$$

Compute w_1^2, w_1^3, w_2^2, w_2^3, and $w_1 w_2$.

*5. (a) Find a formula for the reciprocal of a complex number $z = a + bi$.

(b) Is the reciprocal of the conjugate equal to the conjugate of the reciprocal? In symbols, is the formula

$$\frac{1}{\bar{z}} = \overline{\left(\frac{1}{z}\right)}$$

true?

6. Verify that the following rules are true.

(a) $\overline{z_1 + z_2} = \bar{z}_1 + \bar{z}_2$ (b) $\overline{z_1 z_2} = \bar{z}_1 \cdot \bar{z}_2$

(c) $\overline{(z_1/z_2)} = \bar{z}_1/\bar{z}_2$ (d) $|z_1 z_2| = |z_1||z_2|$

(e) $|\bar{z}| = |z|$

7. (a) Compute i^3, i^4, i^5, i^6, and i^7.

   (b) Give a formula for i^n, where n is a positive integer.

8. Show that the polynomial $x^4 + 4$ has $1 + i$, $1 - i$, $-1 + i$, and $-1 - i$ as roots.

9. Show that the system

$$cx - dy = a$$
$$dx + cy = b$$

has a unique solution unless $c = d = 0$.

10. (a) Solve $\bar{z} = z^2$.   (b) Solve $\bar{z} = z^3$.

11. Show that $|z_1 + z_2|^2 + |z_1 - z_2|^2 = 2(|z_1|^2 + |z_2|^2)$. Interpret this identity geometrically.

12. (a) Show that if z is a complex number, there is an angle θ such that

$$z = r(\cos \theta + i \sin \theta),$$

where $r = |z|$. (This is called the polar representation of z. $\theta = \arg z$ is called an argument of z.) Interpret this representation geometrically.

(b) Show that a complex number can have more than one argument. In addition, show that any two arguments of a nonzero complex number differ by the quantity $2n\pi$, where n in an integer.

(c) If $z = r(\cos \theta + i \sin \theta)$, show that

$$z^2 = r^2(\cos 2\theta + i \sin 2\theta).$$

(d) Show that $\arg z_1 + \arg z_2$ is an argument of $z_1 z_2$. Interpret this result geometrically.

(e) Show that $\arg z_1 - \arg z_2$ is an argument of z_1/z_2.

A.2   SYSTEMS OF EQUATIONS AND THE VECTOR SPACE C^m

In Chapter 1 we discussed systems of equations where all numbers involved were real. If we allow complex numbers to appear in systems of equations, then all the procedures and results of Chapter 1 are still valid. The Gaussian elimination procedure requires that we be able to add, subtract, multiply, and divide the numbers that appear. As we have just seen in Section 1, these operations can still be performed on complex numbers. In addition, all properties of matrices are still valid for matrices with complex entries.

EXAMPLE 1 Let us use elimination to solve the following system.

$$ix_1 + \qquad x_2 = 1$$
$$(1 - i)x_1 + (2 - i)x_2 = 5 - 4i$$

Multiplying the first equation by $-i$ gives us the new first equation $x_1 - ix_2 = -i$. If this is multiplied by $(1 - i)$ and the result subtracted from the old second equation, our system becomes

$$x_1 - ix_2 = 1$$
$$3x_2 = 6 - 3i.$$

The solution, obtained by back substitution, is $x_1 = 1 + i$ and $x_2 = 2 - i$.

Let C^m denote the collection of all m-tuples of complex numbers. Since C^m is closed under the operations of addition and scalar multiplication, it is a vector space. All the results and procedures of Chapter 2 are true for C^m.

EXAMPLE 2 The three vectors $(1, 2, 1 - i)$, $(1, i, 1 + i)$, and $(1, 1 + i, i)$ are linearly independent. We can show this by row-reducing

$$\begin{bmatrix} 1 & 1 & 1 \\ 1 & i & 1 + i \\ 1 & 1 + i & i \end{bmatrix} \quad \text{to} \quad \begin{bmatrix} 1 & 1 & 1 \\ 0 & 1 & (3 - i)/5 \\ 0 & 0 & 1 \end{bmatrix}.$$

The row-reduced matrix is obviously of rank 3.

EXAMPLE 3 The inverses of nonsingular matrices with complex entries can be computed by the method of Section 1.7.

$$\begin{bmatrix} i & 1 & \vdots & 1 & 0 \\ 1 + i & 0 & \vdots & 0 & 1 \end{bmatrix} \quad \text{becomes} \quad \begin{bmatrix} 1 & -i & \vdots & -i & 0 \\ 0 & 1 + i & \vdots & 1 + i & 1 \end{bmatrix},$$

which reduces to

$$\begin{bmatrix} 1 & 0 & \vdots & 0 & (1 + i)/2 \\ 0 & 1 & \vdots & 1 & (1 - i)/2 \end{bmatrix}.$$

This shows that the inverse of

$$\begin{bmatrix} i & 1 \\ 1 + i & 0 \end{bmatrix} \quad \text{is} \quad \begin{bmatrix} 0 & (1 + i)/2 \\ 1 & (1 - i)/2 \end{bmatrix}.$$

PROBLEMS A.2

*1. Show that the vectors $(i, 1)$ and $(1, -i)$ are linearly dependent.

*2. What is the dimension of the vector space spanned by the vectors $(1, 1 + i, -i)$, $(i, -1 + i, 1)$, and $(1 - i, 2, -1 - i)$?

*3. Solve the following systems.

(a) $(1 + i)x_1 + (1 - i)x_2 = 1 - i$
$\quad\quad 2x_1 + \quad\quad ix_2 = 4i$

(b) $(2 + i)x_1 + (3 - i)x_2 = 15$
$\quad\quad x_1 + \quad\quad x_2 = 5$

(c) $(1 + i)x_1 + (2 + i)x_2 = -1 + i$
$\quad\quad ix_1 - (1 + i)x_2 = 2 + 5i$

(d) $\quad\quad x_1 + \quad\quad ix_3 = 0$
$\quad\quad ix_1 + x_2 + (-1 + i)x_3 = -1 + 2i$
$\quad (1 + i)x_1 + x_2 + \quad\quad 2ix_3 = 3i$

*4. Find the inverses of the following matrices.

(a) $\begin{bmatrix} i & 1 \\ 1 - i & 2 - i \end{bmatrix}$

(b) $\begin{bmatrix} 1 & 1 \\ 2 + i & 3 - i \end{bmatrix}$

(c) $\begin{bmatrix} 1 & 0 & i \\ i & 1 & -1 + i \\ 1 + i & 1 & 2i \end{bmatrix}$

(d) $\begin{bmatrix} -i & 1 & 0 \\ 1 + i & 1 & 0 \\ 2i & -1 & 1 \end{bmatrix}$

A.3   THE INNER PRODUCT IN C^m

In Chapter 3 we defined an inner product on R^m which gave R^m the normal geometry of Euclidean m-space. The following definition gives an inner product for C^m.

Definition   If $x = (x_1, \ldots, x_m)$ and $y = (y_1, \ldots, y_m)$ are two vectors in C^m, then the **inner product** of x and y is given by

$$\langle x, y \rangle = x_1\bar{y}_1 + x_2\bar{y}_2 + \cdots + x_m\bar{y}_m.$$

Note that this definition reduces to the definition in Chapter 3 when all components of x and y are real. In Problems 1 and 2 the student is asked to show that this inner product has essentially the same properties as the inner product on R^m and that these properties reduce to the properties given in Theorem 3.1 when all numbers involved are real.

It should come as no surprise that the inner product on C^m gives C^m a geometric structure. Since the inner product $\langle x, x \rangle$ is a nonnegative *real* number we can define a norm on C^m by $\|x\| = \sqrt{\langle x, x \rangle}$. The angle between two nonzero vectors x and y in C^m is given by the formula

$$\cos \theta = \frac{\langle x, y \rangle}{\|x\| \cdot \|y\|}$$

and x and y are orthogonal if and only if $\langle x, y \rangle = 0$. However, a complete interpretation of the inner product on C^m requires an understanding of the complex-valued function $w = \cos z$, where z is a complex variable. This function is studied in complex analysis.

PROBLEMS A.3

1. Let x, y, and z be vectors in C^m and let α be a complex number. Verify the following properties. (Compare with Theorem 3.1.)
   (a) $\langle x, y \rangle = \overline{\langle y, x \rangle}$.
   (b) $\langle x, y + z \rangle = \langle x, y \rangle + \langle x, z \rangle$.
   (c) $\langle \alpha x, y \rangle = \alpha \langle x, y \rangle$.
   (d) For all x, $\langle x, x \rangle$ is a nonnegative *real* number, and $\langle x, x \rangle = 0$ if and only if $x = 0$.

2. (a) Show that the properties given in Problem 1 reduce to the properties given in Theorem 3.1 in the event that all numbers involved are real.
   (b) Show that $\langle x, \alpha y \rangle = \bar{\alpha} \langle x, y \rangle$.
   (c) Why is $\langle x, y \rangle + \langle y, x \rangle = 2 \operatorname{Re}(\langle x, y \rangle)$?

3. (a) Show that the Chauchy–Schwarz Inequality holds in C^m. (*Hint:* Use the proof in Section 3.3 with $\alpha = \langle x, y \rangle / \|y\|^2$.)
   (b) Show that $|\langle x, y \rangle + \langle y, x \rangle| \le 2\|x\| \cdot \|y\|$.
   (c) Show that the triangle inequality holds in C^m.

4. Show that $\|\alpha x\| = |\alpha| \cdot \|x\|$.

5. Let $A = [a_{ij}]$ be an $m \times n$ matrix with complex entries. We define $\bar{A} = [\bar{a}_{ij}]$ to be the **conjugate** of A and $A^* = \bar{A}^T$ to be the **conjugate transpose** of A. Note that these definitions apply to vectors in C^m by letting $n = 1$. Verify the following properties.
   (a) If A has real entries, $A^* = A^T$.
   (b) $\bar{A} \cdot \bar{x} = \overline{Ax}$.
   (c) $A + \bar{A}$ is a real matrix.
   (d) $(AB)^* = B^* A^*$.
   (e) $\overline{\langle x, y \rangle} = x^* y$.
   (f) $\|x\|^2 = x^* x$.
   (g) $\langle Ax, y \rangle = \langle x, A^* y \rangle$.

6. An $m \times m$ matrix A is called **Hermitian** if $A^* = A$.
   (a) Show that a real Hermitian matrix is symmetric.
   (b) Show that $\langle Ax, y \rangle = \langle x, Ay \rangle$ for all x, y in C^m if and only if A is Hermitian.

7. An $m \times m$ matrix U is called **unitary** if its columns form an orthonormal set of vectors in C^m.
   (a) Show that $U^* U = I = UU^*$.
   (b) What properties given in Problem 17 of Section 3.2 remain true for unitary matrices?

A.4   EIGENVALUES AND EIGENVECTORS

An eigenvalue of an $m \times m$ complex matrix A is a root of the polynomial $\det(A - \lambda I)$. This polynomial has exactly m complex roots (counting multiplicities). If λ is any root of this polynomial, we can always find a nonzero (eigen)vector v in C^m such that $Av = \lambda v$.

In Chapter 8 we considered real matrices all of whose eigenvalues were real numbers. In general, the polynomial $\det(A - \lambda I)$ may have nonreal (complex) roots even when A is a real matrix. It is impossible to find an eigenvector in R^m for such nonreal eigenvalues. For example, the eigenvalues of the matrix $\begin{bmatrix} 0 & -1 \\ 1 & 0 \end{bmatrix}$ are i and $-i$. The eigenvectors associated with the eigenvalue i are the nonzero multiples of $(i, 1)$ and the eigenvectors for $-i$ are the nonzero multiples of $(-i, 1)$. These eigenvectors are not in R^2; they are in C^2.

The following example shows that the nonreal complex eigenvalues and complex eigenvectors of a real matrix are important.

EXAMPLE 1   Consider a fish population that has been divided into three age groups. Suppose that the Leslie matrix (see Section 1.8 and Example 3 in Section 8.3) for this population is

$$A = \begin{bmatrix} 0 & 60 & 100 \\ \frac{1}{64} & 0 & 0 \\ 0 & \frac{1}{2} & 0 \end{bmatrix}.$$

Then the eigenvalues of A are the roots of the equation $-\lambda^3 + (60/64)\lambda + (50/64) = 0$, namely $\lambda_1 = 5/4$, $\lambda_2 = (-5 + i\sqrt{15})/8$, and $\lambda_3 = \bar{\lambda}_2$. (Note that the two nonreal complex eigenvalues are complex conjugates of each other. This is always the case when the matrix A is a real matrix. In this case the characteristic polynomial is a polynomial with real coefficients. If λ is a root of a polynomial with real coefficients then $\bar{\lambda}$ is also a root.) Since these eigenvalues are distinct, we can find three linearly independent eigenvectors in C^3. An eigenvector for λ_1 is $v_1 = (400, 5, 2)$. The eigenvector v_2 for λ_2 and the eigenvector v_3 for λ_3 can be chosen so that $v_3 = \bar{v}_2$, because $A\bar{v}_2 = \overline{Av_2} = \overline{Av_2} = \overline{\lambda_2 v_2} = \bar{\lambda}_2 \bar{v}_2$.

Using the notation of Example 3 in Section 8.3, we have $x_k = A^k x_0 = (P \Lambda P^{-1})^k x_0 = P \Lambda^k P^{-1} x_0$, where

$$P = \begin{bmatrix} | & | & | \\ v_1 & v_2 & \bar{v}_2 \\ | & | & | \end{bmatrix} \quad \text{and} \quad \Lambda = \begin{bmatrix} \lambda_1 & & \\ & \lambda_2 & \\ & & \bar{\lambda}_2 \end{bmatrix}.$$

Letting

$$P^{-1} x_0 = \begin{bmatrix} \alpha \\ \beta \\ \gamma \end{bmatrix},$$

we have $x_k = (5/4)^k \alpha v_1 + (\lambda_2)^k \beta v_2 + (\bar{\lambda}_2)^k \gamma \bar{v}_2$. We have expressed the (real) vector x_k as a linear combination of the (complex) eigenvectors.

This last expression enables us to show how the population grows. Consider $\|(\lambda_2)^k \beta v_2\| = |\lambda_2|^k |\beta| \cdot \|v_2\|$. Since $|\lambda_2| < 1$, the number $|\lambda_2|^k$

is very close to 0 when k is large and we can conclude that the vector $(\underline{\lambda}_2)^k \beta v_2$ is very small (i.e., has a very small norm). Similarly, the vector $(\bar{\lambda}_2)^k \gamma \bar{v}_2$ has a small norm when k is large. This says that when k is large, x_k is approximately equal to $(5/4)^k \alpha v_1$. Since α and v_1 are fixed, we can conclude that this population will eventually increase at about 25% per year and that the age distribution will tend toward the distribution given by $v_1 = (400, 5, 2)$.

Similar arguments can be used to investigate the behavior for large k of the solution $x_k = A^k x_0$ of the difference equation $x_k = A x_{k-1}$. When the real matrix A is diagonalizable over the complex numbers and has a real eigenvalue λ whose absolute value is larger than the absolute value of every other eigenvalue, then for large k, the vector $x_k = A^k x_0$ is approximately $\lambda^k \alpha v$ for some real number α, where v is an eigenvector for λ. In the case of Leslie matrices when there is a real eigenvalue λ which is larger in absolute value than all other eigenvalues, then the distribution of the population will usually tend toward the distribution given by an eigenvector of λ.

PROBLEMS A.4

1. The matrix

$$A = \begin{bmatrix} 0 & 0 & 110 \\ 0.05 & 0 & 0 \\ 0 & 0.2 & 0 \end{bmatrix}$$

was introduced in Problem 4 of Section 1.8 as the Leslie matrix of a certain population.
(a) Show that A has three distinct eigenvalues, only one of which is real.
(b) Show that the absolute values of the three eigenvalues are equal.
(c) Discuss the distribution of this population over extended periods of time.

2. (a) Show that the real matrix $A = \begin{bmatrix} 0 & -2 \\ 8 & 0 \end{bmatrix}$ cannot be diagonalized by a real matrix.
(b) Show that A has two complex eigenvalues and find complex eigenvectors associated with the eigenvalues.
(c) Show that A can be diagonalized by a complex matrix.

*3. Find the (complex) eigenvalues and the associated complex eigenvectors for the following complex matrices.
(a) $\begin{bmatrix} 1 & 1-i \\ 1 & -2i \end{bmatrix}$ (b) $\begin{bmatrix} 2-7i & 10+2i \\ 5+i & -1+8i \end{bmatrix}$

4. Let U be a unitary matrix.
(a) If λ is an eigenvalue of U, show that $|\lambda| = 1$. (*Hint:* Consider $\langle Ux, Ux \rangle$ where x is an eigenvector for λ.)
(b) Show that eigenvectors of U belonging to distinct eigenvalues are orthogonal.
(c) The determinant satisfies $\det \bar{A} = \overline{\det A}$. Show that $|\det U| = 1$.

5. (a) Show that all eigenvalues of a Hermitian matrix A are real. (*Hint:* Consider $\langle Ax, x \rangle$ and $\langle x, Ax \rangle$, where x is an eigenvector of A.)
   (b) Show that eigenvectors of A associated with distinct eigenvalues are orthogonal.

6. Hermitian matrices extend the concept of symmetric matrices to the complex situation. In particular, any Hermitian matrix is diagonalizable. Investigate how the results of Section 8.5 can be extended to Hermitian matrices.

Solutions

CHAPTER 1

Section 1.1

EXERCISES

2. $a_{31} = 1$, $a_{21} = 5$, $a_{12} = 2$, $a_{24} = -2$, $b_3 = 2$, $m = 3$, $n = 4$.

3. (b).   4. Any value except $b = 8$.

PROBLEMS 1.1

1. (a), (b), (f).   3. (b), (c), (d).

5. (a), (b), (e).   8. Any value except $b = 3$.

10. For example, $3x_1 + 2x_3 = 5$ or $2x_1 + x_2 - x_3 = 7$.

Section 1.2

PROBLEMS 1.2

1. (a) $2t + 3u = 37$
       $10t + u = 10u + t + 9$.

   (b) $x_1 + x_2 + x_3 = 51$
       $x_1 = 2x_2 + 5$
       $x_2 = 3x_3 + 2$.

(d) $3A - 5B = 16$ (f) $\frac{5}{8}A + \frac{2}{8}B + \frac{3}{8}C = 3$
$\quad\ 6B - 2A = 24.$ $\quad\ \frac{2}{8}A + \frac{5}{8}B + \frac{1}{8}C = 3$
$\qquad\qquad\qquad\qquad \frac{1}{8}A + \frac{1}{8}B + \frac{4}{8}C = 3.$

2. (a) $\quad I_1 - I_2 - I_3 - I_4 = 0$
$\qquad\ -I_1 + I_2 + I_3 + I_4 = 0.$

(c) $\quad I_1 - I_2 - I_3 \qquad\qquad\qquad = 0$ (e) $\quad I_1 - I_2 - I_3 \qquad\qquad\qquad = 0$
$\qquad\qquad I_3 + I_4 \qquad\ - I_6 = 0$ $\qquad\qquad\quad I_3 \qquad\ + I_5 - I_6 = 0$
$\quad -I_1 \qquad\qquad\qquad + I_5 + I_6 = 0$ $\qquad\qquad\qquad\quad -I_4 - I_5 + I_6 = 0$
$\qquad\qquad I_2 \qquad -I_4 - I_5 \qquad = 0.$ $\qquad -I_1 + I_2 \qquad + I_4 \qquad\qquad = 0.$

3. Part (a)

(a) $I_1 + I_2 \qquad\qquad\qquad -1 = 0$ (c) $I_1 + I_2 \qquad\qquad +I_5 \qquad -1 = 0$
$\quad I_1 \qquad + I_3 \qquad\qquad -1 = 0$ $\qquad I_1 + I_2 \qquad + I_4 \quad + I_6 - 1 = 0$
$\quad I_1 \qquad\qquad + I_4 - 1 = 0$ $\qquad I_1 \qquad + I_3 \qquad + I_6 - 1 = 0$
$\qquad I_2 - I_3 \qquad\qquad = 0$ $\qquad I_2 - I_3 + I_4 \qquad\qquad = 0$
$\qquad I_2 \qquad - I_4 \qquad = 0$ $\qquad I_2 - I_3 \qquad + I_5 - I_6 = 0$
$\qquad\qquad I_3 - I_4 \quad = 0.$ $\qquad\qquad\qquad I_4 - I_5 + I_6 = 0.$

Part (b)

(a) $I_1 + 2I_2 \qquad\qquad\qquad -1 = 0$ (c) $I_1 + 2I_2 \qquad\qquad +5I_5 \qquad -1 = 0$
$\quad I_1 \qquad + 3I_3 \qquad\qquad -1 = 0$ $\qquad I_1 + 2I_2 \qquad + 4I_4 \quad + 6I_6 - 1 = 0$
$\quad I_1 \qquad\qquad + 4I_4 - 1 = 0$ $\qquad I_1 \qquad + 3I_3 \qquad + 6I_6 - 1 = 0$
$\qquad 2I_2 - 3I_3 \qquad\qquad = 0$ $\qquad 2I_2 - 3I_3 + 4I_4 \qquad\qquad = 0$
$\qquad 2I_2 \qquad - 4I_4 \quad . \quad = 0$ $\qquad 2I_2 - 3I_3 \qquad + 5I_5 - 6I_6 = 0$
$\qquad\qquad 3I_3 - 4I_4 \quad = 0.$ $\qquad\qquad\qquad 4I_4 - 5I_5 + 6I_6 = 0.$

4. (a) $\frac{1}{2}, \frac{1}{3}, \frac{1}{6}.$

6. $\frac{1}{3}p_1 + \frac{1}{5}p_2 + \frac{1}{6}p_3 + \frac{1}{8}p_4 = p_1$
$\quad \frac{1}{6}p_1 + \frac{2}{5}p_2 + \frac{1}{6}p_3 + \frac{1}{4}p_4 = p_2$
$\quad \frac{1}{3}p_1 + \frac{1}{5}p_2 + \frac{1}{2}p_3 + \frac{1}{8}p_4 = p_3$
$\quad \frac{1}{6}p_1 + \frac{1}{5}p_2 + \frac{1}{6}p_3 + \frac{1}{2}p_4 = p_4.$

10. $a_{1j} + a_{2j} + a_{3j} + \cdots + a_{nj} = 1.$

12. $\qquad\qquad\qquad c = -2$ **13.** $\qquad m + b = -1$
$\qquad a + b + c = 6$ $\qquad\qquad 5m + b = 7$
$\qquad 4a - 2b + c = 0$ $\qquad\qquad -2m + b = -7$
$\qquad 16a + 4b + c = 66.$ $\qquad\qquad\qquad b = -3.$

Section 1.3

EXERCISES

1. $x_1 = 2, x_2 = 0, x_3 = 1, x_4 = 1.$

3. $x_1 = 11 + 2x_4 - 9x_5, x_2 = -1 - x_4 + 3x_5,$
$\quad x_3 = 3 + x_4 - 2x_5.$

5. (a) $3, 2, 1$. (b) x_2. (c) x_1, x_3, x_4.   **7.** $3x_1 + 2x_2 = 17$
(d) $x_1 = 1 - \frac{2}{3}x_2$, x_2 free, $x_3 = 4$, $x_4 = 3$. $x_2 = -2$.

8. (a) $2x_1 + 2x_2 - x_3 = 4$ (b) $x_1 + x_2 - \frac{1}{2}x_3 = 2$
            $x_2 - 2x_3 = 2$ $x_2 - 2x_3 = 2$
                 $3x_3 = 6$. $x_3 = 2$.

PROBLEMS 1.3

2. (b), (e), (f), (g), (h).

3. (a) $x_1 = 2$, $x_2 = 2$.   (d) $x_1 = 6 - 2x_2$, x_2 free, $x_3 = 1$.
(f) $x_1 = 0$, $x_2 = 3$, $x_3 = 1$.
(h) $x_1 = 2 - \frac{3}{2}x_2 + \frac{1}{2}x_5$, x_2 free, $x_3 = 4 - x_5$, $x_4 = 2 + x_5$, x_5 free.
(l) $x_1 = -\frac{1}{2} - \frac{1}{2}x_4 + \frac{1}{2}x_6$, $x_2 = 4 - x_6$, $x_3 = -1$, x_4 free, $x_5 = 4 - x_6$, x_6
    free.
(m) $x_1 = 17 - 2x_2 - 3x_3 - 4x_4 - 5x_5 - 6x_6$, x_2, x_3, x_4, x_5, x_6 free.

Section 1.4

EXERCISES

1. $3x_1 + x_2 + 2x_3 + x_4 + x_5 = 6$
              $x_3 - x_4 + x_5 = 0$
                  $2x_4 - 5x_5 = 2$.

2. $x_1 + 2x_2 + x_3 = 3$ Pivots: 1, 1, and -2.
        $x_2 + 2x_3 = 4$
           $-2x_3 = 1$.

5. $x_1 = -\frac{7}{5} - \frac{9}{5}x_4$, $x_2 = -\frac{17}{5} + \frac{1}{5}x_4$, $x_3 = \frac{9}{5} + \frac{3}{5}x_4$, x_4 free.

PROBLEMS 1.4

1. (a) $2x_1 + 3x_2 = 6$ (d) $x_1 + x_2 + x_3 + x_4 = 1$
           $x_2 = 2$. $x_2 + 2x_3 + 3x_4 = 1$
                           $x_3 + x_4 = 1$
                               $x_4 = 1$.

2. (a) Inconsistent. (c) $x_1 = -4$, $x_2 = -2$, $x_3 = 2$.
(e) $x_1 = 19 + 2x_4$, $x_2 = -2 - 2x_4$, $x_3 = 10 + x_4$, x_4 free.
(g) $x_1 = -\frac{8}{5}$, $x_2 = 0$, $x_3 = -\frac{6}{5}$.
(i) $x_1 = \frac{7}{4} - \frac{1}{2}x_3 + \frac{1}{2}x_4$, $x_2 = -\frac{1}{2}$, x_3, x_4 free.
(k) $x_1 = \frac{3}{19} - \frac{11}{19}x_3$, $x_2 = -\frac{4}{19} + \frac{2}{19}x_3$, x_3 free.
(m) $x_1 = \frac{7}{13}$, $x_2 = \frac{4}{13}$, $x_3 = \frac{2}{13}$.
(o) $x_1 = -2x_2 + 4$, x_2 free, $x_3 = 0$, $x_4 = -1$.
(q) $x_1 = -\frac{1}{3}x_3 + \frac{2}{3}x_5 - x_7$, $x_2, x_3, x_4, x_5, x_6, x_7$ free.

4. (a) $x_1 = 3$, $x_2 = -1$. (d) $x_1 = -1$, $x_2 = -1$, $x_3 = 0$.
(g) $x_1 = 1$, $x_2 = -1$, $x_3 = -2$, $x_4 = 3$.
(i) $x_1 = -1$, $x_2 = 2$, $x_3 = 0$, $x_4 = 1$, $x_5 = -1$, $x_6 = -2$.

Section 1.5

EXERCISES

3. $A - B = \begin{bmatrix} 2 & -4 & 2 \\ 4 & -2 & 0 \end{bmatrix}$, $4A + 3B = \begin{bmatrix} 22 & 12 & 8 \\ 23 & -1 & 21 \end{bmatrix}$,

$B + 2A = \begin{bmatrix} 10 & 4 & 4 \\ 11 & -1 & 9 \end{bmatrix}$.

4. $\begin{bmatrix} -39 \\ 54 \\ 45 \end{bmatrix}$.   **5.** $\begin{bmatrix} 2 & 3 & -1 \\ 1 & 0 & 1 \\ 4 & -3 & 1 \end{bmatrix} \begin{bmatrix} x_1 \\ x_2 \\ x_3 \end{bmatrix} = \begin{bmatrix} 1 \\ 2 \\ 7 \end{bmatrix}$.

6. $3x_1 + x_2 + 4x_3 + 2x_4 = 2$ **7.** $Ax = \begin{bmatrix} 5 \\ 7 \end{bmatrix} = \begin{bmatrix} 1 \\ -1 \end{bmatrix} + 2\begin{bmatrix} 2 \\ 4 \end{bmatrix}$.
$5x_1 - x_2 + 2x_3 + x_4 = 1$
$8x_1 + x_2 + 2x_3 + x_4 = -2$.

10. (a) $\begin{bmatrix} 3 \\ -2 \\ -10 \end{bmatrix}$.

PROBLEMS 1.5

2. $u - v = (4, -2, -5)$, $-3u + 5v = (-18, 6, 33)$,
$u + v - 6w = (-2, -2, -17)$.

4. (a) $(\frac{4}{3}, \frac{7}{3}, -\frac{1}{3}, \frac{16}{3})$. (b) $(-6, -\frac{11}{2}, \frac{5}{2}, -16)$.

6. $A + B = \begin{bmatrix} 5 & 1 & 0 \\ 0 & 3 & 4 \\ -1 & 2 & 10 \end{bmatrix}$, $-2A = \begin{bmatrix} 0 & 0 & 0 \\ 2 & -4 & -6 \\ 0 & -2 & -12 \end{bmatrix}$,

$3A - B = \begin{bmatrix} -5 & -1 & 0 \\ -4 & 5 & 8 \\ 1 & 2 & 14 \end{bmatrix}$.

7. $Au = (0, 18)$, $Av = (-5, 31)$, $Bu = (2, 8)$,
$B(u + v) = (-6, 10)$, $A(3v) = (-15, 93)$.

Section 1.6

EXERCISES

1. $x_1 = -4$, $x_2 = 2$, $x_3 = -1$.

2. $\begin{bmatrix} 2 & 3 & 4 & 1 \\ 0 & 0 & -6 & 6 \\ 0 & 0 & 0 & 9 \end{bmatrix}$.   The pivots are $2, -6, 9$.

PROBLEMS 1.6

1. (a), (c), (d), (f).   **2.** (c), (f).

Section 2.4

PROBLEMS 2.4

1. (a) and (d) are bases for R^2. 2. (a) and (d) are bases for R^2.

3. (a) $(2,3)$. (b) $(1,2), (0,-5)$. (c) $(1,0,2), (0,1,-6)$.

Section 2.5

EXERCISES

1. (a) $(1,1,-1), (0,-1,2), (0,0,5)$.
   (b) $(2,-3,1)$.
2. (a) $(1,2,0), (0,1,5), (-1,3,-2)$.
   (b) $(0,1,1), (3,-1,2)$.
   (c) $(1,1,1), (0,-1,1), (0,2,1)$.

3. (a) $(-9,3,1,0), (-6,1,0,1)$. (b) $(-1,1,0)$.

PROBLEMS 2.5

1. (a) Rank $= 2$; $(3,0,0), (2,1,0)$ is a basis for the column space; $(3,-1,2)$, $(0,0,1)$ is a basis for the row space.
   (b) Rank $= 3$; $(1,0,0), (2,3,0), (3,8,2)$ is a basis for the column space; $(1,2,0,9,3), (0,3,1,-5,8), (0,0,0,0,2)$ is a basis for the row space.
   (c) Rank $= 2$; $(6,0,0,0), (3,2,0,0)$ is a basis for the column space; $(6,3,1,3,1,0), (0,0,0,2,0,2)$ is a basis for the row space.

3. (a) Rank $= 2$; $(2,4),(-1,5)$ is a basis for the column space; $(2,-1,5), (0,7,-2)$ is a basis for the row space.
   (b) Rank $= 2$; $(1,0,3,-1), (6,2,5,1)$ is a basis for the column space; $(1,6,4)$, $(0,2,2)$ is a basis for the row space.
   (c) Rank $= 3$; $(1,1,2,2), (-1,-1,-2,0), (2,2,5,-1)$ is a basis for the column space; $(1,2,-1,1,2), (0,0,2,-1,-5), (0,0,0,0,1)$ is a basis for the row space.

4. (a) Nullity $= 0$. (b) Nullity $= 2$; $(-1,-\frac{3}{4},1,0), (0,\frac{1}{2},0,1)$.
   (c) Nullity $= 2$; $(-\frac{4}{3},1,0), (\frac{7}{3},0,1)$. (d) Nullity $= 0$.
   (e) Nullity $= 1$; $(-1,-1,1,1)$.

6. (a) Dimension $= 2$; $(1,-1,1), (3,5,1)$.
   (b) Dimension $= 3$; $(1,1,1), (0,3,1), (1,0,5)$.
   (c) Dimension $= 2$; $(1,-1,2,0), (0,1,1,1)$.

9. $(1,1,2), (1,0,0), (0,1,0)$. 11. $(1,1,0,0), (1,2,3,1), (1,0,0,0), (0,0,1,0)$.

13. $p = p_3(\frac{1}{2},0,1,0) + p_4(0,1,0,1)$.

Section 2.6

EXERCISES

1. (a) The straight line in R^3 that passes through $(1,1,0)$ and is parallel to the line spanned by $(-\frac{1}{3},\frac{1}{3},1)$.
   (b) The point $(-\frac{5}{4},\frac{9}{4})$.

---

4. (a) $x = (-8,-7,0) + x_3(5,3,1)$.
   (c) Inconsistent.
   (e) $x = (0,0,1,-1,1) + x_2(-1,1,0,0,0)$.

6. (a) $\begin{bmatrix} 1 & 0 & 0 & \vdots & -1 \\ 0 & 1 & 0 & \vdots & 3 \\ 0 & 0 & 1 & \vdots & 0 \end{bmatrix}$.   (b) $\begin{bmatrix} 1 & 2 & 0 & \vdots & 2 \\ 0 & 0 & 1 & \vdots & 4 \end{bmatrix}$.

   (c) $\begin{bmatrix} 1 & 1 & 0 & -3 & \vdots & 1 \\ 0 & 0 & 1 & 1 & \vdots & 3 \end{bmatrix}$.

10. $x = (\frac{5}{9}, -\frac{2}{9}, \frac{3}{9})$,
    $x = (-\frac{4}{9}, -\frac{11}{9}, \frac{12}{9})$,
    $x = (\frac{5}{3}, -\frac{2}{3}, 1)$,
    $x = (-\frac{1}{3}, -\frac{5}{3}, 1)$,
    $x = (\frac{5}{9}, \frac{7}{9}, \frac{12}{9})$.

Section 1.7

EXERCISES

1. $Ab_1 = \begin{bmatrix} 8 \\ 7 \end{bmatrix}$, $Ab_2 = \begin{bmatrix} 12 \\ 5 \end{bmatrix}$, $Ab_3 = \begin{bmatrix} -2 \\ 1 \end{bmatrix}$, $AB = \begin{bmatrix} 8 & 12 & -2 \\ 7 & 5 & 1 \end{bmatrix}$.

2. $AB = \begin{bmatrix} -5 & 10 \\ -11 & 18 \\ -1 & 8 \end{bmatrix}$.   4. $\begin{bmatrix} 2 & 3 & 4 \\ 5 & 6 & -2 \end{bmatrix}$.

5. $k \times m$ and $k \times n$, where k is arbitrary.

7. $\begin{bmatrix} -4 & 3 \\ 3 & -2 \end{bmatrix} = A^{-1}$.

8. $A^T = \begin{bmatrix} 1 & 4 \\ 2 & -1 \\ 3 & 2 \end{bmatrix}$, $B^T = \begin{bmatrix} 1 & 2 & 3 \\ 4 & -3 & -1 \end{bmatrix}$,

   $(AB)^T = B^T A^T = \begin{bmatrix} 14 & 8 \\ -5 & 17 \end{bmatrix}$.

PROBLEMS 1.7

1. $AB = \begin{bmatrix} 7 & -1 \\ 10 & 9 \end{bmatrix}$, $(AB)v = A(Bv) = \begin{bmatrix} 5 \\ 28 \end{bmatrix}$, $Bv = \begin{bmatrix} 8 \\ -2 \\ -3 \end{bmatrix}$.

3. $AB = [1]$, $BA = \begin{bmatrix} 2 & 4 & -2 \\ 0 & 0 & 0 \\ 1 & 2 & -1 \end{bmatrix}$.

4. (a) $AB = \begin{bmatrix} 10 & 3 & 17 \\ -6 & 7 & -8 \end{bmatrix}$, $AC = \begin{bmatrix} 9 & 7 \\ -1 & -13 \end{bmatrix}$,

   $BD = \begin{bmatrix} 9 & 5 & 16 \\ 0 & 1 & 12 \end{bmatrix}$, $BF = \begin{bmatrix} -2 \\ 13 \end{bmatrix}$,

   $EF = [4]$, $FE = \begin{bmatrix} -2 & -1 & -3 \\ 6 & 3 & 9 \\ 2 & 1 & 3 \end{bmatrix}$.

12. $A^{-1} = \frac{1}{8}\begin{bmatrix} 11 & -3 & -1 \\ -9 & 1 & 3 \\ 6 & 2 & -2 \end{bmatrix}$, $x = A^{-1}b = \frac{1}{8}\begin{bmatrix} 11 \\ -1 \\ -2 \end{bmatrix}$.

13. (a) $\begin{bmatrix} \frac{6}{12} & -\frac{3}{12} \\ \frac{2}{12} & \frac{1}{12} \end{bmatrix}$.   (b) Singular.   (d) $\begin{bmatrix} 1 & \frac{1}{2} & \frac{5}{2} & -\frac{23}{2} \\ 0 & \frac{1}{2} & \frac{1}{2} & -\frac{5}{2} \\ 0 & 0 & -1 & 3 \\ 0 & 0 & 0 & 1 \end{bmatrix}$.

Section 1.8

EXERCISES

1. 244.   2. $\begin{bmatrix} + & - \\ - & + \end{bmatrix}$.   3. (937, 1207).   4. $\begin{bmatrix} 0.393 \\ 0.607 \end{bmatrix}$.

PROBLEMS 1.8

2. (c) λ.   (d) $\lambda > 0$.   3. $\begin{bmatrix} 0 & 0 & 0 & 1100 \\ 0.01 & 0 & 0 & 0 \\ 0 & 0.2 & 0 & 0 \\ 0 & 0 & 0.5 & 0 \end{bmatrix}$.   5. (a) 432.   (b) n^3.

CHAPTER 2

Section 2.1

PROBLEMS 2.1

2. (b), (c).

3. (a) Properties 1 and 2.   (b) Property 1.
   (c) Properties 1 and 2.   (d) Property 2.

5. (a) The straight line in R^2 passing through the origin with slope $-\frac{3}{4}$.
   (b) The origin in R^2.
   (c) The whole plane R^2.
   (d) The straight line in R^3 consisting of all multiples of $(4, 3, 1)$.
   (e) The plane in R^3 consisting of all sums of multiples of $(2, 1, 0)$ and $(-2, 0, 1)$.
   (f) The straight line in R^3 consisting of all multiples of $(-1, 2, 1)$.

6. (a) $b_2 - 3b_1 = 0$ and $b_3 - 2b_1 = 0$.
   (b) $-2b_1 - b_2 + b_3 = 0$.
   (c) None.

7. If u and v belong to V and α is any scalar, then $A(u + v) = Au + Av = u + v$ and $A(\alpha u) = \alpha(Au) = \alpha u$; hence $u + v$ and αu belong to V.

Section 2.2

EXERCISES

1. (a) Yes; for example, $b = -11v_1 + 0v_2 + 5v_3$.   (b) No.

2. (a) $(-1, 0, 1, 0)$, $(-2, 1, 0, 1)$.   (b) $(-2, 1, 0, 0)$, $(-1, 0, -1, 1)$.

3. If a_1, a_2, a_3, are the rows of A and r_1, r_2, r_3 are the rows of E, then $a_1 = r_1$ $a_2 = 2r_1 + r_2$, $a_3 = 7r_1 + 2r_2$, and $r_1 = a_1$, $r_2 = -2a_1 + a_2$, $r_3 = 0a_1$.

PROBLEMS 2.2

1. (a) Yes.   (b) No.   (c) Yes.   (d) No.   2. (a), (d).   3. (c), (d).

4. (a) $(\frac{3}{2}, 1)$.   (b) $(-1, 1, 0, 0)$, $(-1, 0, 2, 1)$.
   (c) $(-2, 1, 0, 0, 0)$, $(-1, 0, 1, 1, 0)$, $(-1, 0, 2, 0, 1)$.
   (d) $(0, 0, 1, 0, 0)$, $(1, -1, 0, 1, 0)$.

6. We have $v_1 = u_1$, $v_2 = u_2 - v_1 = u_2 - u_1$, and $v_3 = u_3 - v_1 - v_2 = u_3 - u_1$ $- u_2 + u_1 = u_3 - u_2$. Hence every linear combination of the v_i's is a linear combination of the u_i's,

$$\alpha_1 v_1 + \alpha_2 v_2 + \alpha_3 v_3 = \alpha_1 u_1 + \alpha_2(u_2 - u_1) + \alpha_3(u_3 - u_2)$$
$$= (\alpha_1 - \alpha_2)u_1 + (\alpha_2 - \alpha_3)u_2 + \alpha_3 u_3.$$

Section 2.3

EXERCISES

2. (a), (d).

PROBLEMS 2.3

1. (a), (e), (f).   3. (a) $(1, -3, 1, 0)$, $(2, -4, 0, 1)$.
                         (b) $(-2, 1, 0)$.
                         (c) $(1, 0, 0)$, $(0, -2, 1)$.

5. If $x_k = x_j$, $k \neq j$, then $0 = x_1 v_1 + x_2 v_2 + \cdots + x_n v_n$ has nontrivial solutions; for example, $x_k = 1$, $x_j = -1$, and $x_i = 0$ for $i \neq k, j$.

6. If $0 = x_1(2v_1) + x_2(v_1 + v_2) + x_3(-v_1 + v_3)$, then $0 = (2x_1 + x_2 - x_3)v_1 + x_2 v_2 + x_3 v_3$. Since the v_i's are independent, it follows that $2x_1 + x_2 - x_3 = x_2 = x_3 = 0$. Hence $x_1 = x_2 = x_3 = 0$.

9. If $0 = x_1 Av_1 + x_2 Av_2 + \cdots + x_n Av_n$, then $0 = A(x_1 v_1 + x_2 v_2 + \cdots + x_n v_n)$. Hence $0 = x_1 v_1 + x_2 v_2 + \cdots + x_n v_n$ because the columns of A are independent. Since the v_i's are independent, it follows that $x_1 = x_2 = \cdots = x_n = 0$.

4. (a), (b).   **5.** $\begin{bmatrix} -2 & 1 \\ 17 & -6 \end{bmatrix}^{-1} = \begin{bmatrix} \frac{6}{5} & \frac{1}{5} \\ \frac{17}{5} & \frac{2}{5} \end{bmatrix}$; $\begin{bmatrix} 4 & 9 \\ 8 & 18 \end{bmatrix}$ is singular.

PROBLEMS 2.6

1. (a) The straight line in R^2 that passes through $(\frac{3}{2}, 0)$ with a slope -2.

(b) The straight line in R^3 that passes through $(2, 0, 0)$ and is parallel to the line spanned by $(-2, 1, 1)$.

(c) The plane in R^3 that contains $(9, 0, 0)$ and is parallel to the plane spanned by $(2, 1, 0)$ and $(-4, 0, 1)$.

2. (a), (c), (e), (f).

3. (a) $\begin{bmatrix} \frac{2}{15} & \frac{1}{15} \\ \frac{3}{15} & -\frac{6}{15} \end{bmatrix}$.   (b) Singular.   (c) $\begin{bmatrix} \frac{2}{72} & \frac{8}{72} \\ -\frac{9}{72} & 0 \end{bmatrix}$.

CHAPTER 3

Section 3.1

PROBLEMS 3.1

1. $\begin{bmatrix} 1 & 1 \\ 1 & 2 \\ 1 & 3 \\ 1 & 4 \\ 1 & 5 \end{bmatrix} \begin{bmatrix} d_0 \\ v \end{bmatrix} = \begin{bmatrix} 3 \\ 5 \\ 9 \\ 11 \\ 12 \end{bmatrix}$.   **2.** $\begin{bmatrix} 1 & -1 & 1 \\ 1 & 0 & 0 \\ 1 & 1 & 1 \\ 1 & 2 & 4 \end{bmatrix} \begin{bmatrix} a \\ b \\ c \end{bmatrix} = \begin{bmatrix} -2 \\ -1 \\ 0 \\ 3 \end{bmatrix}$.

Section 3.2

EXERCISES

1. (a) $\sqrt{2}$. (b) $\sqrt{7}$.   **3.** (a) $\sqrt{2}$. (b) $1/2$.

PROBLEMS 3.2

1. (a) $\sqrt{6}$. (b) $(\sqrt{37})/12$. (e) $\sqrt{(1 + \pi^2)}$. (f) $(\sqrt{1 + a^2 b^2})/|a|$.

2. (a) $\sqrt{3}$. (c) $\sqrt{74}/6$.   **3.** (b), (c), and (d).

4. (b) $\langle a, a \rangle + 2\beta \langle a, b \rangle + \beta^2 \langle b, b \rangle$.   **6.** (a), (b).

7. (a) For example, $(-1, 2, 1)$.   **8.** (a) For example, $(1, 1, -2)$, $(-1, 1, 0)$.

9. (a) For example, $(-2, 1, 0, 0)$, $(-1, -2, 5, 0)$.

11. (a) $\|x \pm y\|^2 = \langle x \pm y, x \pm y \rangle = \|x\|^2 \pm 2\langle x, y \rangle + \|y\|^2$. Hence $\|x + y\|^2 = \|x - y\|^2$ if and only if $2\langle x, y \rangle = -2\langle x, y \rangle$, that is, $\langle x, y \rangle = 0$.

14. $A^T = (xx^T)^T = x^{TT}x^T = xx^T = A$.

$A^2 = (xx^T)(xx^T) = x(x^Tx)x^T = xx^T = A$, since $x^Tx = \langle x, x \rangle = 1$.

16. Expand $\|x \pm y\|^2$ as in Problem 11.

17. (a) Let the columns of A be $c_1, c_2, \ldots, c_n$. Then the ijth entry in $A^T A$ is $\langle c_i, c_j \rangle$.
   (b) If A is orthogonal, then by part (a) $AA^T = I$. If $AA^T = I$, then $A^T = A^{-1}$ and by part (a) A is orthogonal.
   (e) $\langle Ax, Ay \rangle = \langle A^T A x, y \rangle = \langle x, y \rangle$.
   (f) Use Problem 14.
   (h) The identity matrices.
   (i) (2) The third column must be $\pm (1/\sqrt{6})(-1, 2, 1)$.

Section 3.3

EXERCISES

1. (a) $\cos \theta = \frac{1}{3}$.   (b) $\cos \theta = \frac{5}{8}$.

PROBLEMS 3.3

1. (a) $\cos \theta = 3/\sqrt{10}$.   (c) $\cos \theta = \sqrt{3}/2$.

2. (a) $(1/2, \sqrt{3}/2), (1/2, -\sqrt{3}/2)$.

3. (a) $(1/\sqrt{2}, 1/\sqrt{2}), (1/\sqrt{2}, -1/\sqrt{2})$.

4. (a) $(1/2, 1/2, 1/\sqrt{2}), (1/2, 1/2, -1/\sqrt{2})$.

6. Use $\cos \theta = \dfrac{\langle x, y \rangle}{\|x\| \, \|y\|} = 1$, or show that $\|x - y\| = 0$.

Section 3.4

EXERCISES

3. $(-1, 1, 0, 0), (-2, 0, 1, 0), (1, 0, 0, 1)$.

PROBLEMS 3.4

1. (b) $(0, 1, 0), (-1, 0, 1)$.   (d) $(-1, 1, 0, 0), (0, 0, 1, 0), (0, 0, 0, 1)$.
   (e) $(5, -5, 0, 2), (2, -2, 1, 0)$.   (f) $(0, -1, 2, 4)$.
   (g) $(-1, 0, 0, 0, 1), (1, 0, 0, 1, 0), (-1, 0, 1, 0, 0), (1, 1, 0, 0, 0)$.
   (h) $(-1, 0, 0, 0, 1), (0, -1, 0, 1, 0), (-1, 0, 1, 0, 0)$.
   (i) $(4, 3, -2, 1, 0), (-3, 1, 2, 0, 2)$.

7. (a) $\dim V^\perp = m - \dim V$ and $\dim V^{\perp\perp} = m - \dim V^\perp = \dim V$. Since V is a subspace of $V^{\perp\perp}$, it follows that $V^{\perp\perp}$ is equal to V.

9. (a) Basis for $C(A)^\perp$: $(2, 1)$; basis for $R(A)^\perp$: $(3, 1)$.
   (b) Basis for $C(A)^\perp$: $(-4, -3, 2)$; $R(A)^\perp = \{0\}$
   (c) $C(A)^\perp = \{0\}$; $R(A)^\perp = \{0\}$.
   (d) Basis for $C(A)^\perp$: $(-4, 0, 0, 1), (-3, 0, 1, 0), (-2, 1, 0, 0)$; $R(A)^\perp = \{0\}$.
   (e) Basis for $C(A)^\perp$: $(-3, -2, 1, 0), (3, 2, 0, 1)$, $R(A)^\perp = \{0\}$.

10. (b) No.
   (d) W is a subset of $V^\perp$. If u is in $V^\perp$, then $u = v + w$ for some v in V and w in W. Hence $0 = \langle u, v \rangle = \langle v, v \rangle + \langle w, v \rangle = \langle v, v \rangle$ and $v = 0$. Therefore, $u = w$, which belongs to W.

Section 3.5

1. $(\frac{4}{3}, \frac{4}{3}, \frac{4}{3})$, $(-\frac{1}{3}, -\frac{1}{3}, \frac{2}{3})$. **2.** $(-\frac{1}{6}, \frac{8}{6}, \frac{17}{6})$, $(\frac{7}{6}, -\frac{14}{6}, \frac{7}{6})$.

4. (a) $(\frac{3}{5}, \frac{6}{5})$. (b) $(\frac{1}{2}, 0, \frac{1}{2})$.

PROBLEMS 3.5

1. (a) $(0, 1)$. (d) $(\frac{10}{3}, -\frac{10}{3}, 0, \frac{10}{3})$.

3. (a) $(\frac{5}{3}, \frac{3}{2}, \frac{5}{3}, \frac{3}{2}, \frac{5}{3})$, $(-\frac{2}{3}, \frac{1}{2}, \frac{4}{3}, -\frac{1}{2}, -\frac{2}{3})$. (b) $\sqrt{114}/6$. (c) $(\frac{5}{3}, \frac{3}{2}, \frac{5}{3}, \frac{3}{2}, \frac{5}{3})$.

5. (a) $(\frac{11}{17}, \frac{8}{17}, \frac{23}{17})$. (b) $\sqrt{153}/17$. (c) $(\frac{11}{17}, \frac{8}{17}, \frac{23}{17})$.

8. (e) $2\sqrt{21}/7$, $(-1/7)(1, -5, -11)$. (f) 1, $(1/3)(4, 8, 7)$.
  (g) $2\sqrt{6}/3$, $(1/3)(3, 1, 4, 1)$. (h) $\sqrt{47}/8$, $(1/8)(2, 17, 9, 7, 9)$.

10. (a) $3/\sqrt{38}$, $(1/38)(61, 44, -47)$. (c) $3/7$, $(1/49)(43, -187, -165)$.
  (e) $\sqrt{3}$, $(1, 2, -2)$. (g) $18/\sqrt{19}$, $(1/19)(35, 19, 37, 35)$.
  (i) $3/\sqrt{7}$, $(1/7)(22, 24, 11, 7, 0, -10)$.

11. (a) $\dfrac{1}{42}\begin{bmatrix} 41 & 2 & 1 & 6 \\ 2 & 38 & -2 & -12 \\ 1 & -2 & 41 & -6 \\ 6 & -12 & -6 & 6 \end{bmatrix}$.

  (b) $\frac{1}{42}(41, 2, 1, 6)$ and $\frac{1}{42}(43, 40, -1, -6)$.

15. (a) $P^2 = A(A^TA)^{-1}A^TA(A^TA)^{-1}A^T = A(A^TA)^{-1}A^T = P$. **16.** (b) $(C(P))^{\perp}$.
  (c) $(I - P)^2 = I - 2P + P^2 = I - P$.
  $(I - P)^T = I - P^T = I - P$.

Section 3.6

1. $\bar{x} = x_3(-\frac{5}{3}, \frac{1}{3}, 1) + (2, 3, 0)$.

PROBLEMS 3.6

1. (a) $\bar{x} = (\frac{5}{6}, \frac{1}{2})$. (b) $\bar{x} = x_3(-4, 2, 1) + (\frac{3}{17}, \frac{3}{17}, 0)$.

3. $a = 1$, $b = c = \frac{3}{2}$.

4. (a) $y = 0.8 + 2.4t$. (b) $y = -\frac{13}{10} + \frac{11}{10}t + \frac{1}{2}t^2$.
  (c) $y = \frac{1}{6}t - t^2 + \frac{5}{6}t^3$. (d) $y = -\frac{5}{3} + \frac{10}{9}t^2$.

5. (a) $y = -\frac{4}{5} - \frac{2}{5}x_1 - \frac{37}{15}x_2$. **8.** $\log \alpha = \frac{7}{6}$, $\log \beta = \frac{11}{2}$.
  (b) $y = -\frac{2}{3}x_1 + \frac{5}{3}x_2$.

Section 3.7

EXERCISES

1. $(1, 2, 3) = 2(1, 1, 1) - \frac{1}{14}(2, -3, 1) - \frac{3}{14}(4, 1, -5)$.

2. Projection $= \frac{1}{10}(8, 11, 2, -1)$.   **3.** $\bar{x} = (\frac{1}{3}, \frac{13}{42})$.

PROBLEMS 3.7

1. (a) $\bar{x} = (\frac{4}{3}, \frac{1}{2})$.   (b) $\bar{x} = (-\frac{5}{2}, \frac{3}{2})$.   (c) $\bar{x} = (\frac{1}{3}, 0)$.
    (d) $\bar{x} = (\frac{1}{3}, \frac{1}{21})$.   (e) $\bar{x} = (\frac{3}{25}, -\frac{6}{25})$.   (f) $\bar{x} = (\frac{9}{25}, -\frac{1}{10})$.
    (g) $\bar{x} = (-\frac{2}{25}, -\frac{2}{25}, \frac{1}{25})$.   (h) $\bar{x} = (3, 2, \frac{1}{2})$.   (i) $\bar{x} = (1, \frac{1}{3}, \frac{13}{12})$.

2. $(6, 5, 0, -7) = (1, 1, 1, 1) + 2(1, -1, 1, -1) + 3(1, 2, -1, -2)$.

4. $y = 1 - t$.

5. (a) Use the fact that $u = \langle u, w_1 \rangle w_1 + \cdots + \langle u, w_m \rangle w_m$.

6. (a) Since the columns of A are orthonormal vectors, $A^T A = I$.

Section 3.8

EXERCISES

1. For example, $(1, 2, 3)$, $(2, -1, 0)$, $(-3, -6, 5)$.   **2.** $(1, -1, 3)$, $(-5, 16, 7)$.

PROBLEMS 3.8

1. $(1, 1, 1, 0)$, $(-1, 1, 0, 1)$, $(1, 0, -1, 1)$, $(0, 1, -1, -1)$.

2. (b) $(1, 1, 1)$, $(-2, 1, 1)$, $(0, -1, 1)$.   (d) $(-1, 2, 0, 2)$, $(0, 0, 1, 0)$, $(0, 1, 0, -1)$.

3. (a) $(1, 1, 0, 0)$, $(-1, 1, 2, 0)$, $(-2, 2, -2, 3)$.

4. $(1, -1, 2)$, $(2, 4, 1)$, $(3, -1, -2)$.

CHAPTER 4

Section 4.1

EXERCISES

1. (a) Function space. (b) Not a function space, not closed under addition, not closed under scalar multiplication.

2. Yes.   **3.** No.

PROBLEMS 4.1

1. (a) Yes. (b) No, not closed under addition nor under scalar multiplication. (c) Yes. (d) Yes. (e) Yes. (f) Yes. (g) No, not closed under scalar multiplication.

4. (a) $x = (-8, -7, 0) + x_3(5, 3, 1)$.

(c) Inconsistent.

(e) $x = (0, 0, 1, -1, 1) + x_2(-1, 1, 0, 0, 0)$.

6. (a) $\begin{bmatrix} 1 & 0 & 0 & \vdots & -1 \\ 0 & 1 & 0 & \vdots & 3 \\ 0 & 0 & 1 & \vdots & 0 \end{bmatrix}$.   (b) $\begin{bmatrix} 1 & 2 & 0 & \vdots & 2 \\ 0 & 0 & 1 & \vdots & 4 \end{bmatrix}$.

(c) $\begin{bmatrix} 1 & 1 & 0 & -3 & \vdots & 1 \\ 0 & 0 & 1 & 1 & \vdots & 3 \end{bmatrix}$.

10. $x = (\frac{5}{9}, -\frac{2}{9}, \frac{3}{9})$,

$x = (-\frac{4}{9}, -\frac{11}{9}, \frac{12}{9})$,

$x = (\frac{5}{3}, -\frac{2}{3}, 1)$,

$x = (-\frac{1}{3}, -\frac{5}{3}, 1)$,

$x = (\frac{5}{9}, \frac{7}{9}, \frac{12}{9})$.

Section 1.7

EXERCISES

1. $Ab_1 = \begin{bmatrix} 8 \\ 7 \end{bmatrix}$, $Ab_2 = \begin{bmatrix} 12 \\ 5 \end{bmatrix}$, $Ab_3 = \begin{bmatrix} -2 \\ 1 \end{bmatrix}$, $AB = \begin{bmatrix} 8 & 12 & -2 \\ 7 & 5 & 1 \end{bmatrix}$.

2. $AB = \begin{bmatrix} -5 & 10 \\ -11 & 18 \\ -1 & 8 \end{bmatrix}$.   **4.** $\begin{bmatrix} 2 & 3 & 4 \\ 5 & 6 & -2 \end{bmatrix}$.

5. $k \times m$ and $k \times n$, where k is arbitrary.

7. $\begin{bmatrix} -4 & 3 \\ 3 & -2 \end{bmatrix} = A^{-1}$.

8. $A^T = \begin{bmatrix} 1 & 4 \\ 2 & -1 \\ 3 & 2 \end{bmatrix}$, $B^T = \begin{bmatrix} 1 & 2 & 3 \\ 4 & -3 & -1 \end{bmatrix}$,

$(AB)^T = B^T A^T = \begin{bmatrix} 14 & 8 \\ -5 & 17 \end{bmatrix}$.

PROBLEMS 1.7

1. $AB = \begin{bmatrix} 7 & -1 \\ 10 & 9 \end{bmatrix}$, $(AB)v = A(Bv) = \begin{bmatrix} 5 \\ 28 \end{bmatrix}$, $Bv = \begin{bmatrix} 8 \\ -2 \\ -3 \end{bmatrix}$.

3. $AB = [1]$, $BA = \begin{bmatrix} 2 & 4 & -2 \\ 0 & 0 & 0 \\ 1 & 2 & -1 \end{bmatrix}$.

4. (a) $AB = \begin{bmatrix} 10 & 3 & 17 \\ -6 & 7 & -8 \end{bmatrix}$, $AC = \begin{bmatrix} 9 & 7 \\ -1 & -13 \end{bmatrix}$,

$BD = \begin{bmatrix} 9 & 5 & 16 \\ 0 & 1 & 12 \end{bmatrix}$, $BF = \begin{bmatrix} -2 \\ 13 \end{bmatrix}$,

$EF = [4]$, $FE = \begin{bmatrix} -2 & -1 & -3 \\ 6 & 3 & 9 \\ 2 & 1 & 3 \end{bmatrix}$.

12. $A^{-1} = \dfrac{1}{8}\begin{bmatrix} 11 & -3 & -1 \\ -9 & 1 & 3 \\ 6 & 2 & -2 \end{bmatrix}$, $x = A^{-1}b = \dfrac{1}{8}\begin{bmatrix} 11 \\ -1 \\ -2 \end{bmatrix}$.

13. (a) $\begin{bmatrix} \frac{6}{12} & -\frac{3}{12} \\ \frac{2}{12} & \frac{1}{12} \end{bmatrix}$.   (b) Singular.   (d) $\begin{bmatrix} 1 & \frac{1}{2} & \frac{5}{2} & -\frac{23}{2} \\ 0 & \frac{1}{2} & \frac{1}{2} & -\frac{5}{2} \\ 0 & 0 & -1 & 3 \\ 0 & 0 & 0 & 1 \end{bmatrix}$.

Section 1.8

EXERCISES

1. 244.   2. $\begin{bmatrix} + & - \\ - & + \end{bmatrix}$.   3. $(937, 1207)$.   4. $\begin{bmatrix} 0.393 \\ 0.607 \end{bmatrix}$.

PROBLEMS 1.8

2. (c) λ. (d) $\lambda > 0$.   3. $\begin{bmatrix} 0 & 0 & 0 & 1100 \\ 0.01 & 0 & 0 & 0 \\ 0 & 0.2 & 0 & 0 \\ 0 & 0 & 0.5 & 0 \end{bmatrix}$.   5. (a) 432. (b) n^3.

CHAPTER 2

Section 2.1

PROBLEMS 2.1

2. (b), (c).

3. (a) Properties 1 and 2. (b) Property 1.
   (c) Properties 1 and 2. (d) Property 2.

5. (a) The straight line in R^2 passing through the origin with slope $-\frac{3}{4}$.
   (b) The origin in R^2.
   (c) The whole plane R^2.
   (d) The straight line in R^3 consisting of all multiples of $(4, 3, 1)$.
   (e) The plane in R^3 consisting of all sums of multiples of $(2, 1, 0)$ and $(-2, 0, 1)$.
   (f) The straight line in R^3 consisting of all multiples of $(-1, 2, 1)$.

6. (a) $b_2 - 3b_1 = 0$ and $b_3 - 2b_1 = 0$.
   (b) $-2b_1 - b_2 + b_3 = 0$.
   (c) None.

7. If u and v belong to V and α is any scalar, then $A(u + v) = Au + Av = u + v$ and $A(\alpha u) = \alpha(Au) = \alpha u$; hence $u + v$ and αu belong to V.

Section 2.2

EXERCISES

1. (a) Yes; for example, $b = -11v_1 + 0v_2 + 5v_3$. (b) No.

2. (a) $(-1, 0, 1, 0)$, $(-2, 1, 0, 1)$. (b) $(-2, 1, 0, 0)$, $(-1, 0, -1, 1)$.

3. If a_1, a_2, a_3, are the rows of A and r_1, r_2, r_3 are the rows of E, then $a_1 = r_1$, $a_2 = 2r_1 + r_2$, $a_3 = 7r_1 + 2r_2$, and $r_1 = a_1$, $r_2 = -2a_1 + a_2$, $r_3 = 0a_1$.

PROBLEMS 2.2

1. (a) Yes. (b) No. (c) Yes. (d) No. 2. (a), (d). 3. (c), (d).

4. (a) $(\frac{3}{2}, 1)$. (b) $(-1, 1, 0, 0)$, $(-1, 0, 2, 1)$.
   (c) $(-2, 1, 0, 0, 0)$, $(-1, 0, 1, 1, 0)$, $(-1, 0, 2, 0, 1)$.
   (d) $(0, 0, 1, 0, 0)$, $(1, -1, 0, 1, 0)$.

6. We have $v_1 = u_1$, $v_2 = u_2 - v_1 = u_2 - u_1$, and $v_3 = u_3 - v_1 - v_2 = u_3 - u_1 - u_2 + u_1 = u_3 - u_2$. Hence every linear combination of the v_i's is a linear combination of the u_i's,

$$\alpha_1 v_1 + \alpha_2 v_2 + \alpha_3 v_3 = \alpha_1 u_1 + \alpha_2 (u_2 - u_1) + \alpha_3 (u_3 - u_2)$$

$$= (\alpha_1 - \alpha_2) u_1 + (\alpha_2 - \alpha_3) u_2 + \alpha_3 u_3.$$

Section 2.3

EXERCISES

2. (a), (d).

PROBLEMS 2.3

1. (a), (e), (f). 3. (a) $(1, -3, 1, 0)$, $(2, -4, 0, 1)$.
   (b) $(-2, 1, 0)$.
   (c) $(1, 0, 0)$, $(0, -2, 1)$.

5. If $x_k = x_j$, $k \neq j$, then $0 = x_1 v_1 + x_2 v_2 + \cdots + x_n v_n$ has nontrivial solutions; for example, $x_k = 1$, $x_j = -1$, and $x_i = 0$ for $i \neq k, j$.

6. If $0 = x_1(2v_1) + x_2(v_1 + v_2) + x_3(-v_1 + v_3)$, then $0 = (2x_1 + x_2 - x_3)v_1 + x_2 v_2 + x_3 v_3$. Since the v_i's are independent, it follows that $2x_1 + x_2 - x_3 = x_2 = x_3 = 0$. Hence $x_1 = x_2 = x_3 = 0$.

9. If $0 = x_1 Av_1 + x_2 Av_2 + \cdots + x_n Av_n$, then $0 = A(x_1 v_1 + x_2 v_2 + \cdots + x_n v_n)$. Hence $0 = x_1 v_1 + x_2 v_2 + \cdots + x_n v_n$ because the columns of A are independent. Since the v_i's are independent, it follows that $x_1 = x_2 = \cdots = x_n = 0$.

Section 2.4

1. (a) and (d) are bases for R^2.   2. (a) and (d) are bases for R^2.

3. (a) $(2, 3)$.   (b) $(1, 2), (0, -5)$.   (c) $(1, 0, 2), (0, 1, -6)$.

Section 2.5

EXERCISES

1. (a) $(1, 1, -1), (0, -1, 2), (0, 0, 5)$.   2. (a) $(1, 2, 0), (0, 1, 5), (-1, 3, -2)$.
   (b) $(2, -3, 1)$.   (b) $(0, 1, 1), (3, -1, 2)$.
                                                             (c) $(1, 1, 1), (0, -1, 1), (0, 2, 1)$.

3. (a) $(-9, 3, 1, 0), (-6, 1, 0, 1)$.   (b) $(-1, 1, 0)$.

PROBLEMS 2.5

1. (a) Rank $= 2$; $(3, 0, 0), (2, 1, 0)$ is a basis for the column space; $(3, -1, 2)$, $(0, 0, 1)$ is a basis for the row space.
   (b) Rank $= 3$; $(1, 0, 0), (2, 3, 0), (3, 8, 2)$ is a basis for the column space; $(1, 2, 0, 9, 3), (0, 3, 1, -5, 8), (0, 0, 0, 0, 2)$ is a basis for the row space.
   (c) Rank $= 2$; $(6, 0, 0, 0), (3, 2, 0, 0)$ is a basis for the column space; $(6, 3, 1, 3, 1, 0), (0, 0, 0, 2, 0, 2)$ is a basis for the row space.

3. (a) Rank $= 2$; $(2, 4), (-1, 5)$ is a basis for the column space; $(2, -1, 5), (0, 7, -2)$ is a basis for the row space.
   (b) Rank $= 2$; $(1, 0, 3, -1), (6, 2, 5, 1)$ is a basis for the column space; $(1, 6, 4)$, $(0, 2, 2)$ is a basis for the row space.
   (c) Rank $= 3$; $(1, 1, 2, 2), (-1, -1, -2, 0), (2, 2, 5, -1)$ is a basis for the column space; $(1, 2, -1, 1, 2), (0, 0, 2, -1, -5), (0, 0, 0, 0, 1)$ is a basis for the row space.

4. (a) Nullity $= 0$.   (b) Nullity $= 2$; $(-1, -\frac{3}{4}, 1, 0), (0, \frac{1}{2}, 0, 1)$.
   (c) Nullity $= 2$; $(-\frac{4}{3}, 1, 0), (\frac{7}{3}, 0, 1)$.   (d) Nullity $= 0$.
   (e) Nullity $= 1$; $(-1, -1, 1, 1)$.

6. (a) Dimension $= 2$; $(1, -1, 1), (3, 5, 1)$.
   (b) Dimension $= 3$; $(1, 1, 1), (0, 3, 1), (1, 0, 5)$.
   (c) Dimension $= 2$; $(1, -1, 2, 0), (0, 1, 1, 1)$.

9. $(1, 1, 2), (1, 0, 0), (0, 1, 0)$.   11. $(1, 1, 0, 0), (1, 2, 3, 1), (1, 0, 0, 0), (0, 0, 1, 0)$.

13. $p = p_3(\frac{1}{2}, 0, 1, 0) + p_4(0, 1, 0, 1)$.

Section 2.6

EXERCISES

1. (a) The straight line in R^3 that passes through $(1, 1, 0)$ and is parallel to the line spanned by $(-\frac{1}{3}, \frac{1}{3}, 1)$.
   (b) The point $(-\frac{5}{4}, \frac{9}{4})$.

4. (a), (b).   **5.** $\begin{bmatrix} -2 & 1 \\ 17 & -6 \end{bmatrix}^{-1} = \begin{bmatrix} \frac{6}{5} & \frac{1}{5} \\ \frac{17}{5} & \frac{2}{5} \end{bmatrix}$; $\begin{bmatrix} 4 & 9 \\ 8 & 18 \end{bmatrix}$ is singular.

PROBLEMS 2.6

1. (a) The straight line in R^2 that passes through $(\frac{3}{2}, 0)$ with a slope -2.
   (b) The straight line in R^3 that passes through $(2, 0, 0)$ and is parallel to the line spanned by $(-2, 1, 1)$.
   (c) The plane in R^3 that contains $(9, 0, 0)$ and is parallel to the plane spanned by $(2, 1, 0)$ and $(-4, 0, 1)$.

2. (a), (c), (e), (f).

3. (a) $\begin{bmatrix} \frac{2}{15} & \frac{1}{15} \\ \frac{3}{15} & -\frac{6}{15} \end{bmatrix}$. (b) Singular. (c) $\begin{bmatrix} \frac{2}{72} & \frac{8}{72} \\ -\frac{9}{72} & 0 \end{bmatrix}$.

CHAPTER 3

Section 3.1

PROBLEMS 3.1

1. $\begin{bmatrix} 1 & 1 \\ 1 & 2 \\ 1 & 3 \\ 1 & 4 \\ 1 & 5 \end{bmatrix} \begin{bmatrix} d_0 \\ v \end{bmatrix} = \begin{bmatrix} 3 \\ 5 \\ 9 \\ 11 \\ 12 \end{bmatrix}$.   2. $\begin{bmatrix} 1 & -1 & 1 \\ 1 & 0 & 0 \\ 1 & 1 & 1 \\ 1 & 2 & 4 \end{bmatrix} \begin{bmatrix} a \\ b \\ c \end{bmatrix} = \begin{bmatrix} -2 \\ -1 \\ 0 \\ 3 \end{bmatrix}$.

Section 3.2

EXERCISES

1. (a) $\sqrt{2}$. (b) $\sqrt{7}$.   3. (a) $\sqrt{2}$. (b) $1/2$.

PROBLEMS 3.2

1. (a) $\sqrt{6}$. (b) $(\sqrt{37})/12$. (e) $\sqrt{(1 + \pi^2)}$. (f) $(\sqrt{1 + a^2 b^2})/|a|$.

2. (a) $\sqrt{3}$. (c) $\sqrt{74}/6$.   3. (b), (c), and (d).

4. (b) $\langle a, a \rangle + 2\beta \langle a, b \rangle + \beta^2 \langle b, b \rangle$.   6. (a), (b).

7. (a) For example, $(-1, 2, 1)$.   8. (a) For example, $(1, 1, -2), (-1, 1, 0)$.

9. (a) For example, $(-2, 1, 0, 0), (-1, -2, 5, 0)$.

11. (a) $\|x \pm y\|^2 = \langle x \pm y, x \pm y \rangle = \|x\|^2 \pm 2\langle x, y \rangle + \|y\|^2$. Hence $\|x + y\|^2 = \|x - y\|^2$ if and only if $2\langle x, y \rangle = -2\langle x, y \rangle$, that is, $\langle x, y \rangle = 0$.

14. $A^T = (xx^T)^T = x^{TT}x^T = xx^T = A$.
    $A^2 = (xx^T)(xx^T) = x(x^Tx)x^T = xx^T = A$, since $x^Tx = \langle x, x \rangle = 1$.

16. Expand $\|x \pm y\|^2$ as in Problem 11.

17. (a) Let the columns of A be $c_1, c_2, \ldots, c_n$. Then the ijth entry in $A^T A$ is $\langle c_i, c_j \rangle$.
   (b) If A is orthogonal, then by part (a) $AA^T = I$. If $AA^T = I$, then $A^T = A^{-1}$ and by part (a) A is orthogonal.
   (e) $\langle Ax, Ay \rangle = \langle A^T Ax, y \rangle = \langle x, y \rangle$.
   (f) Use Problem 14.
   (h) The identity matrices.
   (i) (2) The third column must be $\pm(1/\sqrt{6})(-1, 2, 1)$.

Section 3.3

EXERCISES

1. (a) $\cos \theta = \frac{1}{3}$.   (b) $\cos \theta = \frac{5}{8}$.

PROBLEMS 3.3

1. (a) $\cos \theta = 3/\sqrt{10}$.   (c) $\cos \theta = \sqrt{3}/2$.

2. (a) $(1/2, \sqrt{3}/2), (1/2, -\sqrt{3}/2)$.

3. (a) $(1/\sqrt{2}, 1/\sqrt{2}), (1/\sqrt{2}, -1/\sqrt{2})$.

4. (a) $(1/2, 1/2, 1/\sqrt{2}), (1/2, 1/2, -1/\sqrt{2})$.

6. Use $\cos \theta = \dfrac{\langle x, y \rangle}{\|x\| \, \|y\|} = 1$, or show that $\|x - y\| = 0$.

Section 3.4

EXERCISES

3. $(-1, 1, 0, 0), (-2, 0, 1, 0), (1, 0, 0, 1)$.

PROBLEMS 3.4

1. (b) $(0, 1, 0), (-1, 0, 1)$.   (d) $(-1, 1, 0, 0), (0, 0, 1, 0), (0, 0, 0, 1)$.
   (e) $(5, -5, 0, 2), (2, -2, 1, 0)$.   (f) $(0, -1, 2, 4)$.
   (g) $(-1, 0, 0, 0, 1), (1, 0, 0, 1, 0), (-1, 0, 1, 0, 0), (1, 1, 0, 0, 0)$.
   (h) $(-1, 0, 0, 0, 1), (0, -1, 0, 1, 0), (-1, 0, 1, 0, 0)$.
   (i) $(4, 3, -2, 1, 0), (-3, 1, 2, 0, 2)$.

7. (a) $\dim V^{\perp} = m - \dim V$ and $\dim V^{\perp\perp} = m - \dim V^{\perp} = \dim V$. Since V is a subspace of $V^{\perp\perp}$, it follows that $V^{\perp\perp}$ is equal to V.

9. (a) Basis for $C(A)^{\perp}$: $(2, 1)$; basis for $R(A)^{\perp}$: $(3, 1)$.
   (b) Basis for $C(A)^{\perp}$: $(-4, -3, 2)$; $R(A)^{\perp} = \{0\}$
   (c) $C(A)^{\perp} = \{0\}$; $R(A)^{\perp} = \{0\}$.
   (d) Basis for $C(A)^{\perp}$: $(-4, 0, 0, 1), (-3, 0, 1, 0), (-2, 1, 0, 0)$; $R(A)^{\perp} = \{0\}$.
   (e) Basis for $C(A)^{\perp}$: $(-3, -2, 1, 0), (3, 2, 0, 1)$, $R(A)^{\perp} = \{0\}$.

10. (b) No.
    (d) W is a subset of $V^{\perp}$. If u is in $V^{\perp}$, then $u = v + w$ for some v in V and w in W. Hence $0 = \langle u, v \rangle = \langle v, v \rangle + \langle w, v \rangle = \langle v, v \rangle$ and $v = 0$. Therefore, $u = w$, which belongs to W.

Section 3.5

EXERCISES

1. $(\frac{4}{3}, \frac{4}{3}, \frac{4}{3})$, $(-\frac{1}{3}, -\frac{1}{3}, \frac{2}{3})$. **2.** $(-\frac{1}{6}, \frac{8}{6}, \frac{17}{6})$, $(\frac{7}{6}, -\frac{14}{6}, \frac{7}{6})$.

4. (a) $(\frac{3}{5}, \frac{6}{5})$. (b) $(\frac{1}{2}, 0, \frac{1}{2})$.

PROBLEMS 3.5

1. (a) $(0, 1)$. (d) $(\frac{10}{3}, -\frac{10}{3}, 0, \frac{10}{3})$.

3. (a) $(\frac{5}{3}, \frac{3}{2}, \frac{5}{3}, \frac{3}{2}, \frac{5}{3})$, $(-\frac{2}{3}, \frac{1}{2}, \frac{4}{3}, -\frac{1}{2}, -\frac{2}{3})$. (b) $\sqrt{114}/6$. (c) $(\frac{5}{3}, \frac{3}{2}, \frac{5}{3}, \frac{3}{2}, \frac{5}{3})$.

5. (a) $(\frac{11}{17}, \frac{8}{17}, \frac{23}{17})$. (b) $\sqrt{153}/17$. (c) $(\frac{11}{17}, \frac{8}{17}, \frac{23}{17})$.

8. (e) $2\sqrt{21}/7$, $(-1/7)(1, -5, -11)$. (f) 1, $(1/3)(4, 8, 7)$.
   (g) $2\sqrt{6}/3$, $(1/3)(3, 1, 4, 1)$. (h) $\sqrt{47}/8$, $(1/8)(2, 17, 9, 7, 9)$.

10. (a) $3/\sqrt{38}$, $(1/38)(61, 44, -47)$. (c) $3/7$, $(1/49)(43, -187, -165)$.
   (e) $\sqrt{3}$, $(1, 2, -2)$. (g) $18/\sqrt{19}$, $(1/19)(35, 19, 37, 35)$.
   (i) $3/\sqrt{7}$, $(1/7)(22, 24, 11, 7, 0, -10)$.

11. (a) $\dfrac{1}{42} \begin{bmatrix} 41 & 2 & 1 & 6 \\ 2 & 38 & -2 & -12 \\ 1 & -2 & 41 & -6 \\ 6 & -12 & -6 & 6 \end{bmatrix}$.

   (b) $\frac{1}{42}(41, 2, 1, 6)$ and $\frac{1}{42}(43, 40, -1, -6)$.

15. (a) $P^2 = A(A^TA)^{-1}A^TA(A^TA)^{-1}A^T = A(A^TA)^{-1}A^T = P$.   **16.** (b) $(C(P))^{\perp}$.
   (c) $(I - P)^2 = I - 2P + P^2 = I - P$.
   $(I - P)^T = I - P^T = I - P$.

Section 3.6

EXERCISES

1. $\bar{x} = x_3(-\frac{5}{3}, \frac{1}{3}, 1) + (2, 3, 0)$.

PROBLEMS 3.6

1. (a) $\bar{x} = (\frac{5}{6}, \frac{1}{2})$. (b) $\bar{x} = x_3(-4, 2, 1) + (\frac{3}{17}, \frac{3}{17}, 0)$.

3. $a = 1$, $b = c = \frac{3}{2}$.

4. (a) $y = 0.8 + 2.4t$. (b) $y = -\frac{13}{10} + \frac{11}{10}t + \frac{1}{2}t^2$.
   (c) $y = \frac{1}{6}t - t^2 + \frac{5}{6}t^3$. (d) $y = -\frac{5}{3} + \frac{10}{9}t^2$.

5. (a) $y = -\frac{4}{5} - \frac{2}{5}x_1 - \frac{37}{15}x_2$. **8.** $\log \alpha = \frac{7}{6}$, $\log \beta = \frac{11}{2}$.
   (b) $y = -\frac{2}{3}x_1 + \frac{5}{3}x_2$.

Section 3.7

EXERCISES

1. $(1, 2, 3) = 2(1, 1, 1) - \frac{1}{14}(2, -3, 1) - \frac{3}{14}(4, 1, -5)$.

2. Projection $= \frac{1}{10}(8, 11, 2, -1)$.   3. $\bar{x} = (\frac{1}{3}, \frac{13}{42})$.

PROBLEMS 3.7

1. (a) $\bar{x} = (\frac{4}{3}, \frac{1}{2})$.   (b) $\bar{x} = (-\frac{5}{2}, \frac{3}{2})$.   (c) $\bar{x} = (\frac{1}{3}, 0)$.
   (d) $\bar{x} = (\frac{1}{3}, \frac{1}{21})$.   (e) $\bar{x} = (\frac{3}{25}, -\frac{6}{25})$.   (f) $\bar{x} = (\frac{9}{25}, -\frac{1}{10})$.
   (g) $\bar{x} = (-\frac{2}{25}, -\frac{2}{25}, \frac{1}{25})$.   (h) $\bar{x} = (3, 2, \frac{1}{2})$.   (i) $\bar{x} = (1, \frac{1}{3}, \frac{13}{12})$.

2. $(6, 5, 0, -7) = (1, 1, 1, 1) + 2(1, -1, 1, -1) + 3(1, 2, -1, -2)$.

4. $y = 1 - t$.

5. (a) Use the fact that $u = \langle u, w_1 \rangle w_1 + \cdots + \langle u, w_m \rangle w_m$.

6. (a) Since the columns of A are orthonormal vectors, $A^T A = I$.

Section 3.8

EXERCISES

1. For example, $(1, 2, 3), (2, -1, 0), (-3, -6, 5)$.   2. $(1, -1, 3), (-5, 16, 7)$.

PROBLEMS 3.8

1. $(1, 1, 1, 0), (-1, 1, 0, 1), (1, 0, -1, 1), (0, 1, -1, -1)$.

2. (b) $(1, 1, 1), (-2, 1, 1), (0, -1, 1)$.   (d) $(-1, 2, 0, 2), (0, 0, 1, 0), (0, 1, 0, -1)$.

3. (a) $(1, 1, 0, 0), (-1, 1, 2, 0), (-2, 2, -2, 3)$.

4. $(1, -1, 2), (2, 4, 1), (3, -1, -2)$.

CHAPTER 4

Section 4.1

EXERCISES

1. (a) Function space.   (b) Not a function space, not closed under addition, not closed under scalar multiplication.

2. Yes.   3. No.

PROBLEMS 4.1

1. (a) Yes.   (b) No, not closed under addition nor under scalar multiplication.
   (c) Yes.   (d) Yes.   (e) Yes.   (f) Yes.   (g) No, not closed under scalar multiplication.

2. (a) No, not closed under scalar multiplication.   (b) Yes.
(c) No, not closed under addition nor under scalar multiplication.
(d) No, not closed under addition nor under scalar multiplication.
(e) Yes.   (f) Yes.

5. (a) Yes.   (b) Yes.   (c) No, not closed under addition nor under scalar multiplication.

6. (a) Yes.   (b) Yes.   (c) Yes.   (d) No, not closed under addition nor under scalar multiplication.

Section 4.2

PROBLEMS 4.2

1. (b), (d).   **2.** (a), (b), (c).

4. The constant polynomial 1 is a basis for $P_0(R)$.

5. (a) Dim $= 2$, basis $1, 1 - t$.   (b) Dim $= 3$, basis $t, t^2 + 1, t^2 - 1$.
(c) Dim $= 3$, basis $\sin x, \cos x, \sin 2x$.   (d) Dim $= 3$, basis $\sin x, \cos x, x$.
(e) Dim $= 2$, basis $e^x, \sin x$.

Section 4.3

EXERCISES

1. The length of $\sin x$ is $\sqrt{\pi}/2$, the distance between $\sin x$ and $\cos x$ is $\sqrt{(\pi/2 - 1)}$, and $\sin x$ and $\cos x$ are not orthogonal on $I = [0, \pi/2]$.

PROBLEMS 4.3

1. (a) 1.   (d) $\sqrt{\pi}/2$.   **2.** (a) $\frac{1}{3}$.   (c) $\log \sqrt{2}$.

3. (a) For example, $-2 + 3x, -1 + 2x^2$.   **4.** $1 - 6x + 6x^2$.
(c) For example, $-3 + 4x, -3 + 5x^2$.

7. (a) For example, $1, x^2, -3x + 5x^3$.   (b) For example, $x, -3 + 5x^2, x^3$.
(c) For example, $-1 + 3x^2, -3x + 5x^3$.
(d) For example, $-3 + 5x^2, -3x + 5x^3$.

8. (a) $x^3 = \frac{3}{5}x + \frac{1}{5}(-3x + 5x^3)$,
$1 + x^2 = 1 + x^2$,
$1 + x^3 = 1 + \frac{3}{5}x + \frac{1}{5}(-3x + 5x^3)$.
(b) $x^3 = x^3$,
$1 + x^2 = \frac{8}{3}x^2 - \frac{1}{3}(-3 + 5x^2)$,
$1 + x^3 = \frac{5}{3}x^2 - \frac{1}{3}(-3 + 5x^2) + x^3$.
(c) $x^3 = \frac{3}{5}x + \frac{1}{5}(-3x + 5x^3)$,
$1 + x^2 = \frac{4}{3} + (\frac{1}{3})(-1 + 3x^2)$,
$1 + x^3 = 1 + (\frac{3}{5})x + (\frac{1}{5})(-3x + 5x^3)$.
(d) $x^3 = (\frac{3}{5})x + (\frac{1}{5})(-3x + 5x^3)$,
$1 + x^2 = (\frac{8}{3})x^2 - (\frac{1}{3})(-3 + 5x^2)$,
$1 + x^3 = (\frac{3}{5})x + (\frac{8}{3})x^2 - (\frac{1}{3})(-3 + 5x^2) + (\frac{1}{5})(-3x + 5x^3)$.

Section 4.4

EXERCISES

2. Projection $= 2 \sin x$.   **3.** Projection $= x - \frac{1}{6}$.

4. $h(x) = \dfrac{3}{\pi} x + \dfrac{175}{8} \dfrac{4\pi^2 - 60}{5\pi^3}(x^2 - \frac{3}{5}x)$.

PROBLEMS 4.4

1. (b) $\frac{7}{4}x$.   **2.** $4\pi^2/3 - 4\pi \sin x + 4 \cos x, \frac{1}{2}, 0$.

3. $\pi - \sin 2x, 4\pi^2/3 - 2\pi \sin 2x, 0$.   **4.** (a) $1, x - \frac{1}{2}, x^2 - x + \frac{1}{6}$.
                                                              (c) $x, e^x - 3x$.
                                                              (e) $x, x^2, x^3 - \frac{3}{5}x$.

5. (a) All polynomials $ax^2 + bx + c$ with $a + b + 3c = 0$.

6. $h(x) = \pi, h(x) = 0$.

7. (a) $h(x) = 3x/\pi^2$.   (b) $h(x) = \frac{9}{7}x$.   (c) $h(x) = \frac{3}{5}x$.
     (d) $h(x) = e - 1 + (18 - 6e)(x - \frac{1}{2}) + (35e - 95)(6x^2 - 6x + 1)$.
     (e) $h(x) = (15/2\pi^2) - (45/2\pi^4)x^2$.

9. $(\bar{a}, \bar{b}, \bar{c}) = (0, 3/\pi^2, 0)$.

CHAPTER 5

Section 5.1

EXERCISES

1. $T(-x) = T((-1)x) = (-1)T(x) = -T(x); \quad T(x - y) = T(x + (-y))$
     $= T(x) + T(-y) = T(x) + (-T(y)) = T(x) - T(y)$.

3. If $x = (1, 1)$, then $T(2x) = (4, 2)$ but $2T(x) = (2, 2)$.   **4.** (b), (c).

5. $\begin{bmatrix} 2 & 0 \\ -1 & 3 \end{bmatrix}; \quad \begin{bmatrix} 2 & 0 \\ -1 & 3 \end{bmatrix}\begin{bmatrix} x_1 \\ x_2 \end{bmatrix} = \begin{bmatrix} 2x_1 \\ -x_1 + 3x_2 \end{bmatrix}$.

PROBLEMS 5.1

1. (a) $\begin{bmatrix} 0 & 1 & 0 \\ 1 & 0 & 0 \end{bmatrix}$.   (d) $[2 \quad 3]$.

2. (a) $\begin{bmatrix} 2 & 1 & 0 \\ 0 & 1 & -1 \end{bmatrix}$.   (b) $[1 \quad 2 \quad -1 \quad 5]$.   (c) $\begin{bmatrix} 0 & 0 \\ 0 & 1 \\ 1 & 0 \\ 1 & 1 \end{bmatrix}$.

4. (a) $(\sqrt{3}/2 - 1, \frac{1}{2} + \sqrt{3})$.   (b) $(-\sqrt{3}/2 - 1, \frac{1}{2} + \sqrt{3})$.
     (c) $(-\sqrt{2}/2, 3\sqrt{2}/2)$.   (d) $(3\sqrt{2}/2, -\sqrt{2}/2)$.

5. $(-\frac{14}{5}, \frac{23}{5})$.   6. $\begin{bmatrix} 1 & 0 & 0 \\ 0 & \frac{1}{2} & \frac{1}{2} \\ 0 & \frac{1}{2} & \frac{1}{2} \end{bmatrix}$.

7. $\begin{bmatrix} 1 & 0 & 0 \\ 0 & 0 & 1 \\ 0 & 1 & 0 \end{bmatrix}$, $T(1,2,3) = (1,3,2)$, $T(2,-1,2) = (2,2,-1)$.

8. All vectors x in the plane spanned by $(1,0,0)$ and $(0,1,1)$.

10. $A = 2P - I$, where P is the projection matrix for V. We know that $P = P^T$ and $P^2 = P$ (see Problem 15 in Section 3.5).
   (a) A is symmetric because $A^T = (2P - I)^T = 2P^T - I^T = 2P - I = A$.
   A is orthogonal because $AA^T = AA = (2P - I)^2 = 4P^2 - 4P + I = 4P - 4P + I = I$.
   (b) $\langle T(x), T(y) \rangle = \langle 2Px - x, 2Py - y \rangle$
   $= 4\langle Px, Py \rangle - 2\langle Px, y \rangle - 2\langle x, Py \rangle + \langle x, y \rangle$
   $= 4\langle P^TPx, y \rangle - 2\langle Px, y \rangle - 2\langle P^Tx, y \rangle + \langle x, y \rangle$
   $= 4\langle P^2x, y \rangle - 4\langle Px, y \rangle + \langle x, y \rangle$
   $= 4\langle Px, y \rangle - 4\langle Px, y \rangle + \langle x, y \rangle = \langle x, y \rangle$.
   $\|T(x)\|^2 = \langle T(x), T(x) \rangle = \langle x, x \rangle = \|x\|^2$,
   hence $\|T(x)\| = \|x\|$.

16. (a), (b).

Section 5.2

EXERCISES

1. The square with vertices $(1, -1)$, $(2, -1)$, $(2, -2)$, $(1, -2)$. The image of a rectangle, parallelogram, circle is a rectangle, parallelogram, circle, respectively.

2. Since $T(1,0,0) = T(5,2,-2) = (1,0,2)$, the plane is mapped onto the line spanned by $(1,0,2)$.

3. (a) $(1,0)$, $(1,1)$; $T(R^3) = R^2$.   (b) The line spanned by $(1,1)$.

4. The standard matrix for T is

$$A = \begin{bmatrix} 1 & 1 \\ 1 & -1 \\ 1 & 2 \end{bmatrix}.$$

Since $N(T) = N(A) = \{0\}$, T is one-to-one.

PROBLEMS 5.2

1. (a) $(1,2,0)$, $(1,3,1)$.   (b) $(2,5,1)$, $(0,1,1)$.   (c) $(1,4,2)$, $(1,3,1)$.   (d) $(3,8,2)$.

2. (a) $(1,0)$, $(0,1)$.   (b) $(1,1)$.   (c) $(1,2)$, $(1,1)$.   (d) $(2,1)$.   3. (b), (d).

4. Since A is nonsingular, $N(T) = N(A) = \{0\}$ and $T(R^n) = C(A) = R^n$.

Section 5.3

1. (a) $(2, -1, 3)$.   (b) $(-4, -4, 9)$.   2. $\begin{bmatrix} 1 & 0 \\ -1 & -1 \\ 1 & 2 \end{bmatrix}$.   3. $\begin{bmatrix} 1 & 0 \\ 0 & -1 \end{bmatrix}$.

PROBLEMS 5.3

1. (a) $(1, -2, 3)$.   (b) $(2, -1, 2)$.   2. $\begin{bmatrix} -1 & 0 & -\frac{12}{8} \\ 1 & \frac{14}{8} & \frac{7}{8} \end{bmatrix}$.

3. $\begin{bmatrix} 2 & 3 & 6 \\ -2 & -1 & -1 \end{bmatrix}$.   7. $(1, 2), (-2, 1)$.

CHAPTER 6

Section 6.1

PROBLEMS 6.1

1. Yes.   2. Yes.

3. (b) Let A_{ij} be the $m \times n$ matrix whose entry in the ith row and jth column is 1 and which has all other entries equal to zero. Then the mn matrices A_{ij}, $i = 1, \ldots, m, j = 1, \ldots, n$, are a basis for V. Hence dim $V = mn$.
   (c) No.

4. (a) The matrices A_{ij} [see the solution of Problem 3(b)] with $i \leq j$ are a basis for W. Hence dim $W = 1 + 2 + 3 + \cdots + n = n(n + 1)/2$.
   (b) The matrices A_{ii}, $i = 1, \ldots, n$, are a basis for U. Hence dim $U = n$.
   (c) The dimension is $1 + 2 + 3 + \cdots + n = n(n + 1)/2$.
   (d) The dimension is $1 + 2 + 3 + \cdots + (n - 1) = n(n - 1)/2$.

5. (a), (d).   11. No.

CHAPTER 7

Section 7.1

3. (a) 28.   (b) 83.   (c) -3495.   (d) 66.   (e) 0.

PROBLEMS 7.1

1. (a) -17.   (b) 12.   (c) 4.   (d) 0.   (e) 0.   (f) $-\frac{19}{120}$.   (g) 37.
   (h) -18.   (i) 113.

4. (a) $\frac{1}{2}$.   (b) -45.   (c) 0.   (d) $-\frac{6}{5}$.

6. $\det(AB) = (\det A)(\det B) = (\det B)(\det A) = \det(BA)$.

7. If n is odd, then (see Problem 3) $\det A = \det A^T = \det(-A) = (-1)^n \det A = -\det A$, hence $2 \det A = 0$, that is, $\det A = 0$.
$A = \begin{bmatrix} 0 & 1 \\ -1 & 0 \end{bmatrix}$ is skew-symmetric, but $\det A = 1$.

9. (a) $-ab$. (b) $-abd$. (c) $abdg$.

10. Let r be the remainder obtained by dividing n by 4. Then $\det A = (-1)^k a_{1n} a_{2,n-1} \cdots a_{n-1,2} a_{n1}$, where $k = 0$ if $r = 0, 1$ and $k = 1$ if $r = 2, 3$.

Section 7.2

EXERCISES

1. $A_{11} = 28$, $A_{12} = 20$, $A_{13} = 14$, $A_{21} = -21$, $A_{22} = -15$, $A_{23} = 17$, $A_{31} = 8$, $A_{32} = -10$, $A_{33} = 4$, $\det A = 110$.

PROBLEMS 7.2

1. (a) 206. (b) -6. (c) $\frac{5}{9}$. (d) -3. (e) 4. (f) 60.

3. $\det \begin{bmatrix} 3-\lambda & 1 \\ -6 & -4-\lambda \end{bmatrix} = \lambda^2 + \lambda - 6 = (\lambda - 2)(\lambda + 3); \lambda = 2, \lambda = -3$.

4. (a) No real number λ. (b) $\lambda = 1$.

Section 7.3

EXERCISES

1. $a \times b = (-10, -7, 8)$, $c \times d = (-3, -3, -3)$.

PROBLEMS 7.3

1. (a) $(-3, -3, 5)$, area $= \sqrt{43}$. (b) $(1, 17, -11)$, area $= \sqrt{411}$.
(c) $(13, -5, -4)$, area $= \sqrt{210}$.

2. (a) 9. (b) 7.   **3.** (a) 5. (b) 34.   **4.** (a) 5. (b) $\sqrt{59}$.

Section 7.4

EXERCISES

1. $\dfrac{1}{4} \begin{bmatrix} -12 & -4 & 8 \\ 29 & 10 & -17 \\ 19 & 6 & -11 \end{bmatrix}$.

PROBLEMS 7.4

1. (a) $\dfrac{1}{43} \begin{bmatrix} -3 & 10 & 2 \\ 11 & -8 & 7 \\ 9 & 13 & -6 \end{bmatrix}$. (b) $\begin{bmatrix} 3 & 2 & 1 \\ 5 & 4 & 2 \\ 10 & 7 & 4 \end{bmatrix}$. (c) $\dfrac{1}{10} \begin{bmatrix} 9 & 2 & -6 \\ -15 & 10 & 0 \\ 21 & -2 & 6 \end{bmatrix}$.

$$\textbf{2.} \quad \begin{bmatrix} \dfrac{1}{a} & -\dfrac{b}{ad} & \dfrac{be}{adf} & -\dfrac{c}{af} \\[2mm] 0 & \dfrac{1}{d} & & -\dfrac{e}{df} \\[2mm] 0 & 0 & & \dfrac{1}{f} \end{bmatrix}. \qquad \textbf{4. (a)} \ (\tfrac{23}{37}, -\tfrac{13}{37}). \quad \textbf{(b)} \ (\tfrac{3}{2}, -2, -\tfrac{1}{2}).$$

CHAPTER 8

Section 8.1

EXERCISES

1. $\lambda = 1$.

3. $(5, -1, 1)$ is a basis for $E(4)$. The eigenvectors are $\alpha(5, -1, 1)$, where $\alpha \neq 0$.

PROBLEMS 8.1

3. $2, -3$.   **4. (a)** $-3, 3, 6$.   **(b)** $-1, -1, 2$.

6. (a) $-1, -2, -3, -4$.   **(b)** $1, 1, 2, 2$.   **7.** $A(Av) = A(\lambda v) = \lambda(Av)$.

9. The diagonal entries of the matrix.

12. The only possible eigenvalues of A are 0 and 1.

Section 8.2

EXERCISES

2. $\lambda = 0, -1, 2$. $E(0)$ has basis $(2, -1, 1)$, $E(-1)$ has basis $(-1, 1, 0)$, $E(2)$ has basis $(1, 1, 3)$.

4. $\lambda = 1, -1$. $E(1)$ has basis $(1, 0, 0)$, $(0, 1, 1)$, $E(-1)$ has basis $(1, 1, 0)$.

6. $\lambda = 1, -1$. $E(1)$ has basis $(2, 1, 1)$, $E(-1)$ has basis $(1, 1, 0)$.

PROBLEMS 8.2

1. (a) $\lambda = 2$, basis $(1, 1)$; $\lambda = 3$, basis $(0, 1)$.
  (b) $\lambda = 0$, basis $(2, 1)$; $\lambda = 3$, basis $(-1, 1)$.
  (c) $\lambda = -1$, basis $(1, 1)$; $\lambda = 3$, basis $(-1, 3)$.
  (d) $\lambda = 1$, basis $(0, 1, 0)$;
      $\lambda = -3$, basis $(-4, -3, 16)$.
  (e) $\lambda = 0$, basis $(-1, 1, 2)$; $\lambda = 1$, basis $(2, 0, 1)$;
      $\lambda = 4$, basis $(1, 3, 2)$.
  (f) $\lambda = 2$, basis $(3, 2, 1)$;
      $\lambda = -1 + \sqrt{2}$, basis $(-1 - \sqrt{2}, -1 + \sqrt{2}, 1)$;
      $\lambda = -1 - \sqrt{2}$, basis $(-1 + \sqrt{2}, -1 - \sqrt{2}, 1)$.

(g) $\lambda = 0$, basis $(-1, 0, 1)$;
$\lambda = 1$, basis $(0, 1, 0)$, $(0, 0, 1)$.
(h) $\lambda = 1$, basis $(0, 1, 0)$; $\lambda = 2$, basis $(1, 0, 0)$.
(i) $\lambda = 0$, basis $(1, 0, 0)$; $\lambda = 1$, basis $(3, 2, 1)$.
(j) $\lambda = 0$, basis $(0, 1, 0)$; $\lambda = 2$, basis $(1, 0, 1)$;
$\lambda = -2$, basis $(-1, 0, 1)$.
(k) $\lambda = 0$, basis $(1, 1, 1)$; $\lambda = 1$, basis $(-1, 0, 1)$;
$\lambda = 3$, basis $(1, -2, 1)$.
(l) $\lambda = 1$, basis $(-2, 1, 1)$; $\lambda = 2$, basis $(-1, 0, 1)$;
$\lambda = 3$, basis $(-2, 1, 2)$.
(m) $\lambda = 1$, basis $(1, -2, 2)$; $\lambda = 3$, basis $(1, 0, 0)$.
(n) $\lambda = 2$, basis $(1, 0, 0)$, $(0, 1, 0)$;
$\lambda = -5$, basis $(0, -1, 1)$.
(o) $\lambda = 1$, basis $(-1, 4, 1)$; $\lambda = 3$, basis $(1, 2, 1)$;
$\lambda = -2$, basis $(-1, 1, 1)$.
(p) $\lambda = 1$, basis $(1, 0, 0, 0)$, $(0, 0, 1, 0)$.
(q) $\lambda = 1$, basis $(1, 0, 0, 0)$;
$\lambda = -2$, basis $(0, 0, 1, 0)$, $(0, 0, 0, 1)$.
(r) $\lambda = 2$, basis $(1, 0, 0, 0)$;
$\lambda = -1$, basis $(0, 0, 0, 1)$.

4. A and A^{-1} have the same eigenvectors.

5. $\det(A - \lambda I) = \det((A - \lambda I)^T) = \det(A^T - \lambda I)$. The matrix $\begin{bmatrix} 1 & 0 \\ 1 & 1 \end{bmatrix}$ shows that A and A^T need not have the same eigenvectors.

Section 8.3

EXERCISES

2. (a) $9.96 N_0$. (b) $0.47 N_0$. (c) $15.18 N_0$.

PROBLEMS 8.3

1. (a) $P = \begin{bmatrix} 1 & 0 \\ 1 & 1 \end{bmatrix}$, $\Lambda = \begin{bmatrix} 2 & 0 \\ 0 & 3 \end{bmatrix}$. (b) $P = \begin{bmatrix} 2 & -1 \\ 1 & 1 \end{bmatrix}$, $\Lambda = \begin{bmatrix} 0 & 0 \\ 0 & 3 \end{bmatrix}$.

(c) $P = \begin{bmatrix} 1 & -1 \\ 1 & 3 \end{bmatrix}$, $\Lambda = \begin{bmatrix} -1 & 0 \\ 0 & 3 \end{bmatrix}$. (d) Not diagonalizable.

(e) $P = \begin{bmatrix} -1 & 2 & 1 \\ 1 & 0 & 3 \\ 2 & 1 & 2 \end{bmatrix}$, $\Lambda = \begin{bmatrix} 0 & 0 & 0 \\ 0 & 1 & 0 \\ 0 & 0 & 4 \end{bmatrix}$.

(f) $P = \begin{bmatrix} 3 & -1 - \sqrt{2} & -1 + \sqrt{2} \\ 2 & -1 + \sqrt{2} & -1 - \sqrt{2} \\ 1 & 1 & 1 \end{bmatrix}$, $\Lambda = \begin{bmatrix} 2 & 0 & 0 \\ 0 & -1 + \sqrt{2} & 0 \\ 0 & 0 & -1 - \sqrt{2} \end{bmatrix}$.

(g) $P = \begin{bmatrix} -1 & 0 & 0 \\ 0 & 1 & 0 \\ 1 & 0 & 1 \end{bmatrix}$, $\Lambda = \begin{bmatrix} 0 & 0 & 0 \\ 0 & 1 & 0 \\ 0 & 0 & 1 \end{bmatrix}$.

(h) Not diagonalizable. (i) Not diagonalizable.

(j) $P = \begin{bmatrix} 0 & 1 & -1 \\ 1 & 0 & 0 \\ 0 & 1 & 1 \end{bmatrix}$, $\Lambda = \begin{bmatrix} 0 & 0 & 0 \\ 0 & 2 & 0 \\ 0 & 0 & -2 \end{bmatrix}$.

(k) $P = \begin{bmatrix} 1 & -1 & 1 \\ 1 & 0 & -2 \\ 1 & 1 & 1 \end{bmatrix}$, $\Lambda = \begin{bmatrix} 0 & 0 & 0 \\ 0 & 1 & 0 \\ 0 & 0 & 3 \end{bmatrix}$.

(l) $P = \begin{bmatrix} -2 & -1 & -2 \\ 1 & 0 & 1 \\ 1 & 1 & 2 \end{bmatrix}$, $\Lambda = \begin{bmatrix} 1 & 0 & 0 \\ 0 & 2 & 0 \\ 0 & 0 & 3 \end{bmatrix}$.

(m) Not diagonalizable.

(n) $P = \begin{bmatrix} 1 & 0 & 0 \\ 0 & 1 & -1 \\ 0 & 0 & 1 \end{bmatrix}$, $\Lambda = \begin{bmatrix} 2 & 0 & 0 \\ 0 & 2 & 0 \\ 0 & 0 & -5 \end{bmatrix}$.

(o) $P = \begin{bmatrix} -1 & 1 & -1 \\ 4 & 2 & 1 \\ 1 & 1 & 1 \end{bmatrix}$, $\Lambda = \begin{bmatrix} 1 & 0 & 0 \\ 0 & 3 & 0 \\ 0 & 0 & -2 \end{bmatrix}$.

(p) Not diagonalizable. (q) Not diagonalizable. (r) Not diagonalizable.

2. $\begin{bmatrix} 0 & 1 & 2 \\ 0 & 1 & 2 \\ 0 & 0 & 3 \end{bmatrix}$. 3. No. 5. No. 6. No. 7. $A = \lambda I$.

11. (a) $\det(B - \lambda I) = \det(S^{-1}(B - \lambda I)S) = \det(S^{-1}BS - \lambda S^{-1}IS)$
$= \det(A - \lambda I)$.
(b) No. (c) Yes. (d) Yes. (g) I, the identity matrix.
(h) If $A = S^{-1}BS$ and v is an eigenvector of A, then Sv is an eigenvector of B.

Section 8.4

PROBLEMS 8.4

1. (a) $\det(A - \lambda I) = (2 - \lambda)^3(1 - \lambda)$. $E(2)$ has basis $(1, 0, 0, 0)$, $(0, 0, 1, 0)$, hence dim $E(2) = 2$.
(b) $\lambda = 1$.   .

3. $\det(A - \lambda I) = (4 - \lambda)^2(-1 - \lambda)^3$. $E(4)$ has basis $(1, 0, 0, 0, 0)$, hence dim $E(4) = 1$. $E(-1)$ has basis $(0, 0, 1, 0, 0)$, hence dim $E(-1) = 1$.

Section 8.5

EXERCISES

1. $\lambda = 2$, $E(2)$ has basis $(-1, 0, 1)$, $(0, 1, 0)$.
$\lambda = 4$, $E(4)$ has basis $(1, 0, 1)$.

2. $(-1/\sqrt{2}, 0, 1/\sqrt{2})$, $(0, 1, 0)$, $(1/\sqrt{2}, 0, 1/\sqrt{2})$.

3. The columns of the matrix are the vectors in the answer to Exercise 2.

PROBLEMS 8.5

1. (a) $(1/\sqrt{2})\begin{bmatrix} 1 & 1 \\ -1 & 1 \end{bmatrix}$. (d) $(1/\sqrt{6})\begin{bmatrix} -\sqrt{3} & -1 & \sqrt{2} \\ 0 & 2 & \sqrt{2} \\ \sqrt{3} & -1 & \sqrt{2} \end{bmatrix}$.

(f) $(1/\sqrt{2})\begin{bmatrix} 0 & -1 & 1 \\ 0 & 1 & 1 \\ \sqrt{2} & 0 & 0 \end{bmatrix}$. (k) $(1/\sqrt{2})\begin{bmatrix} 0 & -1 & 0 & 1 \\ -1 & 0 & 1 & 0 \\ 0 & 1 & 0 & 1 \\ 1 & 0 & 1 & 0 \end{bmatrix}$.

2. (b) $\dfrac{1}{9}\begin{bmatrix} 13 & 4 & -2 \\ 4 & 13 & -2 \\ -2 & -2 & 10 \end{bmatrix}$.

3. $A^T = (P\Lambda P^T)^T = P^{TT}\Lambda^T P^T = P\Lambda P^T = A.$

5. $A = P(\lambda I)P^T = \lambda PP^T = \lambda I$, if $P^{-1} = P^T.$

8. If $A^T = A$, then $B^T = (P^TAP)^T = P^TA^TP^{TT} = P^TAP = B$. If $B^T = B$, then $A^T = (PBP^T)^T = P^{TT}B^TP^T = PBP^T = A.$

Section 8.6

EXERCISES

1. $(0, 1, 0), (-1, 1, 1), (2, 1, 2).$

PROBLEMS 8.6

1. (a) Matrix $\begin{bmatrix} 1 & 0 \\ 0 & 6 \end{bmatrix}$, basis $(-2, 1), (1, 2).$

(c) Matrix $\begin{bmatrix} 1 & 0 & 0 \\ 0 & 1 & 0 \\ 0 & 0 & 7 \end{bmatrix}$, basis $(-1, 1, 0), (-1, 0, 1), (1, 2, 3).$

3. The ith column of P is the coordinate vector of w_i with respect to the basis $v_1, v_2, \ldots, v_n.$

Section 8.7

PROBLEMS 8.7

1. $dx/dt = -ax + by + cz$
$dy/dt = dx - ey + fz$
$dz/dt = gx + hy - iz$ where $a, b, c, d, e, f, g, h, i \geq 0.$

Section 8.8

EXERCISES

3. $dx/dt = Ax$, where $A = \begin{bmatrix} 3 & 1 \\ 1 & 3 \end{bmatrix}.$

PROBLEMS 8.8

3. (a) $dx/dt = Ax$, where $x = \begin{bmatrix} y \\ y' \end{bmatrix}$ and $A = \begin{bmatrix} 0 & 1 \\ 1 & 0 \end{bmatrix}$.

(c) $dx/dt = Ax$, where $x = \begin{bmatrix} y \\ y' \\ y'' \end{bmatrix}$ and $A = \begin{bmatrix} 0 & 1 & 0 \\ 0 & 0 & 1 \\ 0 & 2 & -1 \end{bmatrix}$.

(e) $dx/dt = Ax$, where $x = \begin{bmatrix} y \\ y' \\ y'' \\ y''' \end{bmatrix}$ and $A = \begin{bmatrix} 0 & 1 & 0 & 0 \\ 0 & 0 & 1 & 0 \\ 0 & 0 & 0 & 1 \\ -2 & 0 & 1 & 2 \end{bmatrix}$.

Section 8.9

EXERCISES

1. $x(t) = c_1(e^{2t}, -e^{2t}) + c_2(e^{4t}, e^{4t})$.

PROBLEMS 8.9

2. In parts (a), (b), and (c) the given solutions are a basis for the solution space.

CHAPTER 9

Section 9.1

PROBLEMS 9.1

1.

	Multiplications	Additions		Multiplications	Additions
(a)	n	$n - 1$	(b)	n^2	$n^2 - n$
(c)	n^3	$n^3 - n^2$	(d)	n^3	$n^3 - n^2$
(e)	$2n^3$	$2n^3 - 2n^2$	(f)	$2n^3$	$2n^3 - 2n^2$
(g)	$3n^3$	$3n^3 - 3n^2$			

4. (a) $(n^3 - n)/3 + kn^2$.   (b) $(n^3 - n^2)/2 + kn^2$.

6. (a) $n^3 + kn^2$.   (b) Always.

Section 9.2

PROBLEMS 9.2

1.

	Mantissa	Exponent		Mantissa	Exponent
(a)	0.7876	0	(b)	0.7876	3
(c)	0.6250	0	(d)	0.3333	0
(e)	0.1428	0	(f)	0.1428	−1
(g)	0.7000	4	(h)	0.1428	−3

2. 0.0005.   **3.** (a) For each x, $Ax = (0.7500 \cdot 10^7, 0.2496 \cdot 10^7)$.
                          (b) $0.1426 \cdot 10^8$.

Section 9.3

PROBLEMS 9.3

1. (a) $x_1 = 0.6619$, $x_2 = 1.523$, $x_3 = 4.286$.
   (b) $x_1 = -4.182$, $x_2 = 4.658$, $x_3 = 0.6065$.
   (c) $x_1 = 0.5121$, $x_2 = 4.271$, $x_3 = 3.774$.

Section 9.4

PROBLEMS 9.4

1. (a) 2, $(1, 0, 1)$.   (b) 4.14, $(0.64, 1, 0.86)$.   (c) 4, $(0.33, 1, 0.67)$.

3. (a) For A: 50, $(1, -0.286)$; 1, $(-0.333, 1)$.
       For B: 3, $(1, -0.286)$; 2, $(-0.333, 1)$.
   (d) The smaller the value of λ_1/λ_2 the more rapid will be the convergence of the power method.

APPENDIX

Section A.1

PROBLEMS A.1

1. (a) $22 - 4i$.   (b) $3 - 3i$.   (c) $\frac{19}{25} - \frac{8}{25}i$.   (d) 2.   (e) $2i$.   (f) $(-2)^n$.

2. (a) -3.   (b) 4.   (c) 5.   (d) 1.   (e) -3.   (f) -4.   (g) 5.   (h) 1.

3. (a) $\sqrt{2}/2 + (\sqrt{2}/2)i$ and $-\sqrt{2}/2 - (\sqrt{2}/2)i$.
   (b) $-\sqrt{2}/2 + (\sqrt{2}/2)i$ and $\sqrt{2}/2 - (\sqrt{2}/2)i$.

5. (a) $a/(a^2 + b^2) - (b/(a^2 + b^2))i = \bar{z}/|z|$.   (b) Yes.

Section A.2

PROBLEMS A.2

1. $(i, 1) = i(1, -i)$.   **2.** 1.

3. (a) $x = (i, 2)$.   (b) $x = (2 - i, 3 + i)$.
   (c) $x = (3, -2)$.   (d) $x = (1 - i, i, 1 + i)$.

4. (a) $\begin{bmatrix} (-1-2i)/3 & i/3 \\ (1+i)/3 & i/3 \end{bmatrix}$. (b) $\begin{bmatrix} 1+i & (-1-2i)/5 \\ -i & (1+2i)/5 \end{bmatrix}$.

(c) $\begin{bmatrix} 1 & 1 & 0 \\ 1+i & i & 0 \\ 1-i & -i & 1 \end{bmatrix}$. (d) $\begin{bmatrix} 1+i & i & -i \\ 0 & 1+i & -i \\ -1 & -1 & 1 \end{bmatrix}$.

Section A.4

PROBLEMS A.4

3. (a) $-i$, $(i, 1)$; $1-i$, $(1-i, -i)$. (b) $3i$, $(1, i)$; $1-2i$, $(2i, -1)$.

Index